T0318885

Nanotechnology
An Introduction

Nanotechnology
An Introduction
Second Edition

Jeremy J. Ramsden
*Honorary Professor of Nanotechnology
at the University of Buckingham
Research & Technology Director,
Henge Precision Materials Ltd
Fellow of the Institute of Materials,
Minerals and Mining and IUPAC*

AMSTERDAM • BOSTON • HEIDELBERG • LONDON
NEW YORK • OXFORD • PARIS • SAN DIEGO
SAN FRANCISCO • SINGAPORE • SYDNEY • TOKYO
William Andrew Publishers is an Imprint of Elsevier

ELSEVIER

Publisher: Matthew Deans
Acquisition Editor: Simon Holt
Editorial Project Manager: Sabrina Webber
Production Project Manager: Anusha Sambamoorthy
Designer: Greg Harris

William Andrew is an imprint of Elsevier
The Boulevard, Langford Lane, Kidlington, Oxford, OX5 1GB, UK
50 Hampshire Street, 5th Floor, Cambridge, MA 02139, USA

Copyright © 2016, 2011 Elsevier Inc. All rights reserved.

No part of this publication may be reproduced or transmitted in any form or by any means, electronic or mechanical, including photocopying, recording, or any information storage and retrieval system, without permission in writing from the publisher. Details on how to seek permission, further information about the Publisher's permissions policies and our arrangements with organizations such as the Copyright Clearance Center and the Copyright Licensing Agency, can be found at our website: www.elsevier.com/permissions.

This book and the individual contributions contained in it are protected under copyright by the Publisher (other than as may be noted herein).

Notices
Knowledge and best practice in this field are constantly changing. As new research and experience broaden our understanding, changes in research methods, professional practices, or medical treatment may become necessary.

Practitioners and researchers must always rely on their own experience and knowledge in evaluating and using any information, methods, compounds, or experiments described herein. In using such information or methods they should be mindful of their own safety and the safety of others, including parties for whom they have a professional responsibility.

To the fullest extent of the law, neither the Publisher nor the authors, contributors, or editors, assume any liability for any injury and/or damage to persons or property as a matter of products liability, negligence or otherwise, or from any use or operation of any methods, products, instructions, or ideas contained in the material herein.

Library of Congress Control Number: 2011924730

British Library Cataloguing in Publication Data
A catalogue record for this book is available from the British Library

ISBN: 978-0-323-39311-9

For information on all William Andrew publications
visit our website at www.williamandrew.com

**Working together to grow
libraries in developing countries**

www.elsevier.com | www.bookaid.org | www.sabre.org

ELSEVIER   BOOK AID International   Sabre Foundation

...Kicsinyben rejlik a nagy,
Olyan sok a tárgy, s létünk oly rövid

IMRE MADÁCH

Contents

Preface to the Second Edition

Progress in nanotechnology has been so rapid with results so dramatic during the few years that have elapsed since the first edition of this book, it would be tempting to conclude that we are on the cusp of advance. It might, however, be premature to venture such an opinion. At any rate, apart from reflecting this progress, the opportunity to expand this book for the second edition has enabled a number of new topics to be included, which were omitted in the first edition either for reasons of lack of space or because at the time they were not perceived to be sufficiently important.

It is of wider significance that the power of nanotechnology has enabled complexity to be understood at ever smaller scales, which is helping humanity to understand the scientific basis of some of the oldest and most intractable of technologies, such as those involved in food and medicine.

Above all, this book should be considered as a map of the highly multifarious terrain of nanotechnology.

Inevitably, the author of a book of this nature is greatly indebted to countless colleagues and correspondents throughout the world. It would be invidious to mention some without mentioning all, and they are too numerous for the latter, but I hope that in this era of a voluminous research literature their inputs are adequately reflected in the reference list.

I should particularly like to thank the Tokyo Medical and Dental University for hospitality during the summer months, which provided a most congenial environment for completing the new edition.

Jeremy J. Ramsden
Buckingham
September 2015

From the Preface to the First Edition

There are already hundreds of books in print on nanotechnology in English alone, at all levels up to that of the advanced researcher and ranging in coverage from detailed accounts of incremental improvements in existing microtechnologies to far-reaching visionary approaches toward productive nanosystems. Furthermore, not only do many old-established scientific journals in the fields of physics and chemistry now carry nanotechnology sections, but numerous scientific journals dedicated to nanotechnology have appeared and new ones are launched each year. In addition, a flood of commercially-oriented reports about present and future nanotechnology markets is constantly being produced; these are often weighty documents comprising a thousand pages or more, whose reliability is difficult to assess, since even if they make use of publicly available data, the processing of those data in order to arrive at predictions is typically carried out using unrevealed algorithms.

Faced with this huge and burgeoning literature, the newcomer to the field, who may well have a strong background in one of the traditional disciplines such as physics, mechanical or electrical engineering, chemistry or biology, or who may have been working on microelectromechanical systems (MEMS) (also known as microsystems technology, MST), is likely to feel confronted with a somewhat chaotic and colorful scene from which it is often difficult to extract meaning. The goal of this book is to tackle that problem by presenting an overview of the entire field, focusing on key essentials, with which the reader will be able to erect his or her own personal scaffold on which the amazingly diverse plethora of detailed information emerging from countless laboratories all over the world can be structured into some kind of coherent order. The emphasis is therefore on concepts; any attempt to complement this by capturing all the latest specific discoveries and inventions in nanotechnology, other than illustratively, would almost immediately become out of date; the aim of this book might briefly be stated as being "to make sense of nanotechnology"—to explain things thoroughly enough to be intelligible while still remaining introductory.

The book itself is structured around a robust anatomy of the subject. Following a basic introduction (Chapter 1), which includes a brief history of the subject, careful consideration is given to the meaning of the nanoscale (Chapter 2), on which everything else is dependent, since nanotechnology can most simply (but cryptically) be defined as "technology (or engineering) at the nanoscale". This chapter in itself constitutes a succinct summary of the entire field. Chapter 3 is devoted to interfacial forces, which govern key aspects of behavior at the nanoscale. Chapter 4 covers the nano/bio interface, which plays a fundamental rôle in the continuing evolution of nanotechnology, and Chapter 5 deals with the demanding issues of metrology in nanotechnology, which have also strongly influenced nanofabrication technology. In this chapter, the metrology of the nano/bio interface is covered in detail, since this is one of the newest and least familiar parts of the field. Nanomaterials (both nano-objects and nanostructured materials) are covered in Chapter 6—except carbon nanomaterials (and devices), which merit a separate chapter (9). Nanoscale devices of all kinds (except those based on carbon)—mainly information processors and transducers, including sensors—are the topic of Chapter 7, and strategies for their fabrication are covered in Chapter 8, devoted to the three fundamental approaches towards achieving nanoscale manufacture (nanofacture), namely the top–down methods rooted in ultraprecision engineering and semiconductor processing, the bottom-to-bottom approach that is closest to the original concept of nanotechnology (the molecular assembler), and the bottom–up (self-assembly) methods that have been powerfully inspired by processes in the living world. Problems of materials selection, design and so forth are treated in Chapter 10, especially how to deal with vastification; that is, the vast numbers of components in a nanosystem and the almost inevitable occurrence of defective ones. Chapter 11 is devoted to bionanotechnology, defined as the incorporation of biomolecules into nanodevices. The final chapter (12) deals with the impacts of nanotechnology: technical, economic, social, psychological and ethical. Each chapter is provided with a succinct summary at the end as well as suggestions for further reading. A glossary of nanotechnology neologisms is appended, along with a list of the most common abbreviations.

The primary readership is expected to be engineers and scientists who have previously been working in other fields but are considering entering the nano field and wish to rapidly acquire an appreciation of its vocabulary, possibilities and limitations. The secondary readership is anyone curious about nanotechnology, including undergraduates and professionals in other fields. The book should also appeal to those less directly connected with science and engineering, such as insurers and lawyers, whose activities are very likely to be connected with nanotechnology in the future, and traders, commodity brokers and entrepreneurs in general dissatisfied with remaining in ignorance of the technology that they are making use of. It is designed to equip the reader with the ability to cogently appraise the merits or otherwise of any piece of nanotechnology that may be reported in one form or another.

It is a distinct characteristic of nanotechnology that many of its features draw heavily from existing work in chemistry, physical chemistry, physics and biology. Hence, there is relatively little domain-specific knowledge associated with nanotechnology. Most of the themes in this book are covered in great detail in specialist literature that may not exclusively or even overtly be associated with nanotechnology. It seems, therefore, that nanotechnology is most aptly globally characterized as an attitude or mindset, comprising above all the desire both to understand the world at the atomic level and to create objects of beauty and utility by controlling matter at that level. Unlike the science of the subatomic level, however, nanotechnology necessarily concerns itself with superatomic levels as well, since the ultimate objects of its creation must be macroscopic in order to be of use to humanity. Hence, problems of emergent properties also form a part of nanotechnology.

The uniqueness of this book resides in its unifying viewpoint that draws many disparate pieces of knowledge together to create novel technologies. These nanotechnologies are in turn united by the distinctive attitude associated with nanotechnology.

Nanotechnology is inseparably associated with the emergence of qualitatively different behavior when a quantitative difference, namely, increasing smallness, becomes great enough: one might call this a Hegelian viewpoint of nanotechnology. Phenomena that are merely modified *pari passu* with diminishing size without any qualitative change should not, therefore, strictly rank as nanotechnology. This viewpoint avoids the pitfall of having to group practically all of physics, chemistry and biology under nanotechnology because of too vague a definition.

Jeremy J. Ramsden
May 2010

What is nanotechnology?

CHAPTER CONTENTS

INTRODUCTION

Nanotechnology is above all a mindset, a way of thinking about the world that is rooted in atomically precise perception. As such, it represents the apotheosis of man's ceaseless urge to understand the world and use that understanding for practical purposes. Well synonymized as "atomically precise technology" (APT), it encapsulates the vision of building our earthly estate atom-by-atom, controlling architecture, composition and hence physical properties with atomic resolution. "Hard" nanotechnologists promote a future world in which every artifact (and even food) can be constructed atom-by-atom from a feedstock such as acetylene, requiring

Nanotechnology: An Introduction. DOI: 10.1016/B978-0-323-39311-9.00007-8
Copyright © 2016 Elsevier Inc. All rights reserved.

in addition only energy and instructions. A more pragmatic view accepts that there are many intermediate stages in which partially atomically precise construction can improve existing artifacts and create new ones. Similarly, while the resolute aim of "hard" nanotechnologists is to create productive nanosystems (PN) working with atomic precision – the nanoscale assemblers that would execute the instructions and build everything we need from the bottom upwards — a more pragmatic view accepts that while in principle everything can be reproduced and many things imitated via atom-by-atom assembly, in many cases the improvement in properties or performance would be negligible and a hybrid approach will best serve the needs of humanity. This is particularly likely to be the case for large artifacts (such as human dwellings or airplanes) and for relatively complex products such as food that can be quite easily grown naturally.

In this chapter, we shall first look at the basic definitions for nanotechnology, and sketch a concept system ("ontology") for the field. It is also possible to define nanotechnology ostensively, according to what is already generally considered to be nanotechnology, and extended by what is envisaged in the future. A further way of defining it is through its history. We also briefly look at the relation of nanotechnology to biology, which has been a powerful paradigm for convincing engineers that nanotechnology is possible – nanobiotechnology and bionanotechnology form the topics of subsequent Chapters (Chapters 4 and 11, respectively). General motivations for nanotechnology are considered – "Why nanotechnology?" Attention is drawn to the list of neologisms associated with nanotechnology at the end of the book.

1.1 DEFINITIONS AND CONCEPTS
1.1.1 WORKING DEFINITIONS

The simplest definition of nanotechnology is "technology at the nanoscale". The various definitions currently circulating can be reasonably accurately thus para-phrased. Obviously, this definition is not intelligible in the absence of a further definition, namely, that of the nanoscale. Furthermore, definitions of components of nanotechnology, such as "nanofiber", also refer to the nanoscale; indeed every word starting with "nano", which we can generically write as "nanoX", can be defined as "nanoscale X". Therefore, unless we define "nanoscale", we cannot therefore properly define nanotechnology. A rational attempt to do so is made in Chapter 2. Here, we note that provisionally, the nanoscale is considered to cover the range from 1 to 100 nm. Essentially this is a consensus without a strong rational foundation.

A slightly longer but still succinct definition of nanotechnology is simply "engineering with atomic precision", or APT. However, this definition does not explicitly include the aspects of "fundamentally new properties" or "novel" and "unique" that nanotechnologists usually insist upon, wishing to exclude existing artifacts that happen to be small. These aspects are encapsulated by the US National Nanotechnology Initiative's declaration that "the essence of nanotechnology is the ability to work at the molecular level, atom-by-atom, to create large structures with fundamentally new molecular organization ... nanotechnology is concerned with materials and systems whose structures and components exhibit novel and significantly improved physical, chemical, and biological properties, phenomena, and processes due to their nanoscale size" [219]. The US Foresight Institute gives "nanotechnology is a group of emerging technologies in which the structure of matter is controlled at the nanometer scale to produce novel materials and devices that have useful and unique properties". Function is stressed in "the design, synthesis, characterization and application of materials, devices and systems that have a functional organization in at least one dimension on the nanometer scale". This is emphasized even more strongly in "nanotechnology pertains to the processing of materials in which structure of a dimension of less than 100 nm is essential to obtain the required functional performance" [52]. In all such definitions, there is the implicit meaning that, as for any technology, the end result must be of practical use. A dictionary definition of nanotechnology is "the design, characterization, production and application of materials, devices and systems by controlling shape and size in the nanoscale" [1]. An alternative definition from the same dictionary is "the deliberate and controlled manipulation, precision placement, measurement, modeling and production of matter in the nanoscale in order to create materials, devices, and systems with fundamentally new properties and functions" [1]. The emphasis on control is particularly important: it is this that distinguishes nanotechnology from chemistry, with which it is often compared: In the latter, motion is essentially uncontrolled and random, within the constraint that it takes place on the potential energy surface of the atoms and molecules under consideration. In order to achieve the desired control, a special, nonrandom *eutactic* environment needs to be available. How eutactic environments can be practically achieved is still being vigorously discussed. Finally, a definition of nanotechnology attempting to be comprehensive is "the application of scientific knowledge to measure, create, pattern, manipulate, utilize or incorporate materials and components in the nanoscale". This underlines the idea of nanotechnology as the consummation of stage 4 in the sequence of technological revolutions that marks the development of human civilization (Table 12.1 in Chapter 12).

In summary, then, nanotechnology is concerned with observable *objects*, namely, *materials and devices* (as soon as devices become more than very rudimentary tools, they actually become systems) with engineered structure in the nanoscale, and with *processes* capable of manipulating individual atoms or nanoscale blocks (nanoblocks) with ultraprecision.

It is sometimes debated whether one should refer to "nanotechnology" or "nanotechnologies". The argument in favor of the latter is that nanotechnology encompasses many distinctly different kinds of technology. But there seems to be no reason not to use "nanotechnology" in a collective sense, since the different kinds are nevertheless all united by striving for control at the atomic scale. Both terms are, in fact, legitimate. When one wishes to emphasize diverse applications, the plural form is appropriate. The singular term refers above all to the mindset or attitude associated with the technology.

1.1.2 TOWARDS A CONCEPT SYSTEM FOR NANOTECHNOLOGY

Objects are perceived or conceived. The properties of an object (which may be common to a set of objects) are abstracted into characteristics. Essential characteristics (feature specifications) typically falling into different categories (e.g. shape and color) are combined as a set to form a concept; this is how objects are abstracted into concepts, and the set of essential characteristics that come together as a unit to form a concept is called the intension. The set of objects abstracted into a concept is called the extension. Delimiting characteristics distinguish one concept from another. Concepts are described in definitions and represented by designations. The set of designations constitutes the terminology. Concepts are organized into concept systems. A concept system is often called an ontology (which literally means the science of being, but lately is often used in a more restricted sense, namely, that of the study of categories).

Fig. 1.1 shows (part of) an ontology for nanotechnology. To the right to the diagram one has observable *objects* (products) – an axis of tangible things in order of increasing complexity: *materials, devices and systems*. To the left of the diagram one has *processes*. Note the relationships between *metrology* and *fabrication* (nanomanufacturing, usually abbreviated to nanofacture, also called atomically precise manufacturing, APM) and devices. An atomic force microscope (AFM) is used to measure nanoscale features; every measuring instrument is necessarily a device, and pushing nano-objects around with a needle is the basis of bottom-to-bottom fabrication (see Section 8.3).

Especially the leaves of the tree might be associated with some ambiguity regarding their extensions. For example, devices can be characterized by the nature of their working medium: electrons, photons, etc. Yet many devices involve more than one medium: for example, nanoelectro*mechanical* devices are being researched as a way of achieving electronic switching; opto*electronic* control is a popular way of achieving photonic switching; and photochemistry in miniaturized reactors involves both nanophotonics and nanofluidics.

Table 1.1 describes a few of the concepts and their intensions and extensions. At the time of writing, the terminology of nanotechnology is still being debated within national standards organizations as well as supranational bodies such as the Comité européen de normalisation (CEN) and the International Standards Organization (ISO); hence, no attempt has been made to be comprehensive here.

1.2 AN OSTENSIVE DEFINITION OF NANOTECHNOLOGY

An ostensive definition of current nanotechnology can be constructed from the most popular topics (Table 1.2), essentially extensions in the sense of Section 1.1.2.

A number of inferences can be drawn from this table, including the pre-eminence of carbon as a nanomaterial, the nanoparticle as a nano-object, and energy,

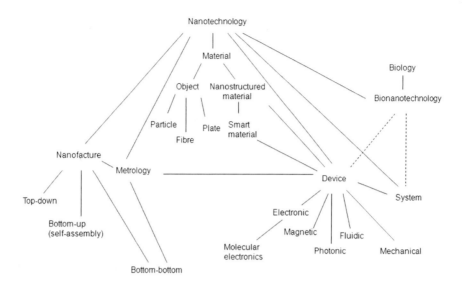

FIGURE 1.1

A concept system (ontology) for nanotechnology. Most of the terms would normally be prefixed by "nano" (e.g. nanometrology and nanodevice). A dashed line merely signifies that if the superordinate concept contributes, then the prefix must indicate that (e.g. bionanodevice and bionanosystem). Biology may also have some input to nanomanufacture (nanofacture), inspiring, especially, self-assembly processes. Not shown on the diagram is what might be called "*conceptual* nanotechnology", or perhaps better (since it is itself a concept), "virtual nanotechnology", which means the (experimental and theoretical) scrutiny of engineering (and other, including biological) processes at the nanoscale in order to understand them better, that is, the mindset or attitude associated with nanotechnology.

Table 1.1 Some Nano-Concepts and their Intensions and Extensions

Intension	Concept	Extension
One or more external dimensions in the nanoscale	Nano-object	Graphene, fullerene
One or more geometrical features in the nanoscale	Nanomaterial	A nanocomposite
Automaton with information storage and/or processing embodiments in the nanoscale	Nanodevice	Single electron transistor

composites (materials) and sensors as applications. Interestingly, "water" features highly on the list. The reasons for this will be apparent from reading Section 3.8.

Table 1.2 Table of the Relative Importance (Ranked by Numbers of Occurrences of Words in the Titles of Papers Presented at the Nano2009 Conference in Houston, Texas) of Nanotechnology Terms and Applications

Rank	Term	Number of Occurrences
1	Carbon, CNT	151
2	Nanoparticle, nanocrystal	138
3	Energy	96
4	(Nano)material	92
5	Nanotube	82
6	(Nano)composite	79
7	(Bio)sensor	55
8	Water	45
9	Device	33
10	Nanowire	33
11	Assembly	31
12	Silicon	30
13	Zinc (oxide)	26
14	Titanium (oxide)	25
15	Quantum	24
16	Silica	21
17	Phage	20
18	Bio	19
19	Photovoltaic	15
20	Nanorod	8
21	Graphene	7
22	Nanopore	7
23	Silver	7

A very simple (albeit privative) ostensive definition of nanotechnology is "If you can see it (including with the aid of an optical microscope), it is not nano", referring to the fact that any object below 100 nm in size is below the Abbe limit for optical resolution using any visible wavelength (Eq. (5.2)). A nanoplate, however, would only be invisible if oriented exactly parallel to the line of sight.

1.3 **A BRIEF HISTORY OF NANOTECHNOLOGY**

Reference is often made to a lecture given by Richard Feynman in 1959 at Caltech [85]. Entitled "There's Plenty of Room at the Bottom", it expounds his vision of machines making the components for smaller machines (a familiar enough operation at the macroscale), which when assembled are themselves capable of making the components for yet smaller machines, and simply continuing the sequence until the atomic realm is reached. Offering a prize of $1000 for the first person to build a working electric motor with an overall size not exceeding 1/64th of an inch, Feynman was dismayed when not long afterwards a student, William McLellan, presented him with a laboriously hand-assembled (i.e. using the technique of the watchmaker) electric motor of conventional design that nevertheless met the specified criteria.

A similar idea was proposed at around the same time by Marvin Minsky: "Clearly it is possible to have complex machines the size of a flea; probably one can have them the size of bacterial cells . . . consider contemporary efforts towards constructing small fast computers. The main line of attack is concentrated on "printing" or evaporation through masks. This is surely attractive; in one operation one can print thousands of elements. But an alternative, equally attractive, has been ignored. Imagine small machines fabricating small elements at kilocycle rates. (The speed of small mechanical devices is extremely high.) Again, one can hope to make thousands of elements per second. But the generality of the mechanical approach is much greater since there are many structures that do not lend themselves easily to laminar mask construction" [209]. One wonders whether Feynman and Minsky had previously read Robert A. Heinlein's short story "Waldo", which introduces this idea (it was published in the August 1942 issue of "Astounding" magazine under the pseudonym Anson MacDonald).

Here, we find the germ of the idea of the assembler, a concept later elaborated by Eric Drexler. The assembler is a universal nanoscale assembling machine, capable not only of making nanostructured materials but also other machines (including copies of itself). The first assembler would have to be laboriously built atom-by-atom, but once it was working its numbers could evidently grow exponentially, and when a large number was extant, universal manufacturing capability, hence the nano-era, would have truly arrived (see also Chapter 8).

However, the idea of a minute device intervening at the level of elementary particles was conceived almost a hundred years earlier by James Clerk Maxwell when he invented his "demon" for selectively allowing molecules to pass through a door, thereby entangling physics with information. Perhaps Maxwell should be considered as the real father of nanotechnology. The demon was described in Maxwell's *Theory of Heat* first published in 1871, but had already been mentioned in earlier correspondence of his.

1.3.1 **ULTRAPRECISION ENGINEERING**

It could well be said that the history of technological advance is the history of ever finer tolerances in machining metals and other materials. A classic example is

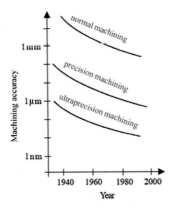

FIGURE 1.2

The evolution of machining accuracy.

-Based on an idea of Norio Taniguchi.

the steam engine: James Watt's high-pressure machine that paved the way for the technology to move from a cumbersome and rather inefficient means of pumping water out of mines to an industrially useful and even self-propelling technology was only possible once machining tolerance had improved to enable pistons to slide within cylinders without leaking.

An approach to the nanoscale seemingly quite different from the Heinlein–Feynman–Minsky–Drexler vision of assemblers starts from the microscopic world of precision engineering, progressively scaling down to ultraprecision engineering (Fig. 1.2). The word "nanotechnology" was itself coined by Norio Taniguchi in 1983 to describe the ultimate limit of this process [278, 279, 280]. He referred to "atomic bit machining".

1.3.2 SEMICONDUCTOR PROCESSING *QUA* MICROTECHNOLOGY

The trend in ultraprecision engineering is mirrored by relentless miniaturization in the semiconductor processing industry. The history of the integrated circuit could perhaps be considered to start in 1904, when Bose patented the galena crystal for receiving electromagnetic waves, followed by Picard's 1906 patent for a silicon crystal. The thermionic valve and the triode were invented, respectively, by Fleming (in 1904) and de Forest (in 1906); it became the basis of logic gates, reaching zenith with the ENIAC, which contained about 20,000 valves. The invention of the point contact transistor in 1947 at Bell Laboratories essentially rendered the thermionic valve obsolete, but the first commercial use of transistors only occurred in 1953 (the Sonotone 1010 hearing aid), the first transistor radio appearing one year later. Meanwhile the idea of an integrated circuit had been proposed by Dummer at the Royal Signals Research Establishment (RSRE) in 1952, but (presumably) he was not allowed to work on it at what was then a government establishment, and the first actual example was realized by Kilby in 1958 at Texas Instruments, closely followed by Noyce at Fairchild in the following year. It is interesting to recall that the Apollo flight computer ("Block II") used for the first moon landing in 1969 was designed in 1964 (the year before Moore's law was first proposed), used resistor–transistor logic (RTL) and had a clock speed of 2 MHz. Intel introduced the first microprocessor, with about 2000 transistors, in 1971, the year in which the pioneering "LE-120A Handy" pocket calculator was launched in Japan. It took another decade before the IBM personal computer appeared (1981); the Apple II had already been launched in 1977. By 2000, we had the Pentium 4 chip with about 1.2×10^6 transistors fabricated with 180 nm process technology. In contrast, the recent dual core Intel Itanium chip has about 1.7×10^9 transistors (occupying an area of about 50×20 mm) with a gate length of 90 nm. A 45 nm transistor can switch 3×10^{11} times per second – this is about 100 GHz. Experimental graphene-based devices achieve more than 1 THz. Despite the extraordinarily high precision fabrication called for in such devices, modern integrated circuits are reliable enough for spacecraft (for example) to use commercial off-the-shelf (COTS) devices. The fabrication plants are not cheap – Intel's 2008 China facility is reputed to have cost 2.5×10^9: a mask alone for a chip made using 180 nm process technology costs about $100,000, rising to one million dollars for 45 nm technology. Despite the huge costs of the plant, cost per chip continues to fall relentlessly: for example, a mobile phone chip cost about $20 in 1997, but only $2 in 2007.

The relentless diminution of feature size, and the concomitant increase of the number of transistors that can be fabricated in parallel on single chip, has been well documented; structures with features a few tens of nanometers in size capable of being examined in an electron microscope were reported as long ago as 1960; device structures with dimensions less than 100 nm were already being reported in 1972, with 25 nm achieved in 1979. Incidentally, the lower size limit for practical semiconductor circuits is considered to be about 20 nm (Section 8.1.1); smaller sizes, hence higher transistor number densities per unit area, will only be achievable using three-dimensional design or quantum logic (Section 7.3). Thus, we see that since the beginning of nanotechnology – identifying this with the conception of Maxwell's demon – nanotechnology has been intimately connected with information science and technology.

1.3.3 NANOPARTICLES

If we define nanotechnology ostensively, we have to concede a very long history: there is evidence that PbS nanocrystals were the goal of a procedure used since Greco-Roman times to color hair black [294]. Nanoparticulate gold has a long history, not least in medicine (see Section 4.2). The Flemish glassmaker John Utynam was granted a patent in 1449 in England for making stained glass incorporating nanoparticulate gold, and the Swiss medical doctor and chemist von Hohenheim (Paracelsus) prepared and administered gold nanoparticles to patients suffering from certain ailments in the early sixteenth century; a modern equivalent is perhaps the magnetic nanoparticles proposed for therapeutic purposes. The secret of the extraordinarily advanced metallurgical features of Damascus swords made more than 400 years ago has recently been found to be carbon nanotubes (CNTs) embedded in the blades [257].

   With such a long history, it is perhaps hardly surprising that at present, nanoparticles represent almost the only part of nanotechnology with commercial significance. The fabrication of different kinds of nanoparticles by chemical means was well established by the middle of the nineteenth century (e.g. Thomas Graham's method for making ferric hydroxide nanoparticles [104]). Wolfgang Ostwald lectured extensively on the topic in the early twentieth century in the USA, and wrote up the lectures in what became a hugely successful book, *Die Welt der vernachlässigten Dimensionen*. Many universities had departments of colloid science (sometimes considered as physics, sometimes as chemistry), at least up to the middle of the twentieth century, when slowly the subject seemed to fall out of fashion, until its recent revival as part of nanotechnology. The field somewhat overlapped that of heterogeneous catalysis, in which it was well known, indeed almost obvious, that specific activity (i.e. per unit mass) tended to increase with increasing fineness of division.

1.4 NANOTECHNOLOGY AS AN EMERGING TECHNOLOGY

The empirical investigation of the development of new technologies shows that they follow an exponential course. This is *a priori* easy to understand: The initial spark presumably starts with one or two people; they make a prototype and deploy it and, according to its success, develop it further. Soon others will notice and start copying it (a large corporation may monopolize the copying by taking out patent protection). The rate of adoption of a new technology is roughly proportional to the number n of people already using it: If we only consider one variable,

$$\frac{dn}{dt} = rn, \tag{1.1}$$

where $r > 0$ is a constant, for which the solution is simple exponential growth,

$$n(t) = n(0)e^{rt}, \tag{1.2}$$

where $n(0)$ is the quantity of n at $t = 0$, typically equal to 1. This is the meaning of "emerging technology"; it is exactly analogous to "r-selection" in ecology (in which r is called the growth rate). The emergence is often painfully slow to the inventors. Their costs usually increase linearly; although the income from the technology (assumed to be proportional to n) will inevitably intersect with the costs at some epoch, the challenge for the inventors is to survive until that happens.

The exponential growth can subsequently manifest itself in different ways. In a large and important industry, such as electronics, there will be an enormous accretion of expertise and many innovations will subsequently occur. A measure different from n may then become appropriate, such as the number of individual circuit elements on a chip, as in Moore's law. Electronics has such wide-ranging applications that many different kinds of adoption display exponential growth; for example, the number of websites; this ultimately depends on the exponentially diminishing cost of computation, which is a direct consequence of Moore's law.

Historically, every technology seems to enter a linear growth phase and then it levels off. Sailing vessels, the steam engine and airships have all suffered that fate. Eq. (1.1) is just the first term of the expansion of the very general

$$\frac{dn}{dt} = \mathcal{F}(n), \tag{1.3}$$

where $\mathcal{F}$ is some unknown function. Expanding further gives

$$\frac{dn}{dt} = rn - \frac{r}{K}n^2, \tag{1.4}$$

where $K > 0$ is another constant (chosen to match the ecological literature). The solution is now

$$n(t) = \frac{K}{1 + e^{-r(t-m)}}, \tag{1.5}$$

the so-called logistic equation, which is sigmoidal with a unique point of inflexion at $t = m, n = K/2$ at which the tangent to the curve is r, and asymptotes $n = 0$ and $n = K$. K is called the carrying capacity in ecology and corresponds to the saturation level of the market. It is emphasized that this result arises from very general considerations; our only nonarbitrary action was to make the coefficient of the term in n^2 negative. There is overwhelming empirical support for this course of events, but a deeper explanation remains elusive. For consumer items, such as a cellphone, if measured by the number of units sold, K must correspond to world population. For other technologies, the relationship may not be so clear. For example, the number of nanoprocessors required to connect every human artifact to the Internet ("the Internet of things") has no clear upper limit because the number of artifacts *per capita* is not really limited. Although the current trend is for this number to increase, some futurists predict a very simple, uncluttered lifestyle in which we can focus almost wholly on higher matters. Even more intriguingly, many technologies have declined after reaching a peak – large sailing boats were replaced by steam, but steam engines themselves subsequently declined.

1.5 *NEC PLUS ULTRA?*

Feynman's lecture was entitled "There's plenty of room *at the bottom*" (emphasis added) and Taniguchi predicted the development of ultraprecision engineering to atomic-scale manipulation. It is indeed far from clear whether subatomic engineering could ever yield a useful mainstream technology (besides which there is still controversy about subatomic structure). In other words, nanotechnology is the apotheosis of man's engineering abilities and once we have mastered it there is nothing beyond. It has never been in man's nature to accept such boundaries, however, and perhaps the ability to handle antimatter would rank as a further challenge. Naturally enough there are plenty of technologies in which subatomic elementary particles such as the electron are handled and controlled, but we cannot *build* anything from them and the instruments of handling and control must perforce be made from atoms.

1.6 BIOLOGY AS PARADIGM

When Feynman delivered his famous lecture [85], it was already well known that the smallest viable unit of life was the cell, which could be less than a micrometer in size. It was surmised that cells contained a great deal of machinery at smaller scales, which has since been abundantly evidenced through the work of molecular biologists. Examples of these machines are molecule carriers (e.g. hemoglobin), enzymes, rotary motors (e.g. those powering bacterial flagella), linear motors (e.g. muscle), pumps (e.g. transmembrane ion channels), and multienzyme complexes carrying out more complicated functions (e.g. the proteasome for degrading proteins, or the ribosome for translating information encoded as a nucleic acid sequence into a polypeptide sequence). When Drexler developed his explicit schema of nanoscale assemblers, allusion to nanoscale biological machinery was explicitly made as a "living proof-of-principle" demonstrating the feasibility of artificial devices constructed at a similar scale [62].

At present, probably the most practically useful manifestation of the biological nanoparadigm is self-assembly. In simple form, self-assembly is well known in the nonliving world (for example, crystallization). This process does not, however, have the potential to become a feasible industrial technology for the general-purpose construction of nanoscale objects, because size limitation is not intrinsic to it. Only when highly regular structures need to be produced (e.g. a nanoporous membrane or a collection of monosized nanoparticles) can the process parameters be set to generate an outcome of a prespecified size. Nature, however, has devised a more sophisticated process, known to engineers as programmable self-assembly, in which every detail of the final structure can be specified in advance by using components that not only interlock in highly specific ways but are also capable of changing their structure upon binding to an assembly partner in order to block or facilitate, respectively, previously possible or impossible interlocking actions. Inspiration for harnessing programmable self-assembly arose from the work of virologists who noticed that preassembled components (head, neck, legs) of bacteriophage viruses would further assemble spontaneously into a functional virus merely upon mixing and shaking in a test-tube. This "shake-and-bake" approach appeared to offer a manufacturing route to nanodevices obviating: (1) the many difficulties involved in making Drexlerian assemblers, which would appear to preclude their realization in the near future; and (2) the great expense of the ultrahigh precision "top–down" approach, whether via UPMT or semiconductor processing. Even if assemblers are ultimately realized, it might be most advantageous to use them to assemble sophisticated "nanoblocks", designed to self-assemble into final macroscopic objects (see Section 8.3.2). In other words, its greatest potential utility will probably arise as a means to bridge the size gap between the nanoscopic products of assemblers and macroscopic artifacts of practical use for humans. Self-assembly is covered in detail in Chapter 8.

1.7 NANO–BIO–INFO–COGNO CONVERGENCE

Other technologies that have emerged relatively recently are biotechnology, information technology (IT), and technologies based on the cognitive sciences. Biotechnology in its primitive form refers to growing microörganisms, most typically bacteria, in fermenters and extracting useful substances from them. For many complicated organic molecules, this route may be more economically attractive than laborious multistep artificial syntheses. In its more advanced form, it refers to the manipulation of the genetic constitution of cells [12]; initially, it was done with microörganisms but currently multicellular eukaryotic organisms are also being worked on. There is a certain similarity between the experimental apparatus needed for such manipulation and those used for manipulating nonliving matter in the nanoscale, which seems to be the origin of the convergence between biotechnology and nanotechnology [13]. As for IT, the gradual miniaturization of top–down semiconductor processing technology [195], which is now well capable of creating nanoscale features, obviously connects nanotechnology with IT; furthermore, any realistic bottom-to-bottom manufacturing technology must be automated and, hence, software driven [189], giving another point of convergence. The connexion with cognitive sciences is more tenuous, beyond the fact that present-day investigations into functioning of neurons often make use of nanometrology tools, and that large-scale attempts to simulate the brain require cutting-edge IT. There is no obvious connexion between biotechnology and the cognitive sciences.

Therefore, although these four are sometimes called "converging emerging technologies", the main feature of the convergence is simply that all four of these technologies are currently advancing very rapidly.

1.8 WHY NANOTECHNOLOGY?

Nanotechnology is associated with at least three distinct advantages:

1. It offers the possibility of creating materials with novel combinations of properties.
2. Devices in the nanoscale need less material to make them, they use less energy and other consumables, their function may be enhanced by reducing the characteristic dimensions, and they may have an extended range of accessibility.

3. It offers a universal fabrication technology, the apotheosis of which is the personal nanofactory.

The burgeoning worldwide activity in nanotechnology cannot be explained purely as a rational attempt to exploit "room at the bottom", however. Two other important human motivations are doubtless also playing a rôle. One is simply "it has not been done before" – the motivation of the mountaineer ascending a peak previously untrodden. The other is the perennial desire to "conquer nature". Opportunities for doing so at the familiar macroscopic scale have become very limited, partly because so much has already been done – in Europe, for example, there are hardly any marshes left to drain or rivers left to dam, historically two of the most typical arenas for "conquering nature" – and partly because the deleterious effects of such "conquest" are now far more widely recognized, and the few remaining undrained marshes and undammed rivers are likely nowadays to be legally protected nature reserves. But the world at the bottom, as Feynman picturesquely called it, is uncontrolled and largely unexplored. On a more prosaic note, nanotechnology may already offer immediate benefit for existing products through substitution or incremental improvement (Fig. 1.3). The space industry has a constant and heavily pressing requirement for making payloads as small and lightweight as possible. Nanotechnology is ideally suited to this end user – provided the nanomaterials, devices and systems can be made sufficiently reliable (see Chapter 10).

1.8.1 NOVEL COMBINATIONS OF PROPERTIES

There are two ways of creating such nanomaterials. One is by adding nano-objects (nanoadditives) to a matrix. For example, organic polymer matrices incorporating carbon nanotubes can be light and very strong, or transparent and electrically conducting. The other is by fabricating materials *de novo*, atom-by-atom.

Since it is usually more expensive to create nanosized rather than microsized matter, one needs to justify the expense of downscaling the additives: As matter is divided ever more finely, certain properties become qualitatively different (see Chapter 2). As examples, the optical absorption spectrum of silicon vapor is quite different from that of a silicon crystal, even though the vapor and crystal are chemically identical; when a crystal becomes very small, the melting point falls and there may be a lattice contraction (i.e. the atoms move closer together) – these are well understood consequences of Laplace's law and may be useful for facilitating a sintering process. If the radius of the crystal is smaller than the Bohr radius of the electron in the bulk solid, the electron is confined and has a higher energy than its bulk counterpart. The optical absorption and fluorescent emission spectra shift to higher energies. Hence, by varying the crystal radius, the optical absorption and emission wavelengths can be tuned. Chemists have long known that heterogeneous catalysts are more active if they are more finely divided. This is a simple consequence of the fact that the reaction takes place at the interface between the solid catalyst and the rest of the reaction medium. Hence, for a given mass the finer the division the greater the surface area. This is not in itself a qualitative change, although in an industrial

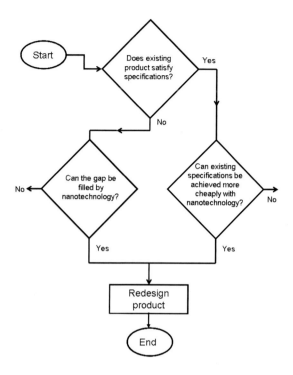

FIGURE 1.3

Flow chart to determine whether nanotechnology should be introduced into a product.

application there may be a qualitative transition from an uneconomic to an economic process.

1.8.2 DEVICE MINIATURIZATION: FUNCTIONAL ENHANCEMENT

The response time of a device usually decreases with decreasing size. Information carriers have less far to diffuse, or travel ballistically, and the resonant frequency of oscillators increases (Section 2.7).

Clearly the quantity of material constituting a device scales roughly as the cube of its linear dimension. Its energy consumption in operation may scale similarly. In the macroscopic world of mechanical engineering, however, if the material costs are disregarded, it is typically more expensive to make something very small; for example, a watch is more expensive than a clock, for equivalent timekeeping precision. On the other hand, when things become very large, as in the case of the Great Westminster Clock familiarly known by the name of its great bell Big Ben, costs again start to rise, because special machinery may be needed to assemble such outsized components, and so on. We shall return to the issue of fabrication in Chapter 8.

Performance (expressed in terms of straightforward input–output relations) may thus be enhanced by reducing the size, although the enhancement does not always continue *ad libitum*: For most microelectromechanical systems (MEMS) devices, such as accelerometers, performance is degraded by downscaling below the microscale (Section 10.8), and the actual size of the devices currently mass-produced for actuating automotive airbags already represents a compromise between economy of material, neither taking up too much space nor weighing two much, and still-acceptable performance. On the other hand, processor speed of VLSI chips is increased through miniaturization, since components are closer together and electrons have to traverse shorter distances.

Miniaturization may enable new domains of application by enhancing device accessibility. In other words, functionality may be enhanced by reducing the size. A typical example is the cellular (mobile) phone. The concept was developed in the 1950s at Bell Labs, but the necessary circuitry would have occupied a large multistorey building using the then current thermionic valve technology and, hence, only became practical with large-scale circuit integration. Similarly, it would not be practicable to equip mass-produced automobiles with macroscopic accelerometers with a volume of about one liter and weighing several kilograms.

Functional enhancement also applies to materials. The breaking strain of a monolithic material typically increases with diminishing size (see, e.g. Section 2.7). This is usually because, for a given probability of occurrence of defects per unit volume, a smaller volume will inevitably contain fewer defects. This advantage may be countered if the surface itself is a source of defects because of the increased preponderance of surface (Section 2.2).

Fabrication procedures may also be enhanced by miniaturization: any moving parts involved in assembly will be able to operate at much higher frequencies than their macroscopic counterparts. New difficulties are, however, created: noise and surfaces. The random thermal motion of atoms plays a much more deleteriously influential rôle than at the macroscale, where it can normally be neglected. The concept of the eutactic environment was introduced partly to cope with this problem. Bottom–up self-assembly, of course, requires noise to drive it; it is amplified up to constructive macroscopic expression by virtue of special features of the components being assembled. The preponderance of surfaces at the nanoscale, which can act as attractors for contamination and catalysts for defect formation, is probably best countered by appropriate energetic engineering of the surfaces (Section 3.2).

1.8.3 A UNIVERSAL FABRICATION TECHNOLOGY

Such a technology is usually considered to be based on nanoscale assemblers (i.e. personal nanofactories). They would enable most artifacts required by humans to be made out of a simple feedstock such as acetylene together with a source of energy (see

Section 8.3) [88]. Note that there is an intermediate level of technological achievement in which objects are constructed with atomic precision, but without the need for construction equipment to be itself in the nanoscale (e.g. using current tip-based ultramicroscope technology and its derivatives). Nanofabrication (nanomanufacture or nanofacture) represents of the ultimate in modularity. Using nanoblocks, purity, the bane of specialty chemical manufacturing, especially pharmaceuticals, is no longer as issue – extraneous molecules are simply ignored by the fabrication system.

1.8.4 ULTIMATE CONTROL OVER NATURE

Those who hope, like Wood et al. [305], that nanotechnology will, finally, allow technology to start exerting a net benign influence on our earthly environment, forget that since man's fundamental nature is that of a predator, he ever seeks to dominate nature [273]. Nanotechnology is therefore attractive because it gives man many new possibilities for increasing his power over nature. This fundamental character of man is very deeply rooted, and it is correspondingly difficult to deflect its desires. For example, it explains why ostensibly very sensible ways of dealing with agricultural pests (which Sir Albert Howard described as "nature's censors", not to be fought but rather to be seen as teachers), such as van den Bosch's biological control, remain unpopular. It is in man's nature to prefer the highly artificial but indeed ingenious method of genetically modifying his crops to allow them to resist glyphosate, and then drenching his fields with the herbicide to kill everything except the crop, instead of laboriously pulling up weeds or changing his crop to a more naturally resistant kind. This artificial method demonstrates his mastery of biotechnology (accomplishing the genetic modification) and chemistry (creating a cheap and effective herbicide). Pride of accomplishment dominates prudence in deploying the technology.

So it is, and doubtless will continue, with nanotechnology: Control of matter in the nanoscale is the ultimate meaningful control and, although there is still much to do to reach this goal, it doubtless represents a kind of apotheosis for mankind. Let us hope that it will be matched by appropriate intelligence and the holding of civilized values when it comes to be widely deployed. Paul the Apostle's dictum of "All things are possible, but not all things are expedient" appears to be very apt.

SUMMARY

Nanotechnology is defined in various ways; a selection of already-published definitions is given, from which it may be perceived that a reasonable consensus even now exists: Nanotechnology is concerned with materials and devices and with processes structured or taking place in the nanoscale. A more formal concept system is developed, in which care is taken to use the terms consistently. Nanotechnology is also defined ostensively (i.e. what objects already in existence are called "nano"?), and by its history. The rôle of biology is introduced as providing a living proof-of-principle for the possibility of nanotechnology; this has been of historical importance and continues to provide inspiration. Motivations for nanotechnology are summarized.

FURTHER READING

1. A.N. Broers, Limits of thin-film microfabrication. Proc. R. Soc. Lond. A 416 (1988) 1–42.
2. K.E. Drexler, Engines of Creation. Anchor Books/Doubleday, New York, 1986.
3. K.E. Drexler, Nanosystems: Molecular Machinery, Manufacturing, and Computation, Wiley-Interscience, 1992.
4. E. Kellenberger, Assembly in biological systems. in: Polymerization in Biological Systems, in: CIBA Foundation Symposium 7 (new series), Elsevier, Amsterdam, 1972.
5. K. Maruyama, F. Nori and V. Vedral, The physics of Maxwell's demon and information, Rev. Modern Phys. 81 (2009) 1–23.
6. W. Ostwald, Die Welt der vernachlässigten Dimensionen, Steinkopff, Dresden, 1914.
7. D.A. Tomalia, In quest of a systematic framework for unifying and defining nanoscience, J. Nanoparticle Res. 11 (2009) 1251–1310.
8. R. Zsigmondy, P.A. Thiessen, Das kolloide Gold, Akademische Verlagsgesellschaft, Leipzig, 1925.

The nanoscale

CHAPTER CONTENTS

INTRODUCTION

Before the advent of nanotechnology, and before the Système International (S.I.) was instituted, the bulk realm was typically referred to as macro and the atomic realm as micro. Scientific consideration of the microworld really began with the invention of the microscope, very appropriately named as it turned out, in view of the subsequent formalization of "micro" as the name of the multiplier 10^{-6}. The Abbe limit of resolution of the optical microscope (Eq. (5.2)) is indeed of the order of one micrometer (the wavelength of visible light). Now that electron microscopes, and even more recently scanning probe microscopes, can resolve features a few nanometers in size, the name microscope has become anachronistic – it would be better to call these instruments ultramicroscopes or nanoscopes. The range intermediate between macro and micro was typically referred to as meso ("mesoscale" simply means "of an intermediate scale"). This had no precise definition (except in meteorology), but typically referred to a realm which, while being indubitably invisible (to the naked eye) and hence "micro", could be adequately modeled without explicitly considering what were by then believed to be the ultimate constituents of matter – electrons, protons, neutrons and so forth – but by taking into account an appropriately coarse-grained discreteness. For example, according to this approach, a protein might be modeled by considering

Nanotechnology: An Introduction. DOI: 10.1016/B978-0-323-39311-9.00008-X
Copyright © 2016 Elsevier Inc. All rights reserved.

each amino acid to be a blob with a characteristic chemical functionality according to its unique side chain ("residue"). One might go further and model an amino acid in a protein as a structureless blob with two integers characterizing, respectively, hydrogen bond (HB)-donating and HB-accepting capacities; nucleotides ("bases") in DNA are commonly modeled as one of A, C, G and T, with the propensities to preferentially pair with, respectively, T, G, C and A. Given that this coarse graining neglects structure below the size of a few nanometers, it might be tempting to simply identify the nanoscale with the mesoscale. Working at the nanoscale would then appear as mainly a matter of computational convenience, since coarse-grained models are much more tractable than more detailed ones (consider Brownian dynamics *versus* molecular dynamics) and often require fewer assumptions. It may well be that a structure characteristic of this scale, such as a lipid bilayer, could be correctly determined from the quantum electrodynamical Hamiltonian, explicitly taking each electron and atomic nucleus into account, but this would be an extraordinarily tedious calculation (and one would need to mobilize the entire computing resources of the world even to reach an approximate solution for any practically useful problem). However, this usage of the term would not imply that there is any particular novelty in the nanoscale, whereas as one makes the transition from typically bulk properties to typically atomic and subatomic properties, one passes through a state that is characteristic neither of individual atoms nor of the bulk, and it would then appear to be reasonable to take the length range over which this transition occurs as the nanoscale.

The current *consensual* definition of the nanoscale is from 1 to 100 nm, although it is still disputed whether these limits should be considered exact or approximate. Finding a consensus is analogous to a multiobjective optimization procedure in which it is not possible to perfectly satisfy every objective considered independently because some of them conflict. A compromise is therefore sought that is as close as possible to all objectives. In many cases, there is a continuum of possible states of each attribute (each attribute corresponding to an objective) and the Pareto front on which the compromise solutions lie is indeed a meaningful solution to the optimization problem. On the other hand, if it is considered that there is only one correct solution for a given objective, anything else being wrong, then the consensus might be Pareto optimal but wrong in every regard – perhaps a little like not a single person in the population having the attributes of Quetelet's "average man". Nevertheless, such a consensual definition might still be useful – at least it has the virtue of simplicity and probably does facilitate discussion of nanotechnology, albeit at a rather basic level, among a diverse range of people.

This chapter goes beyond the current consensus and examines the definition of the nanoscale more deeply. According to one of the definitions of nanotechnology discussed in Chapter 1, "nanotechnology pertains to the processing of materials in which structure of a dimension of less than 100 nm is essential to obtain the required functional performance" [52]. This functional performance is implicitly unattainable at larger sizes; hence, the definition essentially paraphrases "the deliberate and controlled manipulation, precision placement, measurement, modeling and production

Table 2.1 Covalent Radii of Some Atoms

Element	Symbol	Atomic Number	Radius/nm [169]
Carbon	C	6	0.077
Chlorine	Cl	17	0.099
Gold	Au	79	0.150
Hydrogen	H	1	0.031
Silicon	Si	14	0.117
Sodium	Na	11	0.154
Zinc	Zn	30	0.131

of matter in the nanoscale in order to create materials, devices and systems with fundamentally new properties and functions" [1]. The emergence of qualitatively new features as one reduces the size is a key concept associated with nanotechnology. In order to define the nanoscale, we therefore need to search for such qualitative changes.

Given that, in principle, the prefix "nano" can apply to any base unit, not just the meter, one needs to justify the primacy of length in a definition of the nanoscale. Among the different categories of attributes – length, mass, voltage and so on – length does enjoy a certain primacy, perhaps because man first measured things by pacing out their spacial dimension. And, given that the meter roughly corresponds to one human step, we can *prima facie* accept that the nanoscale is based on the nanometer. As nanotechnologists seemingly never tire of relating, the prefix "nano" is derived from the Greek "$\nu\alpha\nuο\sigma$", meaning dwarf, and the meaning of "very small" has been formalized by the S.I. of scientific units adopting "nano" as the name of the multiplier 10^{-9}. Thus, one nanometer (1 nm) means precisely 10^{-9} m, in other words a millionth of a millimeter or a thousandth of a micrometer, which is slightly larger than the size of one atom (Table 2.1).

2.1 THE SIZE OF ATOMS

An atom is important because it designates the ultimate (from a terrestrial viewpoint) particles from which matter is constituted. It was therefore very natural for Richard Feynman to suggest atom-by-atom construction of objects as the lower limit of miniaturization of engineering [85]. It would be highly impracticable to have to start with subatomic particles, such as protons, electrons, neutrons and so forth as building blocks, whereas atomically precise construction, as Feynman rightly perceived and emphasized, is an engineering problem (hence solvable in principle), not a scientific one (requiring the discovery of hitherto unknown laws). Table 2.1 gives the sizes of some atoms.

The scale of the individual atom might be considered as sufficient for a definition of the nanoscale, especially if nanotechnology were to be defined solely in the context of atom-by-atom assembly of objects. But nanotechnology already seems to be much more than that. The definitions of nanotechnology (Chapter 1) emphasize that novel, unique properties emerge at the nanoscale. This implies that merely assembling an otherwise known macro-object atom-by-atom would scarcely warrant the name of nanotechnology.

2.2 MOLECULES AND SURFACES

The dictionary definition of a molecule is typically "the smallest part of a substance that retains all the properties of the substance without losing its chemical identity and is composed of one or more atoms". This combines its etymological meaning as the diminutive of the Latin *moles*, mass (which on its own would make the word essentially synonymous with "particle") with Tyndall's definition as "a group of atoms drawn and held together by what chemists term affinity". This definition is readily applicable to typical covalent molecules such as most organic compounds; a molecule of the carbohydrate called glucose is precisely the particle composed of 6 carbon atoms, 6 oxygen atoms and 12 hydrogen atoms connected in such a way as to make what we know as glucose, and even a single such particle would taste sweet in the characteristic way of glucose, but none of these atoms could be removed without destroying the "glucoseness". Particles of other kinds of substances do not fit the definition so well. A single atom of a metal such as gold, although chemically gold, has a different optical absorption spectrum from that of bulk gold, and the same applies to numerous binary semiconducting compounds such as cadmium sulfide, CdS, which can be prepared as a vapor containing isolated CdS molecules (in the chemical sense). In bulk material, the individual atoms are close enough for the wave functions of their electrons to overlap, but to satisfy Pauli's exclusion principle, their energies must be slightly shifted, forming a band of states instead of a discrete level as in the isolated atom or molecule; see also Section 2.5.

The surface of a particle is qualitatively different from the bulk because it is less connected; the smaller the radius, the greater the proportion of underconnected atoms. Consider a "supersphere", a spherical aggregate of spheres, which we can take to be atoms (Fig. 2.1). By simple geometric considerations, only one of the 19 atoms is not in contact with the surface. If the radii of the atom and the supersphere are r and R, respectively, then the proportion of atoms in the shell must be $1 - [(R - r)/R]^3$. The mean connectivity, and hence cohesive energy, should vary inversely with a fractional power of this quantity. If R is expressed in units of r, then the surface to volume ratio is equal to $3r/R$: in other words, if $R = 3r$, equal numbers of atoms are in the bulk and at the surface. The nanoparticle is, therefore, one that is "all surface". As a result, the melting point T_m of small spheres is lowered relative to bulk material, according to

$$T_m/T_\infty = 1 - (C/R)^n, \tag{2.1}$$

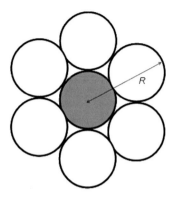

FIGURE 2.1

Cross-section of a spherical nanoparticle consisting of 19 atoms.

where C is a constant, and with the exponent $n = 1$ for metals, as shown by experimental studies. In fact, melting is a rather complex phenomenon and if an approach such as that of Lindemann's criterion is adopted, due account of the difference between the surface and the bulk atoms might be important in determining the temperature of some practical process, such as sintering. Furthermore, it should be noted that if a nanoscale thin film is investigated (i.e. a nanoplate), rather than particles there is no depression of the melting point relative to the bulk [153].

As Wulff has pointed out, for a crystal growing in equilibrium with its vapor or solution, the ratio of the surface tension γ of the growing phase to the distance from the center r should be constant. If the mechanical effect of the surface tension can be reduced to an isotropic pressure, then we have the Laplace law

$$\Delta P = 2\gamma/r, \tag{2.2}$$

where ΔP is the pressure exerted on the crystal compared with zero pressure if the growing phase is an infinite plane. This pressure can do work on the lattice, compressing the crystal. If β is the isothermal compressibility coefficient, then

$$\Delta V/V = -\beta\Delta P = 3\Delta l/l, \tag{2.3}$$

where V is the volume of the unperturbed crystal, and l a characteristic lattice parameter of the unperturbed (bulk) crystal. Conversely, the vapor pressure P of a particle should be inversely dependent on its radius R, as expressed by the Gibbs–Thomson (or Kelvin) law

$$P(R)/P_\infty = e^{2\gamma v/(Rk_\mathrm{B}T)}, \tag{2.4}$$

where P_∞ is the vapor pressure of a surface of infinite radius (i.e. flat); γ and v are, respectively, the surface tension (note that once surface tension is introduced, it should also be borne in mind that surface tension is itself curvature-dependent [268]) and molecular volume of the material; and k_B and T are, respectively, Boltzmann's constant and the absolute temperature. Eq. (2.4) can be used to compare the chemical potentials μ of two spheres of radii R_1 and R_2,

$$\mu_2 - \mu_1 = 2\gamma V \left(\frac{1}{R_2} - \frac{1}{R_1} \right), \tag{2.5}$$

where V is now the molar volume.

These equations do not predict an abrupt discontinuity at the nanoscale. To be sure, there are huge differences between the bulk and the nanoscale (e.g. melting temperature lowered by tens of degrees), but these "unique" properties approach bulk values asymptotically, and the detectable difference therefore depends on the sensitivity of the measuring apparatus, hence is not useful for defining a nanoscale. On the other hand, sometimes there is evidence for qualitative change with diminishing size, as shown, for example, by the appearance of anomalous crystal structures with no bulk equivalent [248].

2.3 NUCLEATION

When a vapor is cooled it will ultimately condense, but if it is very pure it may be greatly supercooled because the appearance of new phase – in this case, a liquid or a solid – is necessarily a discontinuous process and therefore requires a finite nucleus of the new phase to be formed via random fluctuations. At first sight, this seems to be highly improbable, because the formation of the nucleus requires the energetically unfavorable creation of an interface between the vapor and the new condensed phase – indeed, were this not energetically unfavorable the vapor could never be stable. However, if the nucleus is big enough, the energetically favorable contribution of its nonsurface exceeds of the surface contribution, and not only is the nucleus stable, but also grows further, in principle to an unlimited extent (see Fig. 2.2). More precisely, when atoms cluster together to form the new phase, they begin to create an interface between themselves and their surrounding medium, which costs energy $A\gamma$, where A is the area of the cluster's surface, equal to $(4\pi)^{1/3}(3nv)^{2/3}$, where n is the number of atoms in the cluster. At the same time, each atom contributes to the (negative) cohesive energy of the new phase equal to $nv\Delta G$, where ΔG is the difference in the free energies (per unit volume) between the uncondensed and condensed phases, the latter obviously having the lower energy. Summing these two contributions, at first the energy will increase with increasing n, but ultimately the (negative) cohesive energy of the bulk will win. Differentiating with respect to n and looking for maxima yield

$$n^* = -\frac{32\pi\gamma^3}{3v(\Delta G)^3}. \tag{2.6}$$

The critical nucleus size can lay reasonable claim to be considered as the boundary of the nanoscale from the mechano-chemical viewpoint. At least it represents a

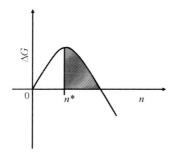

FIGURE 2.2

Sketch of the variation of free energy of a cluster containing n atoms (proportional to the cube of the radius). The maximum corresponds to the critical nucleus size n^*. Clusters that have managed through fluctuations to climb up the free energy slope to reach the critical nucleus size have an equal probability to shrink back and vanish or to grow up to microscopic size. The shaded zone corresponds to a possible nanoscale (see text).

crossover point *qua* discontinuity. The critical nucleus size represents the boundary between "surface dominated" (an unstable state') and "bulk dominated" (a stable state).

The growth of a nucleus depends on the continuing supply of material (i.e. increasing n). If there is only a finite supply of material, sufficient nuclei could be created (by condensation or precipitation) to sink all the uncondensed matter. If the resulting particles were barely larger than the critical nucleus, fluctuations would inevitably cause some of the slightly smaller particles to dissolve and their material would be used by the initially slightly larger ones to grow. However, if slightly more material were provided, possibly added just after the initial nucleation, the particles would be able to grow to a size intermediate between n^* and the point where the free energy becomes negative (the shaded zone in Fig. 2.2), where, although not yet bulk-like, the collection of particles might be almost indefinitely stable.

A corollary of this definition is that nanotechnology is strongly associated with surface technology, since the nanoscale object is necessarily one in which all or almost all of its constituent atoms have the properties of the surface atoms of a bulk piece of what is atomically the same material.

Note that the nucleus, small as it is, has been characterized above using a thermodynamic continuum approach that is perfectly macroscopic in spirit; in other words, it is not apparently necessary to explicitly take atomistic behavior into account to provide a good description of the phenomenon. The excellent agreement between the projections of equations based on (2.7) and experimental results with photographic latent images constituted from only a few silver atoms caused mild surprise when first noticed by Peter Hillson working at the Kodak research laboratories in Harrow, UK. Molecular dynamics simulations also suggest that the continuum approach seems to be valid down to scales of a few atomic radii.

2.4 CHEMICAL REACTIVITY

Consider a heterogeneous reaction A + B → C, where A is a gas or a substance dissolved in a liquid and B is a solid. Only the surface atoms are able to come into contact with the environment; hence, for a given mass of material B, the more finely it is divided the more reactive it will be, in terms of numbers of C produced per unit time.

The above considerations do not imply any discontinuous change upon reaching the nanoscale. Granted, however, that matter is made up of atoms, the atoms situated at the boundary of an object are qualitatively different from those in the bulk (Fig. 2.3). A cluster of seven atoms (in two-dimensional Flatland) has only one bulk atom (see Fig. 2.1), and any smaller cluster is "all surface". This may have a direct impact on chemical reactivity: it is to be expected that the surface atoms are individually more reactive than their bulk neighbors, since they have some free valences (i.e. bonding possibilities). Consideration of chemical reactivity (its enhancement for a given mass, by dividing matter into nanoscale-sized pieces) therefore suggests a discontinuous change when matter becomes "all surface", albeit any observable effect may be very small.

Another implication concerns solubility: The vapor pressure P of a droplet, and by extension the solubility of a nanoparticle, increases with diminishing radius r according to the Kelvin equation (cf. Eq. (2.4))

$$k_B T \ln(P/P_\infty) = 2\gamma v/r. \qquad (2.7)$$

In practice, however, the surface atoms may have already satisfied their bonding requirements by picking up reaction partners from the environment. For example, many metals become spontaneously coated with a film of their oxide when left standing in air, and as a result are chemically more inert than pure material. These films are typically thicker than one atomic layer, implying complex growth processes involving internal transport or diffusion. For example, on silicon the native oxide layer is about 4 nm thick. This implies that a piece of freshly cleaved silicon undergoes some lattice disruption enabling oxygen atoms to effectively penetrate deeper than the topmost layer. Thus, it may happen that if the object is placed in the "wrong" environment, the surface compound may be so stable that the nanoparticles coated with it are actually less reactive than the same mass of bulk matter. A one centimeter cube of sodium taken from its protective fluid (naphtha) and thrown into a

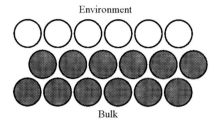

Environment

Bulk

FIGURE 2.3

The boundary of an object shown as a cross-section in two dimensions. The surface atoms (white) are qualitatively different from the bulk atoms (gray), since the latter have six nearest neighbors (in two-dimensional cross-section) of their own kind, whereas the former only have four.

pool of water will act in a lively fashion for some time, but if the sodium were first cut up into micrometer cubes, most of the metallic sodium would have already reacted with moist air before reaching the water.

Consider the prototypical homogeneous reaction

$$A + B \rightarrow C \tag{2.8}$$

and suppose that the reaction rate coefficient k_f is much less than the diffusion-limited rate, that is, $k_f \ll 4\pi(d_A + d_B)(D_A + D_B)$, where d and D are the molecular radii and diffusivities, respectively. Then [259],

$$\frac{dc}{dt} = k_f[\langle a \rangle \langle b \rangle + \Delta^2(\gamma_t)] = k_f \langle ab \rangle, \tag{2.9}$$

where a and b are the numbers (concentrations) of A and B, the angular brackets denote expected numbers and γ_t is the number of C molecules created up to time t. The term $\Delta^2(\gamma_t)$ expresses the fluctuations in γ_t: $\langle \gamma_t^2 \rangle = \langle \gamma_t \rangle^2 + \Delta^2(\gamma_t)$: supposing that γ_t approximates to a Poisson distribution, then $\Delta^2(\gamma_t)$ will be of the same order of magnitude as $\langle \gamma_t \rangle$. The kinetic mass action law (KMAL) putting $\langle a \rangle = a_0 - c(t)$, etc., the subscript 0 denoting initial concentration at $t = 0$, is a first approximation in which $\Delta^2(\gamma_t)$ is supposed negligibly small compared to $\langle a \rangle$ and $\langle b \rangle$, implying that $\langle a \rangle \langle b \rangle = \langle ab \rangle$, whereas strictly speaking it is not since a and b are not independent. Nevertheless, the neglect of $\Delta^2(\gamma_t)$ is justified for molar quantities of starting reagents (except near the end of the process, when $\langle a \rangle$ and $\langle b \rangle$ become very small), but not for reactions in ultrasmall volumes (nanomixers).

These number fluctuations, i.e. the $\Delta^2(\gamma_t)$ term, constantly tend to be eliminated by diffusion. On the other hand, because of the correlation between a and b, initial inhomogeneities in their spacial densities may (depending on the relative rates of diffusion and reaction) lead to the development of zones enriched in either one or the other faster than the enrichment can be eliminated by diffusion. Hence, instead of A disappearing as t^{-1} (when $a_0 = b_0$), it is consumed as $t^{-3/4}$, and in the case of a reversible reaction, equilibrium is approached as $t^{-3/2}$. Deviations from perfect mixing are more pronounced in dimensions lower than three.

As the reaction volume decreases, the relative importance of fluctuations in the concentrations of the various chemical species increases, but this does not in itself constitute a qualitative change. In cases more complicated than the single reaction A + B → C, however, in which additional parallel or sequential reactions are possible, the outcome may actually change. For example, consider what happens if an additional reaction A + C → D is possible. In the usual bulk situation, A-rich regions will rapidly become surrounded by C-rich regions, which in turn will further react with A to form D. If, however, C gets diluted in the ocean of B before it can undergo further reaction with A, D will be formed in very poor yield (see Fig. 2.4). Ways of achieving such rapid dilution could be to somehow divide the reaction space into very small volumes, or to impose a drastically violent mixing régime on the bulk, although friction with the walls of the containing vessel will prevent turbulence from filling the entire volume.

Since nanotechnology is above all the technology of surfaces, it is natural to favor reactions taking place at surfaces, not the heterogeneous reactions in which one agent in bulk three-dimensional space impinges on and reacts with another at a surface (as in electrochemistry), but the situation in which both reagents are confined to a two-dimensional surface. Merely as a consequence of having fewer directional possibilities to move (cf. Pólya's theorem), a reaction of type (2.8) will be vastly accelerated (see [251] for an experimental demonstration).

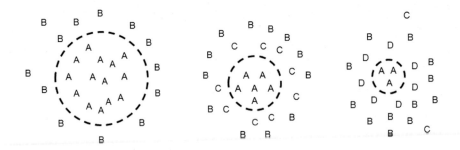

FIGURE 2.4

Illustration (cross-section) of a possible result of an initial mixture of chemical species A and B (on the left) when reactions A + B → C (in the middle) and, additionally, A + C → D (on the right) are possible. Further explanation is given in the text.

2.5 ELECTRONIC AND OPTICAL PROPERTIES

Electronic energy levels. Individual atoms have discrete energy levels and their absorption spectra correspondingly feature sharp individual lines. It is a well-known feature of condensed matter that these discrete levels merge into bands, and the possible emergence of a forbidden zone (band gap) determines whether we have a metal or a dielectric.

Stacking objects with nanoscale sizes in one, two or three dimensions (yielding nanoplates, nanofibers and nanoparticles with, respectively, confinement of carriers in two, one or zero dimensions) constitutes a new class of superlattices or artificial atoms. These are exploited in a variety of nanodevices (Chapter 7). The superlattice gives rise to sub-bands with energies

$$E_n(k) = E_n^{(0)} + \hbar^2 k^2/(2m^*),\qquad(2.10)$$

where $E_n^{(0)}$ is the nth energy level, k the wavenumber and m^* the effective mass of the electron, which depends on the band structure of the material.

Similar phenomena occur in optics, but since the characteristic size of photonic band crystals is in the micrometer range, they are, strictly speaking, beyond the scope of nanotechnology.

Quantum confinement. The general principle is that confinement occurs if a characteristic size (e.g. the thickness of a plate) is less than or equal to the electron coherence length. The shifting optical properties of very small dielectric particles (shift of the fundamental adsorption edge, disappearance of exciton absorption peaks) were first investigated by Berry [20, 21]. Later, Efros and Efros [70], Brus [33] and Banyai and Koch [17] provided explanations in terms of the confinement of charge carriers (electrons, defect electrons or "positive holes" and excitons). In this case, there is a clear criterion for confinement effects to be observable: the actual radius of the particle (assumed spherical) must be less than the charge carrier's Bohr radius r_B [i.e. the smallest possible orbit for the (ground state) electron in hydrogen]. Its magnitude is given by the expression

$$r_B = \frac{4\pi\epsilon_0\hbar^2}{m_e e^2},\qquad(2.11)$$

where $m_e = 0.91 \times 10^{-30}$ kg is the mass of the electron. The Bohr radius of the free electron in vacuum is therefore 53 pm. In a semiconductor, this formula can still be used, but ϵ_0 must be replaced by the actual dielectric constant of the material ϵ_s, and m_e by the effective mass, which depends on the band structure (defined as $m_{\text{eff}} = \hbar^2/[d^2\epsilon/dk^2]$), yielding

$$r_B^* = \frac{\pi\epsilon h^2}{m^* e^2}.\qquad(2.12)$$

This is the natural measure of the nanoscale for electronic properties, especially those involving adsorption and emission of light involving electronic transitions between

Table 2.2 Bohr Radii and Related Parameters of Some Semiconductors [276]

Material	Bandgap/eV[a]	ϵ_s/ϵ_0	$m_{eff}^{electron}/m_e$	m_{eff}^{hole}/m_e	r_B/nm[b]
CdS	2.42	5.4	0.21	0.8	7.5
InSb	0.17	17.7	0.0145	0.40	60
InP	1.35	12.4	0.077	0.64	11
Si	1.12	11.9	0.98	0.16	2.5
ZnO	3.35	9.0	0.27	–	1.7

[a] At 300 K.
[b] Of the electron; see Eq. (2.12); see also [311].

the valence band and the conduction band. Typical values of r_B^* range from a few to a few hundred nanometers (Table 2.2). Therefore, it is practically possible to create particles whose radius r is smaller than the Bohr radius. In this case, the energy levels of the electrons increase (a similar argument applies to defect electrons, i.e. positive holes), and the greater the degree of confinement, the greater the increase. Hence, the band edge of optical adsorption (and band-edge luminescent emission) shifts towards the blue end of the spectrum with decreasing r for $r < r_B$. This is sometimes called a quantum size effect in the scientific literature, and nanoparticles with this property are called quantum dots.

Since r_B varies greatly for different materials (and in some cases the electron and the hole have very different radii), it follows that the nanoscale (considering electronic properties) is material-dependent (see Table 2.2). As pointed out in Section 1.1, until now the nanoscale is often, albeit provisionally, taken as the range between 1 and 100 nm, according to which a nanoparticle would be any particle with a diameter falling in this range. However, a 50 nm CdTe particle would have the optical properties of bulk CdTe, and would not therefore rank as a quantum dot, nor *a fortiori* as a nanoparticle with respect to optical properties since they would not be novel, whereas InSb of the same size would be novel.

The occurrence of impurities is of course size-dependent. Consider a volume V of semiconductor bulk-doped with an impurity at an average number N per unit volume. We suppose that the occurrence of an impurity in a particular place does not depend on the presence of other impurities. The probability p_k that the volume contains k impurity atoms is given by the Poisson distribution as

$$p_k = (VN)^k e^{-VN}/k! \qquad (2.13)$$

and thus the probability that a certain volume has no impurities at all (i.e. $k = 0$) is given by $\exp(-VN)$. There is no discontinuous change as V diminishes, but one could (for example) take the nanoscale as being that length resulting in the volume for which $Vd = 1$. Hence, it will depend on the doping level d.

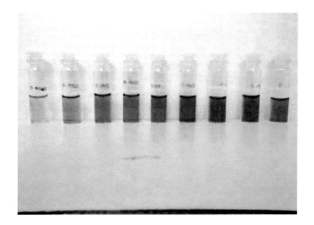

FIGURE 2.5

Gold nanoparticles dispersed in water. The particle size increases from left to right.

-Reproduced with permission from A. Maitra, Nanotechnology and nanobiology – are they children of the same father? Nanotechol. Percept. 6 (2010) 197–204.

Hence, whether an electronic device should be called a nanodevice depends on the type of its construction. If the size of the device falls below the electron mean free path, a novel phenomenon appears (ballistic transport) and, hence, it is appropriately called a nanodevice. If impurities are necessary for the device to function (as in junction devices), then a nanoscaled version of it cannot exist, because performance will not manifest any meaningful discontinuity if there are no impurities at all.

A suspension of metal nanoparticles, with a diameter smaller than the wavelength of light, is typically intensely colored. The optical absorption arises from the surface plasmon, which is excited by the external electromagnetic (optical) field. As the diameter decreases, the surface plasmon resonance (SPR) shifts to the red and its bandwidth greatly increases (Fig. 2.5).

An even simpler criterion for the optical nanoscale could be simply "invisibility", as determined by the Abbe resolution limit (Eq. (5.2)), or the Rayleigh criterion $\alpha \ll 1$, where

$$\alpha = 2\pi r/\lambda, \tag{2.14}$$

where r is the nanoparticle radius and λ the wavelength of the light. Within the Rayleigh régime, the scattering cross-section for nonabsorbing particles of refractive index n is

$$\sigma = \frac{2\lambda^2}{3\pi}\alpha^6 \left(\frac{n^2 - 1}{n^2 + 2}\right)^2 \tag{2.15}$$

and if the incident light has intensity I_0, at a distance d_s from the scattering particle the scattered intensity is, for unpolarized light,

$$I = I_0 \sigma / d_s^2 \qquad (2.16)$$

with a negligible dependence of the angle of observation. A suspension of nanoparticles should not appear turbid.

In order to establish quantitatively how the band-edge luminescent emission varies with particle diameter, consider a one-electron, one-dimensional time-dependent Schrödinger equation with energy eigenvalues E:

$$\left[-\frac{\hbar^2}{2m_e} \frac{d^2}{dx^2} + V(x) \right] \Psi = E\Psi \qquad (2.17)$$

with a potential $V(x) = 0$ for $0 \le x \le 2r$ (within the particle) and $V(x) = \infty$ for $x < 0$ and $x > 2r$. Using a trial wavefunction $\Psi(x) = \sin ax$ for the electron, within the particle we can solve to find $a^2 = 2m_e E/\hbar^2$. Ψ has to vanish at the particle boundaries, which can happen if the phase ax is an integer multiple of π, $ax = n\pi$, whence

$$E_n = \frac{\hbar^2 n^2 \pi^2}{8m_e r^2}, \quad n = 0,1,\dots \qquad (2.18)$$

We can call n the principle quantum number. Note that we have not specified exactly what value of m_e we should use in equation (2.17). It is an open question whether the effective mass of the electron is the same in a very small particle, already known to be subject to lattice contraction and possibly other distortions of the crystal structure (Section 2.2). The experimental verification of these formulas is subjected to a number of difficulties. First, it is quite difficult to prepare monodisperse quantum dots. The Stranski–Krastanov mechanism has been made use of for the preparation of monodisperse GaInAl dots for lasers via semiconductor processing technology (see Section 8.1.2), but most quantum dots are presently made by a chemical method, by reacting two soluble precursors together in solution (Section 6.1.3), with which it is difficult to get particle size distributions with a coefficient of variation much better than 10%. The surface atoms may have different electronic properties from the bulk by virtue of their lower coördination number. This notion was developed by Berry [20, 21] as the main reason for the difference between bulk and nanoparticle optical properties. Particles in contact with another condensed phase, such as a suspending liquid, are likely to have additional electron levels caused by adsorbed impurities; it is extremely unlikely that the particle boundaries correspond to the electron potential jumping abruptly from zero to infinity.

Jellium. Quantum dots are dielectric nanoparticles, typically I–VII, II–VI or III–V semiconductors. Metal clusters can be modeled as jellium [167]. Each metal atom contributes a characteristic number v of electrons to the free electron gas (a corollary of which is that the actual positions of the metal nuclei can be neglected) filling a uniform charged sphere corresponding to the actual cluster containing N metal atoms. The electronic structure is derived by solving the Schrödinger equation for an electron constrained to move within the cluster sphere under the influence of an attractive mean-field potential due to the partially ionized metal atoms. The solution yields energy levels organized in shells and subshells with quantum numbers n and l, respectively, as with atoms. The order of the energy levels in jellium is a little different, however: $1s^2$, $1p^6$, $1d^{10}$, $2s^2$, $1f^{14}$, $2p^6$, etc. – hence even the first valence shell contains s, p, d and f orbitals. The degeneracy of each level is $2(2l + 1)$.

A triumph of the jellium model was the correct prediction of "magic" numbers N_m – corresponding to clusters of exceptional stability. These are clusters with completely filled shells, and the N_m are the values of N fulfilling the condition

$$\sum_{}^{n,l} 2(2l + 1) = nv - q, \qquad (2.19)$$

where q is the charge of the cluster. Hence (for example), uncharged ($q = 0$) Cs ($v = 1$) clusters will have $N_m = 2, 8, 18, 20, 34, 40, 58, \ldots$; Al_{13}^- ($q = 3$) will be stable.

Superatoms. The electron affinity of Al_{13} is comparable to that of a chlorine atom and KAl_{13} is an ionically bound molecule analogous to KCl. Replacing one Al atom with C ($v = 4$) results in an inert species comparable to argon. An extended periodic table can be constructed from superatoms, defined as clusters that are energetically and chemically stable [50]. These entities should be useful as nanoblocks for assembly into useful structures (Section 8.3.2), although this possibility does not yet seem to have been exploited.

2.6 MAGNETIC AND FERROELECTRIC PROPERTIES

Ferromagnetism. In certain elements (e.g. Fe, Co, Ni, Gd) and alloys, exchange interactions between the electrons of neighboring ions lead to a very large coupling between their spins such that, above a certain temperature, the spins spontaneously align with each other. The proliferation of routes for synthesizing nanoparticles of ferromagnetic substances has led to the discovery that when the particles are below a certain size, typically a few tens of nanometers, the substance still has a large magnetic susceptibility in the presence of an external field but lacks the remanent magnetism characteristic of ferromagnetism. This phenomenon is known as superparamagnetism.

Below a certain critical size r_s, the intrinsic spontaneous magnetization of a ferromagnetic material such as iron results in a single domain with all spins in the same direction; above that size the material is divided into domains with different spin directions. Kittel has given a formula to estimate the critical single domain size (diameter) in zero magnetic field,

$$d_s = 1.43\gamma_m/J_s^2, \qquad (2.20)$$

where

$$\gamma_m = 4\sqrt{EK} \qquad (2.21)$$

is the domain wall surface energy density, J_s the spontaneous magnetization, E the exchange energy and K the anisotropy constant. Typical values of d_s are 30 nm for Co, 50 nm for Fe and 70 nm for iron oxides (hematite, Fe_2O_3 and magnetite, Fe_3O_4). These sizes could reasonably be taken to be the upper limit of the nanorealm when magnetism is being considered. Incidentally, the magnitude of the Bohr magneton, the "quantum" of magnetism, is given by $\beta = e\hbar/(2m_e)$.

Within a single domain of volume V, the relaxation time (i.e. the characteristic time taken by a system to return to equilibrium after a disturbance) is

$$\tau = \tau_0 \exp(KV/k_BT), \qquad (2.22)$$

where τ_0 is a constant ($\sim 10^{-11}$ s). Particles are called superparamagnetic if τ is shorter than can be measured; we then still have the high susceptibility of the material, but no remanence. In other words, ferromagnetism becomes zero below the Curie temperature. It is a direct consequence of the mean coördination number of each atom diminishing with diminishing size because there are proportionately more surface atoms, which have fewer neighbors. There is thus a lower limit to the size of the magnetic elements in nanostructured magnetic materials for data storage, typically about 20 nm, below which room temperature thermal energy overcomes the magnetostatic energy of the element, resulting in zero hysteresis and the consequent inability to store magnetization orientation information.

Since the relaxation time varies continuously with size (note that K may also be size-dependent), it is unsuitable for determining the nanoscale; on the other hand r_s could be used. It is known that the shape of a particle also affects its magnetization [229]. A similar set of phenomena occurs with ferroelectricity [41].

2.7 MECHANICAL PROPERTIES

The ultimate tensile strength σ_{ult} of a material can be estimated by using Hooke's law, *ut tensio, sic vis*, to calculate the stress required to separate two layers of atoms from one another, by equating the strain energy (the product of strain, stress and volume) with the energy of the new surfaces created, yielding [237]

$$\sigma_{ult} = (2\gamma Y/x)^{1/2}, \qquad (2.23)$$

where γ is the surface energy (Section 3.2), Y is Young's modulus (i.e. stiffness) and x is the interatomic spacing (i.e. twice the atomic radius, see Table 2.1). A.A. Griffith proposed that the huge discrepancies between this theoretical maximum and actual greatest attainable tensile strengths were due to the nucleation of cracks, which spread and caused failure. Hence, strength not only means stiffness but also must include the concept of toughness, which means resistance to crack propagation. By considering the elastic energy W_σ stored in a crack of length $2l$, namely

$$W_\sigma = \pi l^2 \sigma^2/Y, \qquad (2.24)$$

where σ is the stress, and its surface energy

$$W_\gamma = 4l\gamma \qquad (2.25)$$

(ignoring any local deformation of the material around the crack), a critical *Griffith length* can be defined when $\partial W_\sigma/\partial l$ equals $\partial W_\gamma/\partial \gamma$,

$$l_G = 2\gamma Y/(\pi\sigma^2); \qquad (2.26)$$

if $l < l_G$ the crack will tend to disappear, but if $l > l_G$ then it will rapidly grow. Note the existence of accompanying irreversible processes that accelerate the process, typically leading to microscopic failure, quite possibly with disastrous consequences. By putting $l_G = x$ into formula (2.26), one recovers the estimate of ultimate tensile strength, Eq. (2.23) (neglecting numerical factors of order unity).

If σ is taken to be a typical upper limit of the stress to which a structural component will be subjected in use, one finds values of l_G in the range of nanometers to tens of nanometers. Therefore, as far as mechanical properties of materials are concerned, l_G provides an excellent measure of the nanoscale. If the size of an object is less than l_G, it will have the theoretical strength of a perfect crystal. If it is greater, the maximum tensile strength will diminish $\sim l^{-1}$ (cf. Eq. (2.26)), a result that was demonstrated experimentally by Griffith using glass fibers and by Gordon and others using whiskers of other materials such as silicon and zinc oxide.

In reality, Eq. (2.26) may somewhat underestimate the critical length because of the neglect of material deformation, which causes the work of fracture to exceed that estimated from the surface tension alone. Griffith's criterion is based on Dupré's idea, which completely ignores this deformation. Hence, Bikerman [23] has proposed an alternative hypothesis that a crack propagates when the decrease of strain energy around a critical domain exceeds the increase required to elastically deform this domain to the breaking point.

One consequence of these ideas is the impossibility of comminuting particles below a certain critical size d_{crit} by crushing. Kendall [156] derives, in a rather simple way,

$$d_{\text{crit}} = 32YF/(3\sigma^2), \qquad (2.27)$$

where F is the fracture energy; below this size crack propagation becomes impossible; the material becomes ductile.

Cracks can be arrested by interfaces, and it has long been realized that nature has solved the problem of creating strong (i.e. both stiff and tough) structural materials such as bones, teeth and shells by assembling composite materials consisting of mineral platelets embedded in a protein matrix. The numerous interfaces prevent the propagation of cracks, and provided the stiff components of the composite are not bigger than the critical Griffith length, optimal strength should be obtainable. This has recently been subjected to more careful quantitative scrutiny [150] yielding an expression for the stiffness of the composite:

$$\frac{1}{Y} = \frac{4(1-\theta)}{G_p\theta^2 a^2} + \frac{1}{\theta Y_m}, \qquad (2.28)$$

where G_p and Y_m are, respectively, the shear modulus of the protein and Young's modulus of the mineral, θ is the volume fraction of mineral and a is the aspect ratio of the platelets. It seems that the thickness of the platelets in natural materials (such as various kinds of seashells) roughly corresponds to l_G. By assuming that the protein and the mineral fail at the same time, an expression for the optimal aspect ratio of the mineral platelets can be derived as $\sqrt{\pi Y_m \gamma / l / \tau_p}$, where τ is the shear strength of the protein matrix.

2.8 QUANTUM SMALLNESS

In the preceding sections of this chapter, the various crossover lengths that demarcate the nanorealm from other realms have emerged as upper limits of the nanorealm; above these lengths, the behavior is qualitatively indistinguishable from that of indefinitely large objects, and below them novel properties are observable. But does the nanorealm have a lower limit? This has been implicit in, for example, the discussion of the mesoscale at the beginning of the chapter. It implies that the point where continuum mechanics breaks down and atoms have to be explicitly simulated demarcates the lower limit of the nanoscale. On the other hand, discussion about the possible design of nanoscale assemblers can hardly avoid the explicit consideration of individual atoms. Despite such lively challenges, it is probably fair to say that there has been considerably less discussion about the lower limit of nanotechnology compared with the upper one.

The idea that nanotechnology embodies a qualitative difference occurring at a certain point when one has shrunk things sufficiently links nanotechnology to Hegel, who first formulated the idea that a quantitative difference, if a sufficiently large, can become a qualitative one in his *Wissenschaft der Logik*. In essence nanotechnology is asserting that "less is different". In order to defend condensed matter physics from the criticism that it is somehow less fundamental than elementary particle physics, P.W. Anderson wrote a paper entitled "More is different" [7], in which he developed the idea of qualitative changes emerging when a sufficient number of particles are aggregated together (in other words chemistry is not simply applied physics, nor biology simply applied chemistry, but at each new level, implying simultaneous consideration of a larger number of particles than at the preceding one, new phenomena emerge, the existence of which could not have been predicted even from complete knowledge of the preceding level). Nanotechnology is a further example of the fundamental idea of the emergence of wholly new phenomena at a new level of description, except that now we are not *increasing* the quantity, but *decreasing* it. The various examples gathered in the preceding sections, such as the optical properties of semiconductor nanoparticles already mentioned, or the magnetic properties of nanoparticles, are all concrete manifestations of Hegel's idea of a qualitative change emerging when a quantitative change becomes sufficiently large.

There is, however, an even more fundamental difference that appears when we enter the quantum realm – the world of the absolutely small, as emphasized by Dirac: "It is usually assumed that, by being careful, we may cut down the disturbance accompanying our observations to any desired extent. The concepts of big and small are then purely relative and refer to the gentleness of our means of observation as well as to the object being described. In order to give an absolute meaning to size, such as is required for any theory of the ultimate structure of matter, we have to assume that there is a limit to the fineness of our powers of observation and the smallness of the accompanying disturbance – a limit which is inherent in the nature of things and can never be surpassed by improved technique or increased skill on the part of the observer. If the object under observation is such that the unavoidable limiting disturbance is negligible, then the object is big in the absolute sense and we may apply classical mechanics to it. If, on the other hand, the limiting disturbance is not negligible, then the object is small in the absolute sense and we require a new theory for dealing with it" [60].

Another way in which the quantum realm is absolutely different from the classical one was elaborated upon by Weisskopf [298]. In classical physics, laws predetermine only the general character of phenomena; the laws admit a continuous variety of realizations, and specific features are determined by the initial conditions. On the other hand, in the quantum realm individual atoms have well-defined specific qualities. Furthermore, their identity is unique and immutable. "Two pieces of gold, mined at two different locations and treated in very different ways, cannot be distinguished from one another. All the properties of each individual gold atom are fixed and completely independent of its previous history" [298] (gold has only one stable isotope, it should be noted; this is not the case for all other elements, in which case the isotopic ratio could give a clue to the provenance of a sample). Similarly, all electrons have the same charge, spin and mass. The same identity extends to the crystal structure of a substance (disregarding polymorphism), and indeed to other properties. This existence of well-defined specific qualities is alien to the spirit of classical physics. Furthermore, these "quantum blocks" (atoms, electrons, etc.) are embodied in larger structures assembled from atoms, such as chemical molecules, which also possess quality, specificity and individuality.

The divisibility of *process* is also a main feature of classical physics [298]: every process can be considered as a succession of partial processes. Typically, a reversible change in thermodynamics is effected as a succession of infinitesimal steps. This gives rise to the Boltzmann paradox: since all possible motions in a piece of matter should share in its thermal energy, if there were an endless regression of the kind molecules → atoms → protons etc. → . . . then immense, indeed infinite energy would be needed to heat up matter, but this is evidently not the case. The existence of quanta resolves the Boltzmann paradox: the notion of a succession of infinitesimal steps loses meaning in the quantum realm.

When it comes to the living world, classical features seem to be even more apparent than in the macroscopic inanimate world – human perceptions such as the taste of a foodstuff depend on the recent history of what one has eaten, and the psychological circumstances at the time of eating, for example, and human identity is essentially our life history, in which our actual chemical constitution – anyway constantly changing – plays only a secondary rôle. Yet even life makes use of exquisitely identical macromolecules – nucleic acids (RNA and DNA) and proteins, and in that sense is also alien to the spirit of classical physics.

The ultimate aim of nanotechnology – bottom-to-bottom assembly in a eutactic environment, or programmable assembly at the atomic scale – is far closer to the true quantum world than the classical world. In order to facilitate bridging the gap between nanoscale artifacts and those suitable for human use, "nanoblocks" have been proposed as an intermediate level of object. These nanoblocks would be produced with atomic precision, but their own assembly into larger structures would be easier than working with atoms – self-assembly is likely to be the most convenient general route. Every nanoblock (of a given type) would be identical to every other one, regardless of where it had been produced. Yet each nanoblock would probably contain thousands or tens of thousands of atoms – comparable in size to the proteins of a living cell – and hence would not rank as absolutely small according to Dirac, yet would nevertheless possess quality, specificity and individuality. In this sense, nanotechnology represents a kind of compromise between the classical and quantum realms, an attempt to possess the advantages of both. The advantages can be seen particularly clearly in comparison with chemistry, which attempts to create entities (chemical compounds) possessing quality, specificity and individuality, but in the absence of a eutactic environment the yield of reactions that should lead to a unique product is usually significantly below unity. The difference between chemistry and nanotechnology is analogous to the difference between analog and digital ways of representing information: in the latter the basic entities (e.g. zero and one) have a specific, immutable individuality (even though the voltage representing "one" in a digital computer may actually have a value between, say, 0.6 and 1.5 V), whereas in an analog device the information is directly represented by the actual voltage, which may be subject to some systematic bias as well as to unavoidable fluctuations.

Returning to the issue of measurement, Heisenberg has remarked that the mathematical description of a quantum object does not represent its behavior but rather our knowledge of its behavior. This brings clarity to another classical paradox – in not a single instance is it possible to predict a physical event exactly, since as measurements become more and more accurate, the results fluctuate. The indeterminist school deals with this by asserting that every physical law is of a statistical nature; the opposing school asserts that the laws apply exactly to an idealized world picture, to which an actual measurement can only approximate [235]. Clearly, the latter viewpoint is appropriate to the quantum world, in which we can predict the probability of an event. Nanotechnology in effect creates a simplified version of the world, in which only a finite set of discrete states are available, whose occurrence (e.g. as a result of a nanofacturing process) can be predicted exactly.

Hence, we can say that an ideal nano-object should be small enough to possess quality, specificity and individuality, like a quantum object, but large enough for its state not to be destroyed by measuring one of its attributes, such as its position. Here, we seem to be approaching a fundamental definition of a nano-object, rather than merely a phenomenological one (albeit abstracted to essential characteristics), or an ostensive one.

The quantum limit corresponds to an irreducible lower limit of smallness but, depending on the phenomenon under consideration, it might be way beyond the size scale of a single atom; for example, the ultimate lower length scale is given by the Planck length (defined solely using fundamental constants) $\sqrt{\hbar G/c^3} \approx 1.6 \times 10^{-35}$ m.

SUMMARY

Consideration of what happens to things when we reduce their size reveals two kinds of behavior. In one group of phenomena there is a discontinuous change of properties at a certain size. This change can be very reasonably taken to demarcate the nanoscale. Note that qualitative changes in behavior with size only seem to occur "near the bottom". Hence, there is no need to separate changes occurring at the nanoscale from those occurring at the micrometer, millimeter or even meter scale because there are none of the kind we are talking about. In the other group the properties change gradually without any discontinuous (qualitative) change occurring.

Evidently if nanotechnology is merely a continuation of the trend of ever better machining accuracy, as it was viewed by Taniguchi [278], we do not need to worry about qualitative differences, at least not at the hardware level. In this case, the upper boundary of the nanoscale must be somewhat arbitrary, but one hundred nanometers seems entirely reasonable, from which it follows that it is really immaterial whether this boundary is taken to be approximate or exact. This definition would fit current usage in top-down nanomanufacture (see Chapter 8) – ultraprecision machining and semiconductor processing. In this usage, nano-objects and devices and the processes used to make them are qualitatively the same as at bigger scales. The inclusion of phrases such as "where properties related to size can emerge" in definitions of nanotechnology is in this case superfluous, other than as a warning that something unexpected *might* happen.

In contrast, such unexpected phenomena constitute the essential part of the definition of nanotechnology in the first group, in which the nanoscale reveals itself as property-dependent. Furthermore, the scale will also depend on external variables such as temperature and pressure: what appears indubitably as a nanoscale phenomenon at one temperature might cease to be thus distinguished when hotter or colder. In order to permit comparisons between different sets of phenomena in group 1, the characteristic size parameter can be normalized by the critical nanoscale-defining length of that phenomenon. Thus, for example, when comparing nanoparticles of different materials with respect to their optical properties, their dimensionless size would be given by r/r_B, where r is a characteristic length of the object under consideration, and the nanorealm concerns ratios $r/r_B < 1$. Tables 2.3 and 2.4 summarize some of these nanoscale-defining lengths.

Where there is no discontinuity, we can take the Hegelian concept of quantitative change becoming qualitative if great enough to justify the application of the term "nanotechnology". Usually this demands consideration of function (utility). Thus,

Table 2.3 Summary of Simple Object Based Nanoscale-Defining Lengths

Domain	Defining Length	Formula	Typical Value/nm
Surfaces	(Geometry)		5
Nucleation	Critical nucleus size	(2.6)	5
Optics and electronics	Bohr radius	(2.11)	10
Magnetism	Single domain size	§2.6	50
Mechanics	Griffith length	(2.26)	50

Table 2.4 Summary of Device (System)-Based Nanoscale-defining Lengths

Domain	Defining Length	Formula	Typical Value/nm
Electronic	Electron mean free path		100
Photonic	Evanescent field penetration depth	§5.8.5	100
Fluidic	Debye length	(7.42)	10
Mechanical	Phonon mean free path	[262]	100

even though the circuit elements in the current generation of very large-scale integrated circuits (VLSIs) with features a few tens of nanometers in length work in exactly the same way as their macroscopic counterparts, new function (e.g. practicable personal cellular telephony) emerges upon miniaturization.

FURTHER READING

1. A.I. Ekimov, Al.L. Efros, A.A. Onushchenko, Quantum size effects in semiconductor microcrystals, Solid St. Commun. 56 (1985) 921–924.
2. J.J. Ramsden, Less is different, Nanotechnol. Percept. 6 (2010) 57–60.

Forces at the nanoscale

3

CHAPTER CONTENTS

INTRODUCTION

We have established in the previous chapter that the smallness (of nano-objects) implies preponderance of surfaces over bulk. This implies that interfacial (IF) forces are particularly important in the nanorealm, governing the behavior of nano-objects and determining the performance of nanostructured materials and nanodevices. Of particular importance are the IF forces that are electrostatic in origin: Section 3.2 is mostly about them. The so-called local van der Waals or Casimir force is of unique

Nanotechnology: An Introduction. DOI: 10.1016/B978-0-323-39311-9.00009-1
Copyright © 2016 Elsevier Inc. All rights reserved.

relevance to the nanoscale – it is described first in Section 3.1. As for the other fundamental forces, the gravitational force is so weak at the nanoscale – reckoning either distance or mass – that it can be neglected. Conversely, the range of the strong nuclear force is much smaller than the nanoscale, and hence can also be neglected. The capillary force, which in the past has sometimes been considered as a "fundamental" force, is of course simply a manifestation of the wetting phenomena originating in the IF forces discussed in Section 3.2.

Familiar monolithic substances rely on strong metallic, ionic or covalent bonds to hold their constituent atoms together. These three categories correspond to the "elements" of mercury, sodium chloride and sulfur that Paracelsus added to the classical Greek ones of earth (solid – but possibly also encompassing granular matter), air (gas), fire (plasma) and water (liquid). Presumably mankind first encountered complex multicomponent materials (i.e. composites) in nature – wood is perhaps the best example. While the components individually (in the case of wood, cellulose and lignin) are covalently bonded, bonding at the cellulose–lignin interface depends on the relatively weak forces discussed in Section 3.2. Thus, knowledge of these IF forces is essential for understanding nanocomposites (Section 6.6). They also enable the self-assembly of nano-objects to take place (see Section 8.2.1).

3.1 THE CASIMIR FORCE

A cavity consisting of two mirrors facing each other disturbs the pervasive zero-point electromagnetic field, because only certain wavelengths of the field vibrations can fit exactly into the space between the mirrors. This lowers the zero-point energy density in the region between the mirrors, resulting in an attractive Casimir force between them [38]. The force falls off rapidly (as z^{-4}) with the distance z between the mirrors, and hence is negligible at the microscale and above, but at a separation of 10 nm it is comparable with atmospheric pressure (10^5 N/m), and therefore can be expected to affect the operation of nanoscale mechanical devices (Fig. 3.1), and nanoscale assembly.

3.2 INTERMOLECULAR INTERACTIONS

The forces described in this section are responsible for the relatively weak interactions between two objects, typically in the presence of an intervening medium. They are all electrostatic in origin. As Pilley has pointed out in his book on electricity, it is pervasive in bulk matter but not directly perceptible in everyday life. On the other hand, it is quite likely to be manifested at surfaces. When they carry electrostatic charges (this is sometimes called an electrified interface) – for example, created by rubbing in air, or by ionization in solution – they will manifest Coulombic interaction. Even neutral surfaces will, however, interact via the Lifshitz–van der Waals family of forces. Finally, there are the Lewis acid/base interactions, involving some electron

a   b

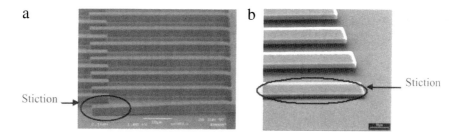

Stiction

Stiction

FIGURE 3.1

An illustration of the practical problems that may arise when micro- or nanocomponents approach each other to within nanoscale distances. (a) Stuck finger on a comb drive; (b) cantilever after release etch adhering to the substrate. These effects are often collectively categorized as manifestations of stiction ("sticking friction"). Condensation of water and the resulting capillary forces also contribute as well as the Casimir force.

-Reproduced with permission from M. Calis, M.P.Y. Desmulliez, Haptic sensing technologies for a novel design methodology in micro/nanotechnology. Nanotechnol. Percept. 1 (2007) 141–158.

exchange between the partners [e.g. the HB], which might be considered as a weak covalent interaction.

3.2.1 THE CONCEPT OF SURFACE TENSION

Surface tension γ is formally defined as the free energy G required to create the extend an interface of area A,

$$\gamma = (\partial G/\partial A)_{T,P}, \tag{3.1}$$

where the practically encountered constant temperature and pressure make the Gibbs free energy the appropriate choice for G. In the SI, the units of γ are N/m, which is the same as energy per unit area (J/m^2). It is customary to refer to γ as a surface tension if the increase of area is reversible, and as a surface energy if it is not.

Generally speaking, work needs to be done to create an interface; it has a higher free energy than the bulk. The work of cohesion of a solid is

$$W^{(\text{coh})} = 2\gamma_1 A = -\Delta G^{(\text{coh})} \tag{3.2}$$

(see Fig. 3.2). On the other hand, the work of adhesion (needed to separate two dissimilar substances 1 and 2) is given by (see Fig. 3.2)

$$W_{12}^{(\text{adh})} = (\gamma_1 + \gamma_2 - \gamma_{12})A = -\Delta G^{(\text{adh})}, \tag{3.3}$$

a formalism introduced in the nineteenth century by Dupré. γ_1 and γ_2 account for the old interfaces lost, and γ_{12} accounts for the new interface gained. Most of the

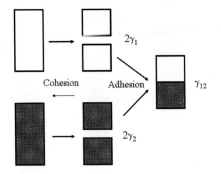

FIGURE 3.2

Cohesion and adhesion of substances 1 (white) and 2 (gray) (see text).

subsequent difficulties experienced by the field of IF interactions have concerned the theoretical calculation (i.e. prediction) of terms involving two (or more) substances such as γ_{12}. The nanoscopic viewpoint is that the "microscopic" surface tension (or energy) γ_{12} depends on specific chemical interactions between the surfaces of the two substances 1 and 2.

Fowkes, Girifalco and Good introduced the reasonable assumption that the tension at the interface of substance 1 against substance 2 is lowered by the presence of substance 2 by an amount equal to the geometric mean of the tensions of the two substances individually, hence equal to $\gamma_1 - (\gamma_1\gamma_2)^{1/2}$, and similarly the tension at the interface of substance 2 against substance 1 is $\gamma_2 - (\gamma_1\gamma_2)^{1/2}$. Summing these two terms, we have

$$\gamma_{12} = \gamma_1 + \gamma_2 - (\gamma_1\gamma_2)^{1/2} = (\sqrt{\gamma_1^{(LW)}} - \sqrt{\gamma_2^{(LW)}})^2 \tag{3.4}$$

called the Girifalco–Good–Fowkes equation. This is equivalent to the work of adhesion being the geometric mean of the works of cohesion, i.e. $W_{12} = (W_{11}W_{22})^{1/2}$. The Dupré equation (3.3) then becomes

$$W_{12}^{(adh)} = 2(\gamma_1\gamma_2)^{1/2}A. \tag{3.5}$$

Fowkes and van Oss developed the idea that the total IF energy is linearly separable into the dispersive (London–van der Waals), dipole-induced dipole (Debye), dipole–dipole (Keesom) and electron donor–acceptor terms, and Lifshitz has pointed out that the London–van der Waals, Debye and Keesom interactions are all of the same type (cf. the Hellman–Feynman theorem), with the same dependence of magnitude on separation between the two interacting substances, and hence

$$\gamma^{(total)} = \gamma^{(LW)} + \gamma^{(ab)}, \tag{3.6}$$

where LW denotes Lifshitz–van der Waals (subsuming all the dispersive terms) and ab denotes (Lewis) acid–base, and *a fortiori*

$$\gamma_{12}^{(total)} = \gamma_{12}^{(LW)} + \gamma_{12}^{(ab)}. \tag{3.7}$$

Whereas the Lifshitz–van der Waals interaction is always attractive, the sign of the Lewis acid–base interaction depends on the relative proportions of Lewis acids and Lewis bases constituting the two interacting substances. Superscript $\ominus$ will be used to denote electron-donating (Lewis base) and superscript $\oplus$ will be used to denote electron-accepting (Lewis acid) moieties; van Oss has proposed that one might again take the geometric mean, namely,

$$\gamma^{ab} = 2(\gamma^{\ominus}\gamma^{\oplus})^{1/2}. \tag{3.8}$$

Two monopolar substances of the same sign will repel each other; attraction depends on the presence of cross-terms. By analogy with Eq. (3.5),

$$W_{12}^{(adh)} = 2[(\gamma_2^{\oplus}\gamma_1^{\ominus})^{1/2} + (\gamma_1^{\oplus}\gamma_2^{\ominus})^{1/2}]A. \tag{3.9}$$

Hence, the ab combining law is

$$\gamma_{12}^{(ab)} = 2[(\gamma_1^{\oplus}\gamma_1^{\ominus})^{1/2} + (\gamma_2^{\oplus}\gamma_2^{\ominus})^{1/2} - (\gamma_1^{\oplus}\gamma_2^{\ominus})^{1/2} - (\gamma_1^{\ominus}\gamma_2^{\oplus})^{1/2}]$$
$$= 2(\sqrt{\gamma_1^{\oplus}} - \sqrt{\gamma_2^{\oplus}})(\sqrt{\gamma_1^{\ominus}} - \sqrt{\gamma_2^{\ominus}}). \tag{3.10}$$

It takes account of the fact that $\ominus$ interacts with $\oplus$, which is why the ab interaction can be either attractive or repulsive. In typical biological and related systems, the Lewis acid–base interaction accounts for 80%–90% of the total interactions. The most familiar manifestation is hydrogen bonding [e.g. double-stranded DNA (the double helix), globular proteins containing α-helices]; π–π interactions (stacking of alternately electron rich and electron-deficient aromatic rings) are frequently encountered in synthetic organic supermolecules.

Let us now consider two solids 1 and 3 in the presence of a liquid medium 2 (e.g. in which a self-assembly process, cf. Section 8.2.1, takes place). ΔG_{123} is the free energy per unit area of materials 1 and 3 interacting in the presence of liquid 2. Using superscript $\parallel$ to denote the IF interaction energies per unit area between infinite parallel planar surfaces,

$$\Delta G_{121}^{\parallel} = -2\gamma_{12} \tag{3.11}$$

and

$$\Delta G_{123}^{\parallel} = \gamma_{13} - \gamma_{12} - \gamma_{23}. \tag{3.12}$$

From the above equations, we can derive

$$\Delta G_{123}^{(LW,ab)\parallel} = \Delta G_{22}^{(LW,ab)\parallel} + \Delta G_{13}^{(LW,ab)\parallel} - \Delta G_{12}^{(LW,ab)\parallel} - \Delta G_{23}^{(LW,ab)\parallel}, \tag{3.13}$$

where ΔG_{13} is the free energy per unit area of materials 1 and 3 interacting directly. It follows that:

• LW forces (anyway weak) tend to cancel out;

Table 3.1 Surface Tension Parameters of Some Solids [224]

Material	$\gamma^{(LW)}$/mJ m^{-2}	$\gamma^{\oplus}$/mJ m^{-2}	$\gamma^{\ominus}$/mJ m^{-2}
Synthetic polymers			
Nylon 6,6	36	0.02	22
PMMA	41	0	13
Polyethylene	33	0	0
Polyethylene oxide	43	0	64
Polystyrene	42	0	1.1
Polyvinylpyrrolidone	43	0	30
PVC	43	0.04	3.5
Teflon	18	0	0
Carbohydrates			
Cellulose	44	1.6	17
Dextran T-150	42	0	55
Metal oxides			
SiO$_2$	39	0.8	41
SnO$_2$	31	2.9	8.5
TiO$_2$	42	0.6	46
ZrO$_2$	35	1.3	3.6

- the so-called "hydrophobic force" is a consequence of the strong cohesion of water ΔG_{22}. Attraction of suspended solids is only prevented by their hydrophilicity. The sign of ΔG_{12} with 2 = water provides an unambiguous measure of hydrophobicity: $\Delta G_{12} < 0 \equiv$ hydrophilic; $\Delta G_{12} > 0 \equiv$ hydrophobic (see also Section 3.2.5).

$\Delta G_{123}^{\parallel}$ can be used to provide a rapid first estimate of whether adhesion between materials 1 and 3 will take place in the presence of medium 2. Tables 3.1 and 3.2 give some typical values (see Section 3.2.3 for information about their determination).

According to the Derjaguin approximation, a sphere of radius r (material 3) interacting with an infinite planar surface (material 1) has the following free energies of interaction as a function of z, the perpendicular distance between the plane and the nearest point of the sphere:

$$\Delta G^{(LW)}(z) = 2\pi \ell_0^2 \Delta G^{(LW)\parallel} r/z, \tag{3.14}$$

where ℓ_0 is the equilibrium contact distance (about 0.15 nm);

$$\Delta G^{(ab)}(z) = 2\pi \chi \Delta G^{(ab)\parallel} \exp[(\ell_0 - z)/\chi]r, \tag{3.15}$$

Table 3.2 Surface Tensions of Some Liquids

Liquid	$\gamma^{(LW)}$/mJ m^{-2}	$\gamma^{\oplus}$/mJ m^{-2}	$\gamma^{\ominus}$/mJ m^{-2}
Water[a]	22	25.5	25.5
Glycerol	34	3.9	57
Ethanol	19	0	68
Chloroform	27	3.8	0
Octane	22	0	0
n-hexadecane	27.5	0	0
Formamide	39	2.3	40
α-bromonaphthalene	44	0	0
Diiodomethane	51	0	0

[a]Absolute values of $\gamma^{\oplus}$ and $\gamma^{\ominus}$ are not known at present; values are arbitrarily assigned to ensure that the known overall γ is correct (Eq. (3.8)).
Data mostly from C.J. van Oss, Forces IFes en milieux aqueux, Masson, Paris, 1996.

where χ is the decay length for the ab interactions; and where electrostatic charges are present,

$$\Delta G^{(el)}(z) = 4\pi\epsilon_0\epsilon\psi_3\psi_1 \ln[1 + \exp(-\kappa z)]r, \quad (3.16)$$

where the ψ are the electrostatic surface potentials of materials 1 and 3 and $1/\kappa$ is the Debye length (inversely proportional to the square root of the ionic strength – this is why electrostatic interactions tend not to be very important in salty aqueous systems, such as many biofluids).

3.2.2 CRITIQUE OF THE SURFACE TENSION FORMALISM

As Bikerman has pointed out [24], there are some difficulties with applying the concept of surface tension to solids. It is generally recognized that most real, machined solid surfaces are undulating as shown in Fig. 3.3. Capillary pressure P_c,

$$P_c = 2\gamma/r, \quad (3.17)$$

where r is the radius of the undulation, would tend to flatten such undulations since the pressure where r is small would engender a strain easily exceeding the yield strength of most metals, yet such undulations persist.

The equation for the three-phase line between a vapor (phase 1) and two liquids (phases 2 and 3) is well known to be

$$\frac{\sin\hat{1}}{\gamma_{23}} = \frac{\sin\hat{2}}{\gamma_{13}} = \frac{\sin\hat{3}}{\gamma_{12}}, \quad (3.18)$$

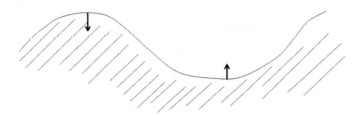

FIGURE 3.3

Sketch of the undulations of a somewhat idealized real machined surface. The arrows indicate the direction of capillary pressure.

where the angles $\hat{1}$, etc., enclose phase 1, etc. The assumption that solids have a surface tension analogous to that of liquids ignores the phenomenon of epitaxy (in which a crystal tends to impose its structure on material deposited it). As an illustration of the absurdities to which the uncritical acceptance of this assumption may lead, consider a small cube of substance 2 grown epitaxially in air (substance 3) on a much larger cube of substance 1 (Fig. 3.4). The angles A and B are both 90° and correspond to $\hat{3}$ and $\hat{2}$, respectively, in Eq. (3.18). It immediately follows that the surface tension of substance 1 equals the IF tension between substances 1 and 2, a highly improbable coincidence. Moreover, since $\hat{1}$ equals 180°, the surface tension of substance 2 must be zero, which also seems highly unlikely.

Given these difficulties in the application of the surface tension concept to rigid solids, it is appropriate to exercise caution when using the approach described in Section 3.2.1. Many of the applications in nanotechnology (especially those concerned with the nano/bio interface) involve, however, soft matter, to which the application of the formalism may be reasonable.

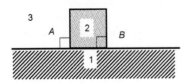

FIGURE 3.4

Illustration of a small cube of substance 2 grown epitaxially on a large cube of substance 1.

-Based on J.J. Bikerman, The criterion of fracture, SPE Trans. 4 (1964) 290–294.

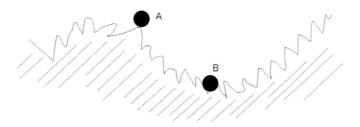

FIGURE 3.5

Cross-section of a real machined surface. Note the two distinct length scales of roughness. A and B are adsorbed impurity molecules.

-Based on J.J. Bikerman, The criterion of fracture, SPE Trans. 4 (1964) 290–294.

3.2.3 EXPERIMENTAL DETERMINATION OF SINGLE-SUBSTANCE SURFACE TENSIONS OF SOLIDS

The general strategy is to measure the advancing contact angles θ on the material 3 whose surface tension is unknown using three appropriate liquids (material 2) with different surface tension components (which are themselves determined, e.g. from hanging drop measurements – Section 3.2.4) [224]. With these values, three simultaneous Young–Dupré equations

$$[\gamma_2^{(LW)}/2 + (\gamma_2^{\oplus}\gamma_2^{\ominus})^{1/2}](1 + \cos\theta) = (\gamma_1^{(LW)}\gamma_2^{(LW)})^{1/2} + (\gamma_1^{\oplus}\gamma_2^{\ominus})^{1/2} + (\gamma_1^{\ominus}\gamma_2^{\oplus})^{1/2} \quad (3.19)$$

can be solved to yield the unknowns $\gamma_1^{(LW)}$, $\gamma_1^{\oplus}$ and $\gamma_1^{\ominus}$. Values of the surface tension parameters for some common materials have been given in Tables 3.1 and 3.2. Methods for computing ψ for ionized solids (e.g. polyions and protonated silica surfaces) are based on the Healy–White ionizable surface group model [120].

A major practical difficulty in experimentally determining surface tensions of solutes is that under nearly all real conditions, surfaces are contaminated in a highly nonuniform fashion. Fig. 3.5 shows how surface roughness can be the origin of this nonuniformity. Adsorbed impurity molecules will experience different IF energies at different locations on the surface due to the different curvatures.

3.2.4 MEASUREMENT OF THE SURFACE TENSION OF LIQUIDS

The determination of the surface tension of liquids by measuring the shape of drops hanging from a rod or tube began with the experiments of Tate, reported in 1864 [281]. Provided the density (specific gravity) of the liquid is known, the surface tension can be computed from certain dimensions of the drop, which can easily be

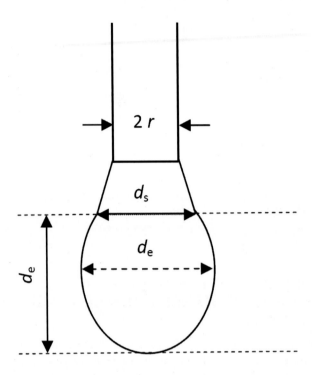

FIGURE 3.6

Cross-section of a pendant drop hanging from a rod of radius r, showing key dimensions used for determining the surface tension of the liquid. d_e is the maximum diameter of the drop.

measured on a photograph of the drop [8, 87, 303, 29], an idealized form of which is shown in Fig. 3.6.

The surface tension is given by

$$\gamma_2 = \Delta \rho g d_e^2 / H, \tag{3.20}$$

where $\Delta \rho$ is the difference in densities of the liquid and the air ($\Delta \rho = \rho_2 - \rho_{air}$), g is the acceleration due to gravity, d_e is the maximum diameter of the drop (see Fig. 3.6) and H is a shape-dependent parameter, very approximately equal to $1/S$, where $S = d_s/d_e$, d_s being the width of the drop at distance d_e above the bottom. Extensive tabulations of $1/H$ in terms of S have been given by Fordham [87].

"Tate's law" (not actually given by Tate [281]) is

$$W = 2\pi r \gamma_2 f, \tag{3.21}$$

where $W = mg$ is the weight of the drop, $m = V\rho$ being its mass and V its volume, r is the radius of the rod from which the drop is hanging and f is a correction factor that may be given by $r/V^{1/3}$.

A hanging drop, especially a "bulging" drop (in which $d_e > 2r$), may not be stable. The remnant (mass m) remaining when a rod is withdrawn from a pool of liquid has been found to be given by [293]

$$t_b = \sqrt{m/\gamma_2},\tag{3.22}$$

where t_b is the "breakup time", defined as the interval between the epoch when connectivity between the rod and the liquid is lost and the epoch when the "ligament" (liquid bridge) becomes quasistatically unstable; interestingly, t_b is independent of γ_2 if the rod is withdrawn at constant acceleration [293].

3.2.5 WETTING AND DEWETTING

Wetting means the spreading of a liquid over a solid surface; dewetting is its converse, the withdrawal of liquid from a surface (the term wetting is used regardless of the nature of the liquid – hence "dry" in this context means not that water is absent, but that all liquid is absent). They are basic processes in countless natural and industrial processes. Although pioneering work in characterizing the IF tensions upon which wetting depends was reported two hundred years ago by Young, the processes are still relatively poorly understood. Wetting is mostly considered from a mesoscopic viewpoint and therefore fits well into the framework of nanotechnology. Few experimental techniques are available for investigating the important solid/liquid interfaces: the contact angle method is simple and probably still the most important, but only a handful of laboratories in the world have shown themselves capable of usefully exploiting it. The history of dewetting, a phenomenon of no less industrial importance than wetting, is much more recent: quantitative experimental work dates from the early 1990s.

It is essentially intuitive to expect that the spreading of a liquid on a solid surrounded by vapor depends on γ_{SV} (S = solid, V = vapor, L = liquid). The quantitative relationship was given by Young in 1805 (cf. Eq. (3.19)),

$$\gamma_{LV} \cos \theta = \gamma_{SV} - \gamma_{SL}.\tag{3.23}$$

The degree of wetting is inversely proportional to the contact angle θ; $\theta = 0$ corresponds to complete wetting. Young's equation (3.23) can be easily derived by noting that the surface tension can be written as a force per unit distance. The IF forces acting on the triple line T, where three phases S, L, V (solid, liquid, vapor) meet must sum to zero in a given direction (x, parallel to the interface) (Fig. 3.7). More formally, it follows from the condition that (at equilibrium) the energies must be invariant with respect to small shifts dx of the position of T. The structure of T may be very complex. For example, for water containing dissolved electrolyte, the local ion composition may differ from that in the bulk; soft solids may be deformed in the vicinity of T. Although Young's equation ignores these details, it provides a remarkably accurate description of contact angles.

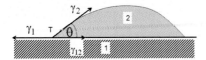

FIGURE 3.7

A drop of a liquid (substance 2) on a solid (substance 1). The vapor is neglected here (i.e. $\gamma_1 \equiv \gamma_{SV}$). T is the triple line.

3.2.6 LENGTH SCALES FOR DEFINING SURFACE TENSION

The region in which deviations from "far-field" quantities occur is known as the core, with radius $r_C \sim 10$ nm. Hence, for typical drops (with a radius $R \sim 1$ mm) used in the contact angle determinations from which surface tension can be calculated, $R \gg r_C$ and the curvature of T may be neglected; atomic scale heterogeneity on the subnanometer scale may also be neglected.

3.2.7 THE WETTING TRANSITION

Complete wetting is characterized by $\theta = 0$, which implies (from Eq. (3.23))

$$\gamma_{LV} = \gamma_{SV} - \gamma_{SL} \tag{3.24}$$

at equilibrium (out of equilibrium, this relation may not hold). The Cooper–Nuttall spreading coefficient S is

$$S = \gamma_{\check{S}V} - \gamma_{SL} - \gamma_{LV}, \tag{3.25}$$

where $\gamma_{\check{S}V}$ is the IF tension of a dry solid. Three régimes can thus be defined:

1. $S > 0$. This corresponds to $\gamma_{\check{S}V} > \gamma_{SV}$, i.e. the wetted surface has a lower energy than the unwetted one. Hence, wetting takes place spontaneously. The thickness h of the film is greater than monomolecular if $S \ll \gamma_{LV}$. The difference $\gamma_{\check{S}V} - \gamma_{SV}$ can be as much as 300 mJ/m^2 for water on metal oxides. Such systems, therefore, show enormous hysteresis between advancing and receding contact angles. Other sources of hysteresis include chemical and morphological inhomogeneity (contamination and roughness).
2. $S = 0$. Occurs if $\gamma_{\check{S}V}$ practically equals γ_{SV}, as is typically the case for organic liquids on molecular solids.
3. $S < 0$. Partial wetting. Films thinner than a certain critical value, usually ~ 1 mm, break up spontaneously into droplets (cf. Section 8.1.2).

3.3 **CAPILLARY FORCE**

It follows from the previous considerations that an attractive interaction between a liquid of density ρ and a solid will cause that liquid to rise to a height h within a vertical tube of diameter r made from the solid emerging from a reservoir of the liquid until the gravitational pull on the liquid column equals the IF attraction

$$h = \frac{1}{2\gamma_{LV}\cos\theta_{SL}}\rho g r, \qquad (3.26)$$

where g is the acceleration due to gravity. For a small diameter tube, the rising of the liquid can be appreciable and easily visible to the naked eye. This is the classical manifestation of the capillary force. Since water is both a strong HB donor and a strong HB acceptor, it is able to interact attractively with a great many different substances, especially minerals, which tend to be oxidized.

Consequences for nano-objects. It is part of everyday experience that a human emerging from a bath carries with him a film of water – depending on the hydrophobicity of his skin – that may be of the order of 100 μm thick, hence weighing a few hundred grams – negligible compared with the weight of its bearer. A wet small furry creature like a shrew has to carry about its own weight in water. A wet fly would have a surplus weight many times its own, which is presumably why most insects use a long proboscis to drink; bringing their bodies into contact with a sheet of water usually has disastrous consequences. The inadvertent introduction of water (e.g. by condensation) into microsystems may completely degrade their performance (cf. Fig. 3.1); nanosystems are even more vulnerable.

The force environment of nano-objects and nanodevices. Apart from the capillary force referred to above, we should also consider objects wholly immersed in a liquid. In this environment, viscosity, rather than inertia, dominates. In contrast to the Newtonian mechanics appropriate to describing the solar system, the movement of a bacterium swimming in water is governed by a Langevin equation comprising frictional force and random fluctuations (Brownian motion). This also encompasses the realm of soft matter, a good definition of which is matter whose behavior is dominated by Brownian motion (fluctuations).

3.4 **HETEROGENEOUS SURFACES**

The expressions given in the preceding sections have assumed that the surfaces are perfectly flat and chemically homogeneous. Most manufactured surfaces are, however, rough and may be chemically inhomogeneous, either because of impurities present from the beginning, or because impurities acquired from the environment have segregated due to roughness (Fig. 3.5).

3.4.1 WETTING ON ROUGH AND CHEMICALLY INHOMOGENEOUS SURFACES

If the roughness is in the nanorange – a few nanometers to a micrometer – the (subnanometer-sized) molecules of the liquid interact locally with planar segments of the surface, while still yielding a unique contact angle, denoted by θ^*, supposed different from the contact angle θ of the same liquid with a perfectly smooth planar surface. Now suppose that the liquid drop on the rough surface is perturbed by a small horizontal displacement by a distance dx in the plane of the surface. Because of the roughness, characterized by the ratio r of actual to apparent surface areas, the real distance is $r dx$, $r > 1$. According to Young's law (3.23), the work done is

$$dW = (\gamma_{SL} - \gamma_{SV}) r dx + \gamma_{LV} \cos \theta^* dx. \tag{3.27}$$

The drop should then find an equilibrium state at which $dW/dx = 0$, yielding

$$(\gamma_{SV} - \gamma_{SL}) r = \gamma_{LV} \cos \theta^*. \tag{3.28}$$

Comparison with (3.23) yields Wentzel's law,

$$\cos \theta^* = r \cos \theta. \tag{3.29}$$

In other words, since $r \geq 1$, roughness will always make a hydrophobic surface more hydrophobic and a hydrophilic surface more hydrophilic. This is the basis of technologies to manufacture superhydrophobic and superhydrophilic surfaces. Typically, natural superhydrophobic materials such as the leaves of the lupin (Fig. 3.8) have roughnesses at multiple length scales.

A similar approach has been used to derive the Cassie–Baxter law for surfaces chemically inhomogeneous (at the nanoscale). Suppose that such a surface is constituted from fractions f_i of N different materials, that individually in pure form yield contact angles θ_i with the liquid under test. Then,

$$\cos \theta^\dagger = \sum_{i=1}^{N} \cos \theta_i, \tag{3.30}$$

where $\theta^\dagger$ is the effective contact angle. The advent of nanofabrication has yielded intriguing situations of drops resting on the tops of arrays of nano- or micropillars, let us suppose with a height r and covering a fraction f of the surface. If the drop remains on the top of the pillars (a situation that has been called the fakir effect), the surface would presumably be sensed as smooth (with the air–liquid interface between the pillars remaining parallel to the mean plane of the substrate), but chemically heterogeneous, and the Cassie–Baxter law would be applicable, simplifying to $\cos \theta^\dagger = f(1 + \cos \theta) - 1$. If on the other hand the drop completely wets both the horizontal and vertical surfaces, the surface is sensed as rough and the Wentzel law would be applicable. As the "intrinsic" contact angle varies, the drop can minimize its energy by either obeying the Wentzel law (for a hydrophilic material) or the Cassie–Baxter law (i.e. displaying the fakir effect, for hydrophobic surfaces). The crossover point between the two régimes is given by $\cos \theta = (f - 1)/(r - f)$.

FIGURE 3.8

Drops of rainfall on the leaves of the lupin showing the superhydrophobicity.

3.4.2 **THREE-BODY INTERACTIONS**

The main structural unit of proteins is the alpha helix. Simple polyamino acids such as polyalanine will fold up spontaneously in water into a single alpha helix. Many proteins, such as the paradigmatical myoglobin, can be adequately described as a bundle of alpha-helical cylinders joined together by short connecting loops. The alpha helix is held together by HBs between the ith and $(i + 4)$th amino acid backbones. Denatured myoglobin has no difficulty in refolding itself spontaneously in water, correctly reforming the alpha helices, but one might wonder why the process is so robust when the denatured polypeptide chain is surrounded by water molecules, each of which is able to donate and accept two HBs (albeit that at room temperature, 80%–90% of the possible HBs between water molecules are already formed – see Section 3.8). The paradox is that although refolding is an intramolecular process, the overwhelming molar excess of water should predominate, ensuring that the backbone HBs are always fully solvated by the surrounding water and, hence, ineffectual for creating protein structure.

One of the intriguing features of natural amino acid sequences of proteins is their blend of polar and apolar residues. It was formerly believed that the protein folding problem involved simply trying to pack as many of the apolar residues into the protein interior as possible, in order to minimize the unfavorable free energy of interaction between water and the apolar residues. Nevertheless, a significant number (of the order of 50% of the total, for a folded globular protein of moderate size) of apolar residues remain on the folded protein surface.

The presence of an apolar residue in the vicinity of a HB is a highly effective way of desolvating it. Folding success actually involves the juxtaposition of appropriate apolar residues with backbone HBs. The effectiveness of desolvation of a HB can be computed by simply counting the number of apolar residues within a sphere of about 7 Å radius centered midway between the HB donor and the HB acceptor [83]. This approach, which can be carried out automatically using the atomic coördinates in the protein data bank (PDB), reveals the presence of *dehydrons*, underdesolvated (or "underwrapped") HBs. Clusters of dehydrons constitute especially effective sticky patches on proteins.

3.5 WEAK COMPETING INTERACTIONS

In any assembly process starting from a random arrangement, it is very likely that some of the initial connexions between the objects being assembled are merely opportunistic and at a certain later stage will need to be broken in order to allow the overall process to continue. For this reason, it is preferable for the connexions to be weak (i.e. the types considered in Section 3.2, e.g. HBs) in order to enable them to be broken if necessary. The impetus for breakage comes from the multiplicity of competing potential connexions that is inevitable in any even moderately large system.

One example is provided by superspheres (see Section 8.2.9). Another is biological self-assembly (e.g. of compact RNA structures, Section 8.2.11) – bonds must be broken and reformed before the final structure is achieved. This is particularly apparent because these molecules are synthesized as a linear polymer, which already starts to spontaneously fold (which means forming connexions between parts of the polymer that are distant from each other along the linear chain) as soon as a few tens of monomers have been connected. As the chain becomes longer, some of these earlier bonds must be broken to allow connexions between points more distant along the chain to be made. Since, as HBs, they have only about one tenth of the strength of ordinary covalent bonds, they have an appreciable probability of being melted (broken) even at room temperature. Furthermore, the polymer is surrounded by water, each molecule of which is potentially able to participate in four HBs (although at room temperature only about 10% of the maximum possible number of HBs in water are broken – see also Section 3.8). Hence, there is ceaseless competition between the intramolecular and intermolecular HBs.

3.6 COÖPERATIVITY

When considering the interactions (e.g. between precursors of a self-assembly process), it is typically tacitly assumed that every binding event is independent. Similarly, when considering conformational switches, it has been assumed that each molecule switches independently. This assumption is, however, often not justified: switching or binding of one facilitates the switching or binding of neighbors, whereupon we have coöperativity (if it hinders rather than facilitates, then it is called anticoöperativity). A coöperative processes can be conceptualized as two subprocesses, nucleation and growth. Let our system exist in one of the two states (e.g. bound or unbound, conformation A or B), which we shall label 0 and 1. We have [264]

$$\text{nucleation:} \qquad \cdots 000 \cdots \overset{\sigma S}{\rightleftharpoons} \cdots 010 \cdots \qquad (3.31)$$

and

$$\text{growth:} \qquad \cdots 001 \cdots \overset{S}{\rightleftharpoons} \cdots 011 \cdots , \qquad (3.32)$$

where σS and S are equilibrium constants and the parameter σ the degree of coöperativity. Let $\{1\} = \theta$ denote the probability of finding a "1"; we have therefore $\{0\} = 1 - \theta$. The parameter λ^{-1} is *defined* as the conditional probability of "00" given that we have a "0", written as (00) and equal to $\{00\}/\{0\}$. It follows that $(01) = 1 - (00) = (\lambda - 1)/\lambda$. According to the mass action law (MAL), for growth we have

$$S = \frac{\{011\}}{\{001\}} = \frac{(11)}{(00)} \qquad (3.33)$$

from which we derive $(11) = S/\lambda$ and, hence, $(01) = 1 - (00) = (\lambda - 1)/\lambda$ and $(10) = 1 - (11) = (\lambda - S)/\lambda$. Similarly, for nucleation,

$$\sigma S = \frac{(01)(10)}{(00)^2} = (\lambda - 1)(\lambda - S). \qquad (3.34)$$

Solving for λ gives

$$\lambda = [1 + S + \sqrt{(1 - S)^2 + 4\sigma S}]/2. \qquad (3.35)$$

To obtain the sought-for relation between θ and S, we note that $\theta = \{01\} + \{11\} = \{0\}(01) + \{1\}(11)$, which can be solved to yield

$$\theta = \frac{1 + (S - 1)/\sqrt{(1 - S)^2 + 4\sigma S}}{2}. \qquad (3.36)$$

Coöperativity provides the basis for programmable self-assembly (Section 8.2.8), and for the widespread biological phenomenon of "induced fit", which occurs when two molecules meet and "recognize" each other, whereupon their affinity is increased.

3.7 PERCOLATION

Percolation can be considered as a formalization of gelation. Let us consider the following. Initially, in a flask, we have isolated sol particles, which are gradually connected to each other in the nearest-neighbor fashion until the gel is formed. Although the particles can be placed anywhere in the medium subject only to the constraint of hard-body exclusion, it is convenient to consider them placed on the squares of a two-dimensional square lattice (imagine a chess- or checkerboard with all squares initially white and just the occupied squares shaded black). Two particles are considered to be connected if they share a common side (this is called site percolation). Alternatively, the particles are placed at the intersections of the lines making up the lattice (imagine a Gō board); neighboring lattice points are connected if they are bonded together (bond percolation). In principle, the lattice is infinite but in reality it may merely be very large. Percolation occurs if one can trace the continuous path of connexions from one side to the other. Initially, all the particles are unconnected. In site percolation, the lattice is initially considered to be empty, and particles are added. In bond percolation, initially all particles are unconnected and bonds are added. The problem is to determine what fraction of sites must be occupied, or how many bonds must be added, in order for percolation to occur. In the remainder of this subsection, we shall consider site percolation. Let the probability of a site being occupied be p (and of being empty, $q = 1 - p$). The average number of singlets per site is $n_1(p) = pq^4$ for the square lattice, since each site is surrounded by four shared sides. The average number of doublets per site is $n_2(p) = 2p^2q^6$, since there are two possible orientations. A triplet can occur in two shapes, straight or bent, and so on. Generalizing,

$$n_s(p) = \sum_t g(s,t)p^s q^t, \tag{3.37}$$

where $g(s,t)$ is the number of independent ways that an s-tuplet can be put on the lattice, and t counts the different shapes. If there is no "infinite" cluster (i.e. one spanning the lattice from side to side), then

$$\sum_s sn_s(p) = p. \tag{3.38}$$

The first moment of this distribution gives the mean cluster size

$$S(p) = \frac{\sum s^2 n_s(p)}{p}. \tag{3.39}$$

Writing this as a polynomial in p using Eq. (3.37), it will be noticed that for $p < p_c$ the value of the series converges, but for $p > p_c$ the value of the series diverges. The "critical" value $p = p_c$ corresponds to the formation of the "infinite" cluster. For site percolation on a square lattice, $p = p_c = 0.5$; the universal Galam–Mauger formula [92]

$$p_c = a[(d - 1)(q - 1)]^{-b} \tag{3.40}$$

with $a = 1.2868$ and $b = 0.6160$ predicts p_c, with less than 1% error, for all known lattices of connectivity q embedded in a space of dimension d. The larger the lattice, the sharper the transition from not percolating to percolating. For a 3×3 lattice, there can be no percolation for two particles or less, but the probability of randomly placing three particles in a straight line from edge to edge is evidently 1/3.

3.8 THE STRUCTURE OF WATER

So much nanotechnology takes place in water, including microfluidic-enabled reactors, many self-assembly processes, as well as nanobiotechnology and bionanotechnology, that it is important to recall some of the salient features of this remarkable liquid.

A water molecule – H–O–H – can participate in four HBs. The two electron lone pairs (LPs) on the oxygen atom are electron donors, hence HB acceptors. The two hydrogens at the ends of the hydroxyl groups (OH) are HB donors, hence electron acceptors. The equilibrium

$$H_2O_{\text{fully bonded}} \rightleftharpoons OH_{\text{free}} + LP_{\text{free}} \tag{3.41}$$

is balanced such that at room temperature about 10% of the OHs and LPs are nonbonded, i.e. free. It is especially noteworthy that the concentrations of these two free species are 7–8 orders of magnitude greater than the concentrations of the perhaps more familiar entities H^+ and OH^-, and their chemical significance is correspondingly greater.

The OH moiety has a unique infrared absorption spectrum, different according to whether it is hydrogen-bonded or free, which can therefore be used to investigate reaction (3.41). A striking example of how the equilibrium can be controlled is given by the spectroscopic consequences of the addition of cosolutes. If sodium chloride is added to water, the Na^+ and Cl^- ions can, respectively, accept and donate electron density to form quasihydrogen bonds to appropriate donors and acceptors in roughly equal measure, and the intensity of the electron-accepting OH band in the infrared spectrum does not change. If sodium tetraphenylborate is added, the borate ion is prevented by its bulky phenyl ligands from interacting with the water, resulting in fewer bonded OH groups; hence, in order to maintain equilibrium (3.41), the concentration of free LPs must diminish. Conversely, if tetrabutylammonium chloride is added, there will be an excess of bonded OH.

On the basis of extensive experiments on hydrogels, Philippa Wiggins has proposed that two kinds of water, low and high density with, respectively, more and less hydrogen bonding, can be created by surfaces in contact with aqueous solutions. Although the theoretical interpretation of the phenomenology is as yet incomplete, whatever its origins it must inevitably have profound effects on nanoscale fluidic circuits, in which essentially all of the fluid phase is in the vicinity of solid surfaces.

SUMMARY

The large ratio of surface to volume characteristic of nanoscale objects and devices places IF forces in a prominent position in governing their behavior. The surface tension formalism allows the magnitudes of these forces between objects made from different materials in the presence of different liquids or vapors to be quickly estimated from tables of experimentally derived single-substance surface tensions.

In aqueous systems, Lewis acid–base interactions, most notably hydrogen bonding, typically dominate the IF forces.

Real, that is morphologically and chemically heterogeneous, surfaces require some modification to the basic theory. In some cases, a simple mean-field correction may be adequate; in others nanostructure must explicitly be taken into account. This is especially strikingly shown in protein interactions, which are actually three-body in nature and depend on the subtle interplay of solvation and desolvation.

Multiple forces of different strength and range may be operating simultaneously. This provides the basis for programmability. The spontaneous assembly of objects into constructions of definite size and shape is only possible if programmability is incorporated, and in the nanoscale this can typically only be achieved by judicious design at the level of the constituent atoms and groups of atoms of the objects.

Multibody connexions provide the basis for coöperativity, an essential attribute of many "smart" devices. Another collective phenomenon is percolation, which is a paradigm for the assembly process known as gelation.

FURTHER READING

1. J.J. Bikerman, Surface energy of solids. Phys. Status Solidi 10 (1965) 3–26.
2. C. Binns, Prodding the cosmic fabric with nanotechnology. Nanotechnol. Percept. 3 (2007) 97–105.
3. M.G. Cacace, E.M. Landau and J.J. Ramsden, The Hofmeister series: salt and solvent effects on interfacial phenomena. Q. Rev. Biophys. 30 (1997) 241–278.
4. R.H. French, et al., Long range interactions in nanoscale science. Rev. Modern Phys. 82 (2010) 1887–1994.
5. P.G. de Gennes, Wetting: statics and dynamics. Rev. Modern Phys. 57 (1985) 827–863.
6. J.S. Rowlinson and B. Widom, Molecular Theory of Capillarity, Clarendon Press, Oxford, 1982.

The nano/bio interface

4

CHAPTER CONTENTS

Nanotechnology: An Introduction. DOI: 10.1016/B978-0-323-39311-9.00010-8
Copyright © 2016 Elsevier Inc. All rights reserved.

INTRODUCTION

As nanotechnology becomes more widely written and talked about, the term "nano/bio interface", as well as several variants such as the nano–bio and bio–nano interfaces, has emerged. This chapter gives an extended definition of the nano/bio interface.

One of the nanotechnology's protagonists, Eric Drexler, has robustly countered skepticism that Feynman's concept of making machines that make smaller machines could be continued down to the atomic level (Section 1.3) by adducing the many examples of biological mechanisms operating at the nanoscale [62]. This biological proof-of-principle has since been vastly reinforced by the acquisition of ever more detailed knowledge of the assembly, structure and mechanisms of these biological machines. Apart from this rather abstract meaning, the nano/bio interface clearly must signify the physical interface between a nanostructured nonliving domain and the living domain. The focus of attention is often how a living cell interacts with a nanostructured substratum, or how a biomacromolecule adsorbs to such a surface. These are important practical matters, but the nano/bio interface has a broader meaning. It constitutes a special case of the nonliving/living interface, in which the scale of salient features on the nonliving side is constrained to fall within a certain range. Nothing is, however, specified regarding the scale of the living side. It could be the very largest scale, that of an entire ecosystem or of human society. Below that, it might be useful to consider (multicellular) organisms, organs, tissues, cells and biomolecules. Clearly very different phenomena enter at each scale.

The distinction between nano–bio and bio–nano interfaces can be made in terms of *information* flowing from the nano domain to the bio domain (nano–bio) and vice versa (bio–nano). These two situations are denoted, respectively, as the nano–bio and bio–nano interfaces. The meanings of the two can be quite different. For example, considering the interface between nanotechnology and society, the nano–bio interface denotes the impact nanotechnology has on society; for example, how it might be completely transformed by the advent of molecular manufacturing. Conversely, the bio–nano interface denotes the introduction of a regulatory framework for nanotechnology. Considering our general environment, the nano–bio interface encompasses the response of soil microbes to nanoparticles added for remediation purposes, and conversely the bio–nano interface implies the destruction of engineered nanoparticles (perhaps added for remediation purposes) by soil microbes. At the scale of an organism (e.g. a human being), the nano–bio interface signifies the technology of scaling atomic-scale assembly up or out to provide human-sized artifacts; conversely, the bio–nano interface corresponds to the man–machine interface that a human being would use to control a nanoscale assembly process. Insofar as digitally encoded control is likely to be used, there should be no difference, in principle, between the interface for controlling a macroscale process and one for controlling a nanoscale process.

FIGURE 4.1

Left hand: a sharply curved "finger" of a protein abuts an inorganic surface. Right hand: a sharply curved inorganic nanoparticle abuts the membrane of a living cell.

Feynman mentions Albert R. Hibbs's suggestion of a miniature "mechanical surgeon" (nowadays referred to as a nanoscale robot or nanobot) able to circulate within the bloodstream and carry out repairs *in situ* [85]. Hogg conceives such a device as being about the size of a bacterium, namely, a few hundred nanometers in diameter [129]. Here, the nano/bio interface is delineated by the outer surfaces of the device and its nanoscale appendages; that is, the zone between the device and its living host. More generally, this meaning of the nano/bio interface refers to any situation in which a nanomaterial or a nanodevice is in contact with living matter.

If the radius of the "bio" side of the interface is less than that of the "nano" side (Fig. 4.1), we can refer to bio–nano; conversely, if the radius of the "nano" side of the interface is less than that of the "bio" side, we can refer to nano–bio.

A further meaning of "nano/bio interface" is the means with which humans interact with a nanodevice. While the protagonists of nanobots typically envisage a fully autonomous device, appraising its surroundings, processing the information internally and carrying out appropriate tasks accordingly, Hibbs presumed that information about its surroundings would be transmitted to a human surgeon, who upon analysing the data would then direct the repair work, much in the way that certain operations are already carried out today by remotely controlling tools fitted to the end of an endoscope, for example. In this case, the requisite nano/bio interface will be the same as the man–machine interface made familiar through digital information-processing technology: The man–machine interface has been a preoccupation of computer scientists ever since the inception of IT. At present, the issue scarcely arises for nanotechnology, since we do not yet have sophisticated nanosystems that need to be interfaced. The closest current realizations of assemblers operating with atomic precision are tip-based scanning probe devices that are digitally controlled; hence, their nano/bio interface is indeed a standard IT man–machine interface. This kind of control will, however, be far too cumbersome for assemblers that are themselves nanosized – the lags in regulation would tend to generate chaos – hence the assemblers will need to operate with a great deal of autonomy. Although the current generation of screen-based graphical user interfaces (GUIs) might be slightly more convenient than punched tape or cards, the laborious letter-by-letter entry of instructions or data via a keyboard remains slow, frustrating and error-prone. While hardware continues to advance exponentially (Moore's law), software and the man–machine interface continue to lag behind and limit human exploitation of IT.

Also encompassed within this meaning of the nano/bio interface is how a nanomachine can produce at the human scale. For example, a single nanostructured microsystem may synthesize only attograms of a valuable medicinal drug, too little even for a single dose administered to a patient. The solution envisaged is scaleout, i.e. massive parallelization – the chemical reactor equivalent of parallel computing. This has already been realized, albeit on a modest scale involving less than a dozen microreactors.

Possibly, much modern genetic engineering should also be included here. In essence, a great deal of molecular biology actually falls into nanobiotechnology.

4.1 THE "PHYSICAL" NANO/BIO INTERFACE

From the biological viewpoint, the nano/bio interface can be considered at three scales at least (examples given in parentheses):

1. the organismal scale (e.g. wearing clothing made from a nanomaterial);
2. the cellular scale (e.g. cell adhesion to a nanomaterial);
3. the molecular scale (e.g. protein adsorption to a nanomaterial).

There is an additional interface above the organismal scale, namely between society and nanotechnology, which comprises several unique features and will be dealt with in a separate section. To reiterate, the nano/bio interface can be considered a special case of the living–nonliving (or bio–nonbio) interface, in which the nonliving side has nanoscale features. If it is a nanomaterial (a nanostructured material or a nano-object), then many of the issues have been dealt with in a general way under the rubric of biocompatibility. If it is a nanodevice, then the issues are likely to belong to metrology (Chapter 5) or to bionanotechnology (Chapter 11). If it is a nanosystem, then effectively it will be interacting with a system on the living side, and we are again in the realm of the interface between society and nanotechnology (Chapter 12). The focus in the following sections is how the presence of nanoscale features introduces unique behavior not present otherwise.

4.1.1 ORGANISMS

It is a moot point whether nanostructure is perceptible by organisms. Certainly it is not directly visible, given that the organs of sight use visible light (cf. Eq. (5.2)). Could we otherwise become aware of (e.g. feel) a difference between a textile woven from a nanostructured fiber and one that was not, for example? Indirectly there might be no difficulty; with the help of a little water, a cravat whose fibers are nanostructured to make them water-repellent would be immediately distinguishable using the naked eye from an ordinary one. Natural fibers of course have structure at the nanoscale, but they are not nanostructured in the sense of being deliberately engineered at that scale. Similar considerations apply to foodstuffs: two aliments of the same overall composition may well feel different when first placed in the mouth if they have different structures at the nanoscale, but if this structure has not been deliberately engineered then we are not talking about nanotechnology according to the definitions (Section 1.1). Experiments to determine whether deliberately differently engineered structures are differently perceived do not yet appear to have been carried out in a sufficiently systematic fashion to be useful; indeed this aspect of the nano/bio interface is, as yet, practically unexplored.

The study of the deleterious biological consequences of nano-objects, especially nanoparticles, penetrating into the body constitutes the field of nanotoxicology (see Section 4.4). Particles may gain access to the body by inhalation, ingestion, through the skin, or following the introduction of material in some medical procedure such as implantation. If they are introduced into the circulation they are likely to be taken up by the macrophage system (see also Section 4.1.3). Toxic, or at any rate inflammatory, effects may arise indirectly through the adsorption of blood proteins on the surface of a particle and their subsequent denaturation (see Section 4.1.4). In other words, the effects on the overall organism depend on processes involving suborganismal scales down to that of molecules.

4.1.2 TISSUES

The tissue–substratum interface has been intensively investigated for many decades, starting long before the era of nanotechnology, in the search to develop more biocompatible orthopedic and dental implant materials. It has long been realized that a rough prosthesis will integrate better into existing bone than a smooth one. Contrariwise, an ultrasmooth stent is known to be less prone to adsorb proteins from the bloodstream than a rough one. This knowledge has not, however, yet been systematically refined by correlating nanoscale morphological features with the rate and quality of assimilation. Although surfaces have been treated so as to incorporate certain molecules into them, this has been done in a (bio)chemical fashion rather than in the spirit of atomically precise engineering, and there has been no comprehensive systematic study of nanoscale features evoking a response not seen otherwise, and disappearing if the features are expanded to the microscale.

The tissular nano/bio interface is of acute interest to the designers of nanoparticles for drug delivery. The main general challenge is achieving long circulation times of the nanoparticles in the bloodstream, which is in turn achieved by evading opsonization — a problem of protein adsorption, dealt with in detail in Section 4.1.4. There is also the question of how the nanoparticles are eliminated after they have done their work. Since solubility increases with increasing curvature according to the Gibbs–Thomson law (2.7), the smaller the particle radius, the faster it will dissolve. The designers of surgical instruments are often concerned with tribology aspects. Although naïvely it might seem that a perfectly smooth surface would have the lowest coefficient of friction, asperity serves to diminish the contact area and, hence, may even be advantageous. Nano-engineered asperity has, as yet, been incompletely explored as a route to control the friction experienced by objects (e.g. a surgical knife) in contact with tissue. An interesting research direction seeks to harness biological lubricants, notably the ubiquitous glycoprotein mucin, in order to control friction [308]. A further aspect of the tissue–nanomaterial interface is biomineralization. The tough, hard shells of marine organisms such as the abalone have a composite structure essentially consisting of mineral nanoplates cemented together by a protein matrix [150]. The exact mechanism of their formation has yet to be elucidated but presumably is essentially an extracellular process. Bone is apparently constructed on similar principles, although it has a more sophisticated hierarchical structure.

4.1.3 **CELLS**

This section considers two themes: (1) a nano-object small enough to penetrate inside a cell; and (2) the response of a supported cell to nanostructured features of its substratum (support).

Penetration of nano-objects into the cell's interior

It has been a long-standing frustration of electrophysiologists that the electrodes typically used to probe electrical activity in the interior of the brain are gigantic in comparison with the size of neurons and dendrites; hence, observations not only imply gross damage due to initial installation of the electrode but also that signals are recorded simultaneously from dozens, hundreds or thousands of neurons. The availability of carbon nanotubes (Section 9.4), which can be excellent electrical conductors, offers the possibility of creating an electrode considerably smaller than a single cell. This has the potential to revolutionize experimental neurophysiology. The main practical obstacle is the difficulty of manipulating the nanotubes: they must be attached to some kind of back-plate.

If the nano-object is freely floating, such as the drug-delivery nanoparticle already referred to, the main problem is controlling uptake. A very small molecule or nano-object (<10 nm) can enter (and leave) a cell directly by diffusion across the cell membrane. Particles in the range 10–100 nm are taken up by the cell by pinocytosis (a clathrin-assisted mode of endocytosis). They cannot leave the cell by this mode. Particles bigger than 200 nm are usually phagocytosed by phagocytes, a category of cells that includes macrophages, neutrophils, etc. Invagination of the particle results in a vesicle called a phagosome, which then fuses with one or more lysosomes containing hydrolytic enzymes, which break down the particle.

The penetration of a drug-bearing nanoparticle into the cell's interior may be precisely the goal of the particle designer, but nanoparticles used in other applications with a nonbiological purpose; for example, as ultraviolet-absorbing sunscreen applied to the skin, may, by similar mechanisms, also end up inside the cell where they may have toxic consequences. The investigation of these consequences constitutes what is currently the most active part of the field of nanotoxicology (Section 4.4). A related problem is the crossing of the blood–brain barrier by nano-objects.

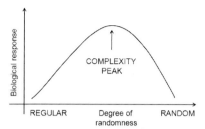

FIGURE 4.2

Proposed relationship between environmental complexity and biological response.

The response of a cell to external nanostructured features

The fundamental premiss engendering interest in living cells interacting with nanostructured surfaces is that there is strong evidence that cell surfaces are themselves heterogeneous at the nanoscale, whence the hypothesis that by matching artificial substratum patterns with natural cell surface ones, a degree of control over the cell can be achieved. "Patterns" are here to be understood as denoting statistical regularity only. Perhaps the degree of regularity of cell surface features determining the binding to substrata is intermediate between wholly regular and random; that is, maximally complex (Fig. 4.2).

While some responses may be of a purely physical nature, at least in part [the characteristic spreading (i.e. the transformation from sphere to segment) shown by some cell types placed on a planar surface could result merely from cell viscoelasticity opposing adhesion to the surface], others involve cellular intelligence: it is an adaptive response, the mechanism of which involves the reception of environmental information by the cell surface and the transmission of that information to the cell's genome, following which action in the form of activating or deactivating the synthesis of certain proteins (i.e. changes in gene expression) results. These proteins then engender a certain response. Sometimes it is the unique history of cell–environment interactions that determines cell behavior. Table 4.1 summarizes cell–environment responses.

Eukaryotic cells

The above hypothesis of maximum complexity originated in many observations that the behavior of cells depends on the nature of the basement membrane supporting them. A classic example is the different patterns of neurite outgrowths from neurons supported on different extracellular matrix materials such as laminin and tenascin. An early example of cells brought into contact with artificial materials was the experiments of Carturan et al. on immobilizing yeast within inorganic gels [40]. Eukaryotic cells are typically a few to several tens of μm in diameter. They are enveloped by a lipid bilayer (the plasmalemma) and the shape is controlled by the

Table 4.1 Cell Environmental Responses in Roughly Increasing Or
der of Complexity. See Section 5.8 for the Techniques Used to Measure
the Responses

Response	Level	Timescale	Technique[a]
Adhesion	Energetic	s	1,2
Spreading (morphology)	Energetic	min	1,2
Growth alignment	Energetic?	h	1
Microexudate secretion	Gene expression	min	2,3
Growth factor secretion	Gene expression	min	3
Alteration of metabolism	?	?	3
Differentiation	Gene expression	days	1,2,3,4
Speciation (i.e. cancer)	Chromosome rearrangement	years	1,2,3,4

[a]*Techniques useful for determining responses are 1, microscopy (usually optical, but may include scanning probe techniques); 2, nonimaging IF techniques (e.g. optical waveguide lightmode spectroscopy); 3, biochemical techniques (e.g. immunocytochemistry); 4, nucleic acid arrays.*

cell itself. In suspension, they tend to adopt the shape of the lowest surface:volume ratio (viz., a sphere) from purely mechanical considerations but on a solid surface tend to spread (i.e. transform into a segment). There is already a vast literature on the interaction of individual living cells with *microstructured* surfaces, defined as having features in the range 100 nm–100 μm. The main result from this large body of work is that the cells tend to align themselves with microscale grooves. The reader appraising this literature should be warned that sometimes these surfaces are referred to as "nano" without any real justification (since the features are larger than nanoscale). The goal of investigations of the nano/bio interface could be to demonstrate that unique features of cell response arise as feature sizes fall below 100 nm (taking the consensual definition of the upper limit of the nanoscale). Surprisingly, this has hitherto been done in only a very small number of cases. One of a few examples is the work by Teixeira et al. [282], who showed that when the ridges of a grooved pattern were shrunk to 70 nm, keratocyte alignment was significantly poorer than on ridges 10 times wider (whereas epithelial cell alignment was unchanged). Alternatively, following the spirit of Chapter 2, one could define one or more "biological response nanoscales" by presenting cells with a systematically diminishing set of structural features and observing at what scale the cells started to respond (cf. Fig. 4.2). Alignment is of course only one, perhaps the most trivial, of the huge variety of measurable responses. Table 4.1 summarizes some of the others.

A rather diverse set of observations may be systematized by noting that the fundamental response of a cell to an environmental stimulus is an adaptive one. Following Sommerhoff [270], one can distinguish three temporal domains of adaptation: behavioral (short term); ontogenic (medium term) and phylogenetic

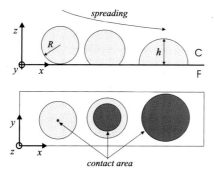

FIGURE 4.3

Idealized cell (initial radius R) spreading on a substratum F in the presence of medium C. Upper panel: cross-sections, showing the transition from sphere to segment (h is the ultimate height of the cell). Lower panel: plans, as might, for example, be seen in a microscope (the dark region shows the actual contact area).

(long term). Although Sommerhoff was discussing organisms, these three domains have their cellular counterparts. At the cellular level, behavioral adaptation is largely represented by energetically driven phenomena [174], perhaps most notably the spreading transition characteristic of many cells adhering to substrata (Fig. 4.3). Morphological changes have been shown [159], and the rate of spreading has been quantitatively shown, to depend on nanoscale features of the substratum [9].

The secretion of microexudates, growth factors, etc. would appear to involve the reception of chemical and/or morphological signals by the cell surface and their transmission to the nucleus in order to express hitherto silent genes. Indeed, this secretion appears to be part of an adaptive response to condition an uncongenial surface. Molecular biology has shown that the amino acid triplet arginine–glycine–aspartic acid (RGD), characteristically present in extracellular matrix proteins such as fibronectin, is a ligand for integrins (large transmembrane receptor molecules present in the cell membrane), binding to which triggers changes depending on cell type and circumstances. Such phenotypic changes fall into the category of ontogenic adaptation, also represented by stem cell differentiation. Recently reported work has shown that 30-nm diameter titania nanotubes promoted human mesenchymal stem cell adhesion without noticeable differentiation, whereas 70–100-nm diameter tubes caused a tenfold cell elongation, which in turn induced cytoskeletal stress and resulted in differentiation into osteoblast-like cells [223]. Phylogenetic adaptation involves a change of genetic constitution and would correspond to, for example, the transformation of a normal cell to a cancerous one. There are well-established examples of such transformations induced by certain nanoparticles. Two important criteria for deciding whether a nanoparticle might be carcinogenic seem to be as follows:

1. Whether the particle is insoluble (this does not, of course, apply to nanoparticles made from known soluble carcinogens active when dispersed molecularly);
2. Whether the particle is acicular and significantly longer than macrophages (see Fig. 4.11).

If both these criteria are fulfilled, the macrophages ceaselessly try to engulf the particles and they become permanent sites of inflation, which may be the indirect cause of tumorigenesis. Nevertheless, given the long induction periods (typically decades) that elapse between exposure to the particles and the development of a tumor, elucidation of the exact mechanism remains a challenge. Incidentally, there appear to be no known examples of nanostructured substrata that directly induce cancer through contact. However, it should be kept in mind that the mechanism for any influence may be indirect (as in [223]; note also that an extracellular matrix protein might bind to a certain nanotextured substratum, change its conformation – see Section 4.1.4 – and expose RGD such that it becomes accessible to a cell approaching the substratum, triggering ontogenic changes that would not have been triggered by the protein-free substratum) and is generally unknown. The secretion of macromolecular, proteinaceous substances that adsorb on the substratum is an especially complicating aspect of the nano/bio interface, not least because any special nanostructured arrangement initially present is likely to be rapidly eliminated thereby.

Prokaryotic cells

In contrast to eukaryotes, prokaryotes (archaea and bacteria) are typically spherical or spherocylindrical, smaller (diameters are usually a few hundred nm) and enveloped by a relatively rigid cell wall predominantly constituted from polysaccharides, which tends to maintain their shape. Prokaryotes have received rather less attention than eukaryotes, although the interaction of bacteria with surfaces is also a topic of great importance for the biocompatibility of medical devices, which is normally viewed in terms of adhesion rather than adaptation. Nevertheless, if one considers bacterial communities, the formation of a biofilm, based on a complex mixture of exudates, and which usually has extremely deleterious consequences for humans, should be considered as an adaptive response at least at the ontogenic level, since the expression pattern of the genome changes significantly. A worthy (and hitherto unreached) goal of investigation, therefore, is whether one can prevent biofilm formation by nanostructuring a substratum.

4.1.4 BIOMOLECULES

A protein is comparable in size to or smaller than the nanoscale features that can nowadays be fabricated artificially. The problem of proteins adsorbing to nonliving interfaces has been studied for almost 200 years and an immense body of literature has been accumulated, much of it belonging to the realm of physical chemistry and without any specific nanoscale ingredient. A good deal of the phenomenology can be satisfactorily interpreted on the basis of the Young–Dupré equations (Chapter 3), which allow one to link experimentally accessible single-substance surface tensions [e.g. via contact angle measurements on protein adsorbents (substrata), and on thin films carefully assembled from proteins] to IF energies. Wetting is a typical mesoscale phenomenon with a characteristic length ~30 nm, thus averaging out much molecular detail. This approach has allowed the systematization of a great deal of data for different proteins adsorbing on different substrata in the presence of different liquid media, rationalizing the interaction in terms of the IF free energy ΔG_{123}, the subscripts 1, 2 and 3 denoting adsorbent, liquid medium and adsorbate, respectively. As schematized in Fig. 4.4, the adsorption process involves first surmounting a repulsive energy barrier of height ΔG_a (the profile of which determines the adsorption kinetics, see Chapter 3), followed by residence at the interface in a potential well of depth ΔG_b. As complex objects, proteins typically undergo changes during residence on the surface, such as dehydration of their zone of contact with the substratum and/or denaturation, with consequential changes in the energetics of interaction.

The model represented by Fig. 4.4 was developed from thermodynamic principles without special assumptions regarding the structures of adsorbent, adsorbate and intervening medium. A clue that this model is too simplistic to represent reality was actually already discovered over a hundred years ago by Hofmeister. A nanoscale approach to the proteinaceous nano/bio interface takes cognizance of the following:

1. Details of the molecular structure on the interface, with explicit recognition of the solvent (water);
2. The density of dehydrons [83];
3. The surface tensions of highly curved features (eg, nanoscale ridges) differ from the values associated with planar surfaces;
4. Proteins typically have highly heterogeneous surfaces at the nanoscale [39].

This cognizance may be termed the (bio)physical chemistry of the nano/bio interface. Item (1) leads to the Hofmeister effect [56]; the full implication of item (2) has not yet been worked out; items (3) and (4) may lead to a different balance of forces and, as with living cells, it has been hypothesized (and demonstrated) that matching protein heterogeneity with artificial substratum nanoscale heterogeneity leads to anomalous behavior in protein adsorption [2]. At present, there is very little theoretical prediction of phenomenology at the nano/bio interface. Although the behavior of a single protein approaching a substratum can nowadays be considered to be reasonably well understood and predictable, real biomedical problems involve a multiplicity of proteins. It is well known that substrata exposed to blood experience a succession of dominant adsorbed proteins (the Vroman effect); until now this has not been comprehensively investigated using nanostructured substrata, and indeed to do so purely empirically without any guiding theory would be a daunting task. As part of this research direction, one should include the phenomenon of the protein "corona" hydrodynamically associated with a nano-object suspended in a proteinaceous medium (e.g. blood), or with a surface exposed to such a medium.

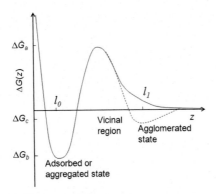

FIGURE 4.4

Sketch of the IF interaction potential $\Delta G_{123}(z)$ experienced by a protein (or other nano-object) approaching a substratum. The potential is the sum of different contributions, individually varying smoothly and monotonically, and the actual shape depends on their relative magnitudes and decay lengths. In this hypothetical (but typical) example, at moderate distances z from the substratum the net interaction is repulsive, dominated by long-range hydrophilic repulsion (at low ionic strength, electrostatic repulsion might be dominant). Sometimes (as shown by the dashed portion of the curve), a secondary minimum appears; low-energy objects unable to surmount the barrier ΔG_a may reside at a separation ℓ_1. At short distances, the attractive Lifshitz–van der Waals interaction dominates; adsorbed objects reside at a separation ℓ_0. At very short distances, the Born repulsion dominates. Further explanation is given in the text.

This corona can be expected to shift its composition as initially adsorbed proteins are exchanged for others. The mechanism of such *protein exchange* processes, in particular, their dependence on the IF free energies, is still very imperfectly understood, but it seems (at least, according to present knowledge, there is no reason to think otherwise) that by appropriately modifying the surface tensions and taking

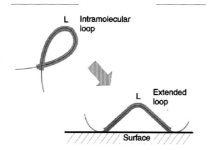

FIGURE 4.5

Surface-induced protein denaturation, showing the substitution of the intramolecular contacts in the native conformation by substratum–protein contacts and the concomitant transition from a compact globular conformation to an extended denatured one.

into account the heterogeneities of both adsorbate and adsorbent (possibly by simply summing all possible combinations of pairwise interactions) one would be able to correctly predict the entire phenomenology, including toxicology aspects. Once in residence on the surface, the protein may exchange its intramolecular contacts for protein–substratum contacts (Fig. 4.5), without necessarily any change of enthalpy, but the entropy inevitably increases because the extended, denatured conformation occupies a much larger proportion of the Ramachandran map than the compact, native conformation (Fig. 4.6).

4.2 NANOMEDICINE

One of the most important manifestations of the nano/bio interface is the application of nanotechnology to that branch of applied biology known as medicine. Mention has already been made of Feynman's inclusion of "microscopic surgeons" in his vision of what came to be called nanotechnology [85]. The dictionary definition of medicine is "the science and art concerned with the cure, alleviation and prevention of disease, and with the restoration and preservation of health". As one of the oldest of human activities accessory to survival, it has of course made enormous strides during the millennia of human civilization. Formally, it was well captured by the dictum "Primum nil nocere" (often ascribed to Hippocrates although not found in his oath); during the past few hundred years, and especially during the past few decades, it has been characterized by an enormous technization, and the concomitant enormous expansion of its possibilities for curing bodily malfunction. The application of nanotechnology,

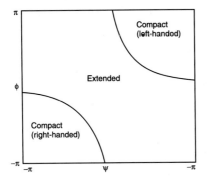

FIGURE 4.6

Sketch of the partition of the Ramachandran map showing the allowed topologies for an L-alanyl-like residue as a function of the two dihedral angles ϕ and ψ. The precise location of the separatrices (solid lines) is currently unknown, but the extended basin invariably has a larger area than the compact regions.

the latest scientific–technical revolution, is a natural continuation of this trend; the definition of nanomedicine is the application of nanotechnology to medicine.

4.2.1 A CONCEPT SYSTEM FOR NANOMEDICINE

Fig. 4.7 shows part of a concept system for nanomedicine. The primary division is between "auxiliary" and "direct", the latter signifying materials and devices that come into direct contact with living organisms. The "auxiliary" concept of medical diagnosis is part of nanotechnology by virtue of its heavy dependence on computation, as is molecular modeling for drug discovery. Automated synthesis also, of course, depends on software control.

As nanoparticles for introducing into the body become more sophisticated, they become devices (as soon they have information-processing capability). Theranostic devices combine sensing for diagnosis with drug release for therapy, but the implantable sensors currently under consideration are at least one or two orders of magnitude bigger than those with which a theranostic particle must be equipped, hence the concepts not been connected.

4.2.2 AUXILIARY APPLICATIONS

The three main activities encompassed by the auxiliary or indirect applications are

1. Any application of nanotechnology assisting drug discovery;

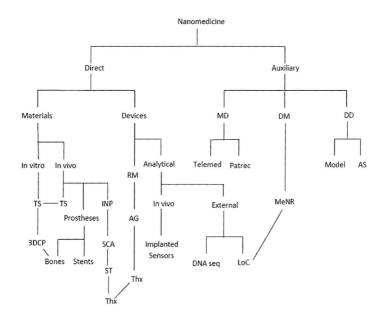

FIGURE 4.7

A concept system for nanomedicine. Abbreviations (from left to right, and top to bottom): MD, medical diagnosis; DM, drug manufacture; DD, drug discovery; Telemed, telemedicine; Patrec, pattern recognition; Model, molecular modeling; AS, automated synthesis; RM, responsive materials; TS, tissue scaffolds; INP, injectable nanoparticles; MeNR, microsystem-enabled nanoreactors; SCA, steerable contrast agents; AG, artificial glands; 3DCP, three-dimensional cell printing; ST, self-targeting (nanoparticles); DNA seq, DNA sequencing; LoC, lab-on-at-chip; Thx, theranostic nanoparticles/devices.

2. Any application of nanotechnology assisting drug manufacture; and

3. The indirect application of nanotechnology to medical diagnostics.

The main part of activity 3 is taken up by the enhanced processing power of nanoscale digital information processors, applied to areas such as pattern recognition and telemedicine, which might be defined as diagnosis and treatment by a medical doctor located geographically remotely from the patient. Activity (1) covers a very broad spectrum, including superior digital data processing capability [which impacts laboratory information management systems (LIMSs) and laboratory automation] as well as nanobiology. Activity (2) is mainly concerned with miniature – albeit micro rather than nano – chemical mixers and reactors; it is hoped that the microsystems paradigms for chemical synthesis will make it feasible to generate quasicustomized versions of drugs, if not for individual patients at least for groups of them. It would be appropriate to include under this heading the advanced materials, many of which are nanocomposites with superior barrier properties, used for packaging drugs.

4.2.3 DIRECT APPLICATIONS

The "direct" division encompasses *in vitro* and *in vivo* applications. Most important *in vitro* applications are materials ("tissue scaffolds") for helping to grow, in culture, tissues destined for implanting in human body (in so far as this depends on superior understanding of the behavior of living cells, it can be said to be part of nanobiology) and "labs-on-chips" – miniature mixers, reactors, etc., for carrying out analyses of clinical fluids. A typical lab-on-a-chip belongs to microsystems rather than nanotechnology, however, and is therefore out of the scope of this book, although there may be nanotechnology incorporated within the microfluidic channels (e.g. a nanoparticulate catalytic surface lining a tube), in which case one can speak of micro-enabled nanotechnology.

Advances in microtechnology have enabled biological cells to be printed using essentially the same technology as additive manufacturing (Section 8.3), which can be thought of as bottom-to-bottom fabrication not miniaturized right down to the nanoscale. As well as engineering tissue, the printing technique is also useful for preparing biosensors (e.g. microcantilevers bearing living cells) and cell research (e.g. preparing pure populations of a particular cell) [75].

4.2.4 DNA SEQUENCING

A very interesting area is the development of miniature DNA sequencing devices based on the physical, rather than the chemical, properties of the nucleotides [319]. The four different DNA "bases" (or nucleotides, symbolized as A,C,G,T) differ not only in their chemical nature, but also in their physical nature, most significantly as regards size and shape. An early hope was to use that quintessentially "nano" metrology device, the atomic force microscope (Fig. 5.1), to determine the base sequence of DNA by directly imaging the polymer, at sufficient resolution to be able to distinguish the four different bases by shape. This has turned out to be far more difficult than originally envisaged, but other methods based on examining a single or double DNA chain are being invented and examined.

The favored scheme is to pass the DNA strand through a nanopore while measuring ionic conductance (of the electrolyte solution in which the DNA is dissolved), either along or across the pore, with the resolution of a single base [77]. The different nucleotides can be thus distinguished, but it is difficult to capture the DNA and drive it through the pore. Both artificial and biological pores are being investigated. An interesting development would be to use the pores assembled from oligopeptides in an electric field-dependent manner [107]. This does not yet appear to have been investigated.

In vivo applications encompass nanostructured materials, such as materials for regenerative medicine (which might well be rather similar to the tissue scaffolds used *in vitro*) and materials for implantable devices, which might include prostheses for replacing bones and stents for keeping blood vessels open [256]. Nanotechnology is attractive for implants because, through creating novel combinations of material properties, it should be possible to achieve performances superior to those available using traditional materials.

Additive manufacturing has now advanced to the point at which living cells can be assembled in a controlled fashion ("bioprinting") [75]. This is developing to a degree that it can be used for creating complete vascularized systems, ready for implementation into the body, or to create living bones as an alternative to metal or ceramic implants.

Some of these materials are of great sophistication, such as advanced polymers for self-regulated insulin delivery [202]. Such stimulus-responsive materials fall into the category of sensorial materials [176], but go beyond them by also having an actuation function. Another useful application is a nanosponge designed to absorb toxins released by infectious bacteria at wound sites [295]. One of the functions of such toxins is to defend the bacteria from attack by the body's immune system. By removing the toxins, the body can itself deal with the infection without requiring the use of antibiotics.

4.2.5 *IN VIVO* APPLICATIONS

Direct *in vivo* applications of nanotechnology also encompass the implantation of devices for sensing biochemicals (physiological markers). The smaller the device, the more readily it can be implanted. The technology involved is basically the same as that being developed for nanobots (Section 4.2.7). The sensors may simply be functionalized nanoparticles [148]. Having a gold core, they can be readily detected using various kinds of sensor. At present, such materials are probably most useful for analysing biopsies. Indeed, a very large category of devices introduced into the body is that of medicinal nanoparticles, as will be explored in the next section.

4.2.6 MEDICINAL NANOPARTICLES

The most active current field of *in vivo* nanomedicine is the use of nanoparticles of progressively increasing sophistication. The two main classes of applications are as contrast agents, mainly for imaging purposes but more generally for diagnostics and as drug delivery agents. In both cases, the particles are administered systemically. The major challenge is to ensure that they circulate long enough before opsonization to be able to accumulate sufficiently in the target tissue. If the particles are superparamagnetic (see Section 2.6), they can be steered by external electromagnetic fields, accelerating their accumulation at the target. Since cancer cells typically have a significantly faster metabolic rate than that of host tissue, substances will tend to be taken up preferentially by cancer cells, assuming that there is a pathway for uptake. In other cases, and also to enhance the uptake differential between cancerous and noncancerous tissue, the nanoparticles are "functionalized"; that is, the surface is specially treated in some way (e.g. by coating the surface with a ligand for a receptor known to exist in the target cells and not elsewhere). For example, this can be used to target metastases, which are extremely difficult to identify by conventional medical means [37].

The simplest medicinal nanoparticles are those made of a single substance (e.g. magnetite) which, when taken up by the tissue, can be subjected to an external electromagnetic field such that they will absorb its energy, become hot and transfer the heat to the cells surrounding them, thereby killing them (necrosis). The particles can themselves be considered as the drug, which requires the external field for activation. Other examples of single-substance drug particles are gold and silver. Nanoparticulate (potable) gold was already introduced as a therapeutic agent by Paracelsus about 500 years ago. Nanoparticulate silver is nowadays widely used (e.g. in toothpaste) as a general antibacterial agent.

At the next level of sophistication, the particles constitute a reservoir for a small organic-molecular drug. Prior to use the particles (which may contain numerous pores of an appropriate nature) are saturated with the drug, which starts to dissociate once the particles are introduced into the human body. At the current level of the technology, this approach works best if the characteristic dissociation time is long compared with time needed to become accumulated at the target. Hemoglobin is the paradigm for this type of particle but is still far from having been successfully imitated by an artificial device (see Chapter 11). Novel materials for drug delivery are still being actively sought. Metal–organic framework (MOF) molecules (Section 6.5) are among the latest candidates to be investigated for this purpose [132].

Slightly more sophisticated are particles in which the reservoir is initially closed, the closure being designed to resist opening under conditions likely to be encountered before final arrival at the target. A simple example is a calcium carbonate nanoshell enclosing a stomach drug designed to be administered orally. The calcium carbonate will dissolve in the acidic conditions of the stomach, thereby releasing the drug. This is an example of a responsive particle that might even be called adaptive under a highly restrictive set of conditions. The ideal "smart" adaptive nanoparticle will sense its surroundings for the presence of its target; bind to the target when it reaches it; sense its surroundings for the local concentrations of the therapeutic agent it carries; and release its burden if the surrounding concentration is low. At present, most research in the field is concerned with devising novel adaptive materials for drug delivery (nanostructure is especially useful here because it allows multiple attributes to be combined in a single material).

Any system based on a reservoir has a finite capacity and hence therapeutic lifetime. For many therapeutic regimens, this may not be a disadvantage, since many drugs should only be administered for a limited duration. For long-term therapy, the goal is to create nano-objects able to manufacture the drug from substances they can gather from their immediate environment (i.e. veritable nanofactories).

4.2.7 MEDICINAL NANODEVICES

A responsive particle is already capable of information processing and is, therefore, a device. Particles combining the ability to concentrate themselves at a target with delivery of an appropriate therapeutic agent actually combine diagnostics with therapy, a combination that is sometimes called theranostics. This particular technological

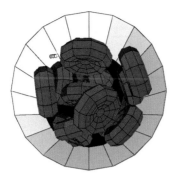

FIGURE 4.8

Sketch of the likely form of a future nanobot (the small bacterium-sized cylinder in the upper left quadrant) drawn to scale in a blood vessel containing erythrocytes.

-Reproduced with permission from T. Hogg, Evaluating microscopic robots for medical diagnosis and treatment, Nanotechnol. Percept. 3 (2007) 63–73.

development has, however, met with some opposition because in traditional medicine, after a diagnosis is made a discussion between the physician and the patient usually ensues in order to decide whether to proceed with therapy and if so what kind, whereas a theranostic device acts autonomously without the explicit authorization of the patient.

The ultimate in sophistication of the nano-object *qua* device is the nanoscale robot or "nanobot". Microscopic or nanoscopic robots are an extension of existing ingestible devices that can move through the gastrointestinal tract and gather information (mainly images) during their passage. As pointed out by Hogg [129], minimal capabilities of such devices are: (chemical) sensing; communication (receiving information from, and transmitting information to, outside the body, and communication with other nanobots); locomotion – operating at very low Reynolds numbers (around 10^{-3}), implying that viscosity dominates inertia; computation (e.g. recognizing a biomarker would typically involve comparing sensor output to some preset threshold value; due to the tiny volumes available, highly miniaturized molecular electronics would be required for constructing on-board logic circuits of practically useful data processing power); and of course power – it is estimated that picowatts would be necessary for propelling a nanobot at a speed of around 1 mm/s. Such nanobots could also incorporate one or more drug reservoirs, or miniature devices for synthesizing therapeutic substances using materials found in their environment. It is very likely that to be effective, these nanobots would have to operate in large swarms of billions or trillions. Fig. 4.8 sketches a putative nanobot.

It may well be that the most important future rôle for nanobots will be to carry out tricky repairs that at present require major surgery; for example, clearing plaque from the walls of blood vessels, or renewing calcified heart valves. The body has wonderful repair mechanisms of its own but for generally presently unknown reasons certain actions appear to be impossible.

4.2.8 IMPROVING THE HOSPITAL ENVIRONMENT

The photocatalytic (photoelectrochemical) effect was discovered in 1972 by Fujishima and Honda using a macroscopic titanium dioxide electrode [90]. Subsequently, the same effect was discovered in titanium dioxide nanoparticles [66]. Indeed, because the photocatalysis takes place at the surface of the material, the use of nanoparticles greatly enhances the effect, because of their immense specific surface area. The mechanism is as follows: the absorption of a photon by the semiconductor (titanium dioxide) generates an electron in the conduction band and a "positive hole" (whence the electron came) in the valence band. This process is illustrated in Fig. 4.9. The two charged entities (electron and positive hole) rapidly migrate to the surface of the particle where they can react with substances adsorbed onto its surface or in its close vicinity [246]. In fact, the primary mechanism of charge separation (without which the photogenerated electron and positive hole would simply recombine) is different in the microscopic electrodes and in the nanoparticles. In the former, the charges are separated by being generated within the space charge layer in the vicinity of the electrode surface. This typically drives the electrons to the interior of the electrode and the positive holes to the surface. Hence, only oxidation reactions can take place at the surface. In contrast, the nanoparticle is too small for there to be any significant space charge; the charge carriers are statistically separated because both migrate to different locations on the surface faster than they can recombine.

The electron is a powerful reducing agent and the positive hole is a powerful oxidizing agent. Most organic compounds are mineralized to carbon dioxide in the presence of illuminated titanium dioxide [122]. Water is an essential reagent in most of these mineralization processes. In 1985, it was reported that bacteria could be photocatalytically sterilized using titania particles [203].

The efficacity of any given semiconductor to photocatalytically destroy contaminants in the environment depends, first, on the energies of the conduction and valence bands relative to the reduction and oxidation (redox) potentials of the contaminants. The band energies of titanium dioxide are favorably positioned for a broad spectrum of contaminant destruction. It is also important that the illuminated semiconductor does not destroy itself (via the process called photocorrosion). Titanium dioxide happens to be remarkably stable in this respect.

It would therefore appear that if the surfaces (walls, ceilings, etc., as well as textiles, e.g. curtains) within a hospital are coated with titanium dioxide nanoparticles, the interior is then continuously decontaminated and sterilized. Furthermore, titanium dioxide is an abundant, relatively cheap material that is nontoxic to humans (titanium is the most commonly used material for implants in human body and the metal is

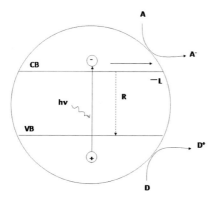

FIGURE 4.9

Diagram of the photocatalytic effect in a semiconductor nanoparticle, showing the excitation of an electron–hole pair by a light quantum ($h\nu$) followed by the migration of the charge carriers to the particle surface, where the electron carries out reduction of acceptor A and the hole oxidation of donor D.

natively coated with a thin film of oxide). Hence, by coating the interior surfaces, one is not introducing any toxic material into the hospital.

Although it has been pointed out that "the level of evidence supporting different disinfection and cleaning procedures performed in healthcare settings worldwide is low" [58], there seems to be little doubt that cleaning does lower the microbial burden within a hospital. The photocatalytic approach compares very favorably with traditional disinfectants because:

1. it operates continuously without attention, other than ensuring that it is illuminated;
2. there appears to be no mechanism whereby microörganisms could acquire resistance against the photocatalytic action;
3. the use of microbiocidal chemicals, which are also unpleasant for humans to handle, is avoided, hence eliminating the incidence of skin and other allergies among cleaning staff.

4.2.9 WIDER IMPLICATIONS

It is customary nowadays to take a global view of things, and in assessing the likely impact of nanotechnology on medicine this is very necessary. Nanotechnology is often viewed to be the key to far-reaching social changes, and once we admit this link then we really have to consider the gamut of major current challenges to human civilization, such as demographic trends (overpopulation, aging), climate change, pollution, exhaustion of natural resources (including fuels) and so forth. Nanotechnology is likely to influence many of these, and all of them have some implications for human health. Turning again to the dictionary, medicine is also defined as "the art of restoring and preserving health by means of remedial substances *and the regulation of diet, habits, etc.*" With this in mind, it would be woefully inadequate if the impact of nanotechnology on medicine were restricted to consideration of the development of the topics discussed in the preceding parts of this section. One may even pause to consider the validity and implications of Illich's view that modern medicine seeks to achieve bodily perfection and immortality, both of which are actually unattainable and, as he argues, not even desirable. Nanomedicine increases the focus on the technical capabilities of medical practice to the detriment of proper consideration of its ethical aspects. Awareness of this potential imbalance is the first step towards redressing it.

4.3 NANOPARTICLES IN THE ENVIRONMENT

The main environmental use of engineered nano-objects is for the remediation of polluted air, water and soil. The main contribution to clean air lies in the incorporation of catalysts in the exhaust stream of combustion devices in order to eliminate a variety of noxious substances, including carbon monoxide and nitrogen oxides. Applications range from physically small – motor vehicle exhausts – to very large – the flue stacks of fossil fuel-burning electricity generating stations. Typically, the catalyst is pressed into hollow honeycomb-like shapes; hence, even if the original material is nanoparticulate, beyond assisting the sintering process through the lowering of the melting point (Section 2.2), the nanoscale of division is no longer required in the final product. It has been demonstrated that cerium oxide is a combustion catalyst (it appears to work by activating oxygen on its surface) and can be added as nanoparticles to automotive fuel, improving combustion efficiency.

Most of the interest in nanoparticles for water remediation center on the extensive work of many researchers on the photocatalytic oxidation and reduction of organic pollutants using semiconductors, among which titanium dioxide is favored because of its stability with respect to photocorrosion. Its band gap is, however, relatively wide; hence, it is only activated by light from the violet end of the spectrum and ultraviolet radiation. The oxidizing and reducing entities are generated on the surface of the semiconductor; hence, the pollutant molecules must be adsorbed on the surface in order to react. The advantage of using nanoparticulate material is obvious since the surface area becomes very large. It should be noted that the mechanism of charge separation of the electrons and positive holes generated by band gap illumination is fundamentally different in a nanoparticle compared with bulk material (cf. Section 7.12.3) [66].

There are, however, many practical difficulties involved in using nanoparticles for water remediation, mostly connected with the problem of separating the water from the particles after the pollutants have been destroyed. If the particles are somehow immobilized, as in the catalysts used for combustion exhaust remediation, then the advantage of the huge surface area is lost. One attractive approach is to create core–shell nanoparticles with a magnetic core (e.g. made from magnetite) and a photocatalytic shell (e.g. made from titania). After remediation, a powerful external magnetic field can be used to collect the particles, which can also then be reused.

The interest in soil remediation stems from the known ability of iron to decompose organochlorine molecules upon contact. Scrap metal has already been used to remove such pollutants from contaminated sites. Again, finely dividing the material increases the surface area and, hence, reactivity. In the case of iron, the main drawback is that the reactivity to atmospheric oxygen also increases and initially zerovalent particles may end up being wholly oxidized. Surprisingly, little seems to be known about the ability of iron oxides to catalyse the decomposition of organochlorines.

Furthermore, the nanoparticles are likely to end up being incorporated in the soil, which is a very complex multiphase material [261], and inactivated by being encapsulated in humic acids. If they remain free, they may be taken up by the numerous microörganisms essential for the efficient operation of soil ecosystems, with largely unknown effects. For this reason, many countries have introduced a moratorium on the deliberate introduction of engineered nanoparticles into the soil, even in the case of such apparently benign materials as the iron oxides.

4.4 NANOTOXICOLOGY

There is currently great concern about the potential adverse impacts of nano-objects on human health. Nanotoxicology is defined as the study of the toxicology of nano-materials; while allergic haptic reactions to certain morphologies and combinations of atoms cannot be ruled out (i.e. triggered by skin contact with the surface of a bulk nanostructured material), overwhelming attention is currently given to nano-objects inhaled or ingested or otherwise entering the body.

Contact with nano-objects occurs under three types of circumstance: occupational, in the premises involved in manufacturing the nano-objects; clinical, in which nano-objects are deliberately introduced into the body for the purposes of diagnosis, therapy or both; and accidental, in which members of the public may be exposed to nano-objects released from an industrial container broken in transit, for example.

Nano-objects used in cosmetics and other personal care products (e.g. ultraviolet-absorbing semiconductor nanoparticles in sunscreen; bactericidal nanosilver particles in toothpaste) rank essentially as medicines in the above categorization; contact with the nanomaterial can be avoided simply by renouncing the use of the product. It should be noted that most regulatory régimes impose far less stringent requirements on cosmetics and personal care products than on pharmaceuticals.

The occupational *hazard* from a nano-object is generally the same as the hazards encountered in other contexts; *exposure* is, however, likely to be significantly higher;

hence, greater attention needs to be paid to the risks. Much of nanotoxicology is presently concerned with assessing hazard rather than exposure. Nevertheless, there are encouraging signs that exposure is also being looked at more carefully [67].

4.4.1 ENVIRONMENTAL NANOPARTICLES

Any consideration of the risk from engineered nanoparticles must include a sensible assessment of the background exposure to natural nanoparticles present in the environment, to which mankind is presumably adapted. Ubiquitous natural nanoparticles include the ultrafine component of desert sand, which is often transported over great distances by the wind, salt aerosols produced by the evaporation of sea spray, volcanic ash, the products of combustion and so forth. Nevertheless, it should be pointed out that there has been an enormous increase in the amount of combustion on the Earth, concomitantly with the exploitation of fossil fuels. Urban atmospheres have become quite sooty due to vehicle exhaust emissions and domestic heating. The construction industry also contributes a great deal of fine dust from demolition activities. Although these nanoparticles are omnipresent, they are not necessarily healthy. It should also be borne in mind that ever more people are living in an urban environment. Furthermore, some deserts are not natural but have been created by deliberate, large-scale geoengineering. The former Aral Sea is an example, the dry bed of which is now a terrible source of dust.

Apart from these more or less natural nanoparticles, a key issue is whether our human defense mechanisms can adapt to cope with new nano-objects emanating from the nanotechnology industry. It is clear that they cannot in the case of long insoluble fibers (e.g. amphibole asbestos – which are of course natural – and carbon nanotubes). In the case of other environmental nano-objects such as smoke (from fires and tobacco) and mineral particles, the dose (concentration, duration, intermittency, etc.) and accompanying lifestyle seem to play a primordial rôle.

Apart from nanoparticles given therapeutically, wear of implanted prostheses, such as hip replacements, generates micro- and nanoparticles in abundance (e.g. Fig. 4.10). In fact, before the advent of the nano-industry, much of the knowledge about the biological effects of nanoparticles came from the investigation of the inflammation caused by joint prostheses.

4.4.2 ASSESSING HAZARD FROM NANO-OBJECTS

The toxicity of nano-objects falls into two categories: (1) they can be considered as chemicals with enhanced reactivity due to their small size (the smaller the object the more reactive it is because of its curvature – see Section 2.4); hence, if the substance from which they are made is toxic or can become toxic in the body, the nano-object is likely to be toxic – as an example consider mercury (Hg), which if ingested as millimeter-sized droplets typically passes through the body unmetabolized, but which in nanoparticulate form is likely to be readily oxidized to highly toxic Hg^+ or Hg^{2+}; or (2) they are not toxic *per se* but initiate an immune response, ranging from mild temporary inflammation to persistent chronic inflammation and maybe ultimately

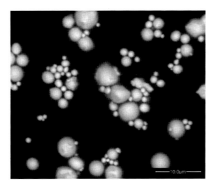

FIGURE 4.10

Scanning electron micrograph of CoCr particles within tissues adjacent to a human implant, retrieved at revision surgery.

-Reproduced with permission from P.A. Revell, The biological effects of nanoparticles, Nanotechnol. Percept. 2 (2006) 283–298.

carcinogenesis. Many metal oxide nanoparticles readily catalyse the production of reactive oxygen species, which can be quite damaging to living matter.

The mechanism of the immune response evoked by many nano-objects is denaturation of proteins adsorbed onto their surface from their environment (see Fig. 4.5), which may be the cytoplasm, intracellular space or a fluid such as blood (whether denaturation actually takes place depends on the chemical nature of the nano-object's surface [226]). The denatured protein-coated nano-object is then identified as nonself by the immune system, which initiates the processes that should lead to elimination of the foreign object.

4.4.3 ENTRY OF NANO-OBJECTS INTO CELLS

Very small nano-objects (less than 10 nm) can usually enter a cell directly by diffusion across the cell membrane. Particles of size 10–100 nm are taken up via a clathrin-assisted mode of endocytosis called pinocytosis (Section 4.1.3); particles greater than 200 nm are usually consumed by macrophages [193].

4.4.4 ELIMINATION OF NANO-OBJECTS

A nanoparticle that is soluble in the cellular or extracellular milieu in which it finds itself will evidently be readily eliminated, without harm to the organism unless the product of solubilization is toxic.

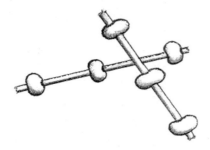

FIGURE 4.11

Sketch of a typical scanning electron micrograph of two amphibole asbestos needles
within the pleural cavity occupied by several
macrophages.

-Reproduced with permission from C.J. van Oss, R.F.
Giese, Properties of two species of deadly nano-needles,
Nanotechnol. Percept. 5 (2009) 147–150.

The most powerful weapon possessed by the macrophages, namely, the lysosomes
containing hydrolytic enzymes that fuse with the phagosome formed by invagination
of the nano-object, is largely useless against the inorganic semiconductors from
which nanoparticles are most commonly made. The entire phagocyte can, however,
be eliminated along with the nanoparticles it has ingested.

One of the chief problems is caused by nanofibers. If insoluble and excessively
elongated, as is blue asbestos and most carbon nanotubes, they can be neither
solubilized nor ingested by the macrophages (Fig. 4.11), which nevertheless persist
indefinitely in a futile attempt to destroy them, engendering sickness, typically via
inflammation.

The case of nanofibers well illustrates the importance of shape, as well as size
and chemical composition, in determining the uptake, metabolism and excretion of
nanoparticles.

Not surprisingly, given the intricate internal structure of the kidney [286], renal
clearance is particularly liable to be hindered if a nanoparticle is too large. Note that
the acquisition of a protein corona (Sections 4.1.4 and 5.8.1) may increase the size
and switch a particle from excretable to retained. A particular challenge is that surface
treatments to prevent opsonization will also make clearance more difficult.

Developers of nanoparticle-based therapeutics need to respect the doctrine,
conventionally ascribed to Hippocrates, of "Primum nil nocere" and consider what
happens to the millions or billions of systemically introduced nanoparticles after they
have accomplished their therapeutic task.

4.4.5 ASSESSING EXPOSURE TO NANO-OBJECTS

Exposure via inhalation is, rightly, given the most attention because it is generally the most efficient way of introducing foreign substances into the body. Inhalation exposure is especially relevant to occupational nanotoxicology. Dermal exposure may occur intentionally if sunscreen containing nanoparticles is applied to the skin, and accidentally if a suspension of nano-objects is spilt or splashed. Therapeutic nanoparticles are almost exclusively administered intraperitoneally.

The degree to which nano-objects penetrate into the respiratory tract not surprisingly depends on the size of the objects. Particles bigger than about 10 μm are trapped in the nasopharyngeal compartment, but particles smaller than about 100 nm penetrate deep into the alveolar region. Within the respiratory tract, objects are eliminated partly by a mechanical process, the mucociliary "escalator". Particles deep in the alveoli may be eliminated by the blood circulation, implying that they have to traverse one or more cell membranes by the mechanisms already discussed.

4.4.6 HANDLING NANO-OBJECTS

If a canister filled with nano-objects is opened, the contents will immediately start to fly round the room, likely posing a considerable hazard. For this reason, nano-objects made on an industrial scale are usually deliberately agglomerated into objects that may be hundreds of nanometers in diameter, before they leave the factory. Some of the earliest nano-objects manufactured on a large scale, such as the well-known P25 (titania), are irreversibly aggregated. They may not then rank as a true nanomaterial, but at least they pose a much diminished hazard.

4.4.7 ENVIRONMENTAL NANOTOXICOLOGY

This field, also known as nano-ecotoxicology, is concerned with the hazard, fate and effects of nano-objects in the environment. The general principles are the same as for regular nanotoxicology: living species are exposed to nano-objects and the effects are recorded (e.g. [30]). One important difference is that inhalation is unlikely to be a significant route of entry in most nonhuman ecosystems. The species may be higher animals (fish are often chosen) or smaller organisms such as the common water flea *Daphnia magna*. The latter are often used in mortality studies in which the concentration of a toxin at which 50% of the fleas die within 48 h is determined (and called EC_{50}). The former are, of course, more troublesome, especially if comparable statistical power is to be achieved, but they offer the opportunity to observe a much vaster range of consequences than mere mortality, especially more or less subtle behavioral effects. The most recent focus is on the prokaryotic denizens of the soil.

Unlike occupational nanotoxicology, in which any exposure is likely to be pristine nano-objects, in environmental nanotoxicology due cognizance must be taken of the fact that our planet has an oxidizing atmosphere, and has had one probably for at least 2000 million years, mainly due to the action of plants [185]. This implies that most metals, other than gold, platinum and so forth (the noble metals), will be oxidized. Hence, many kinds of metallic nanoparticles will not be stable in nature.

4.4.8 PROSPECTS

Although a great deal of work has been and continues to be done on the biological effects of nano-objects, the present effort seems both uncoördinated (with much duplication of effort worldwide alongside the persistent presence of significant gaps in knowledge) and misguided. Firm general principles as adumbrated above already seem to have been established, and there is no real need to verify them *ad nauseam* by testing every conceivable variant of nano-object. Such tests are easy to do at a basic level (the typical procedure seems to be as follows: introduce a vast quantity of nano-objects into the test species, observe behavior, physiological variables, etc., and finally establish in which cells or tissues the nano-objects were accumulated, if not excreted). To arrive at meaningful results, however, far more attention needs to be paid to chronic exposure under realistic conditions. Furthermore, given that we already know that the occurrence of tumors traceable to exposure to nano-objects typically manifests itself after a delay of several decades, tests need to be long-term. The increasingly dominant mode of funding scientific research in the developed world, namely, project grants lasting for two or three years, is not conducive to the careful, in-depth studies that are required to achieve real advances in understanding (neither in nanotoxicology nor, it may be remarked, in other fields of scientific endeavor); rather, this mode encourages superficial, pedestrian work with easily foreseen outcomes. Hopefully at least a few organizations like the National Institutes of Health in the USA will continue to enable a scientist to devote a lifetime of truly exploratory work to his or her chosen topic.

SUMMARY

The nano/bio interface comprises three meanings. First is the conceptual one: the "living proof of principle" that nanoscale mechanisms (the subcellular molecular machinery inside a living cell) exist and can function. Within this meaning there is also an inspirational aspect: living cells are known to distinguish between natural structures differing from one another at the nanoscale, suggesting that artificial mimics can be used to invoke specific living responses. Second is the man–machine interface aspect: how can humans control atomic-scale assembly? Conversely, how can atomic-scale assembly be scaled up to provide artifacts of human dimensions? Third is the literal physical boundary between a living organism and a nanomaterial, device or system. This applies both to nanobiotechnology (the application of nanotechnology to biology, e.g. implantable medical devices) and bionanotechnology (the use of biological molecules in nanodevices, i.e. the topic of Chapter 11). This "bio-physical" interface has several characteristic scales from the biological viewpoint: organismal, cellular and biomolecular. Each scale is examined, considering the nano/bio as a special case of the general problem of the nonbio–bio (nonliving–living) interface.

Perhaps the most important manifestation of nanobiotechnology is nanomedicine. This is defined and discussed, concluding with the toxicology of nano-objects.

FURTHER READING

1. R. Arshady and K. Kono (eds), Smart Nanoparticles in Nanomedicine, Kentus Books, London, 2006.
2. G. Hunt and M. Riediker, Building expert consensus on problems of uncertainty and complexity in nanomaterial safety, Nanotechnol. Percept. 7 (2011) 82–98.
3. M. Longmire, P.L. Choyke and H. Kobayashi, Clearance properties of nano-sized particles and molecules as imaging agents: Considerations and caveats, Nanomedicine (Lond.) 3 (2008) 703–717.
4. G. Oberdörster, E. Oberdörster and J. Oberdörster, Nanotoxicology: an emerging discipline evolving from studies of ultrafine particles, Environ. Health Perspect. 113 (2005) 823–839.
5. J.J. Ramsden, D.M. Allen, D.J. Stephenson, J.R. Alcock, G.N. Peggs, G. Fuller and G. Goch, The design and manufacture of biomedical surfaces, Ann. CIRP 56/2 (2007) 687–711.
6. P.A. Revell, The biological effects of nanoparticles, Nanotechnol. Percept. 2 (2006) 283–298.
7. D. Rickerby and M. Morrison, Prospects for environmental nanotechnologies, Nanotechnol. Percept. 3 (2007) 193–207.

CHAPTER

Nanometrology

5

CHAPTER CONTENTS

Nanotechnology: An Introduction. DOI: 10.1016/B978-0-323-39311-9.00011-X
Copyright © 2016 Elsevier Inc. All rights reserved.

INTRODUCTION

The ultimate goal of nanometrology is to provide the coördinates and identity of every constituent atom in a nano-object, nanostructured material or nanodevice. This goal raises a number of problems, not least in the actual representation and storage of the data. Thanks to techniques such as X-ray diffraction (XRD), one can readily determine atomic spacings to a resolution of the order of 0.1 nm and this information, together with the external dimensions of an object, effectively achieves our goal provided the substance from which the object is made is large, monolithic and *regular*. This achievement must, however, be considered as only the beginning since nanomaterials (and nanodevices considered as complex nanomaterials) may be fabricated from a multiplicity of substances, each one present as a domain of unique and irregular shape. Techniques such as XRD require averaging over a considerable volume to achieve an adequate signal to noise ratio for the finest resolution – hence the stipulations that the object being examined be large and regular. As an illustration, consider the challenge of reverse engineering a VLSI on a "chip" – bearing in mind that this is essentially a two-dimensional structure. XRD of the chip could not yield useful information for the purpose of making an exact replica of the circuit. The challenge of nanometrology is to achieve atomic-scale resolution for such arbitrary structures (the chip is not of course truly arbitrary – the structures form a functional circuit, but knowledge of the function alone would not suffice to help one reconstruct the details, not least because there are likely to be several hardware routes to achieve a given function).

Any successful manufacturing technology requires appropriate metrology, and the atomic-scale precision implicit in nanotechnology places new demands on measuring instruments. Reciprocally, the development of the quintessential nanometrology tool, namely, the scanning probe microscope (SPM, better called an ultramicroscope or nanoscope) has powerfully boosted nanotechnology itself, since these instruments have become the backbone of efforts to develop bottom-to-bottom manufacturing procedures (Section 8.3). The focus of this chapter is the nanometrology of surfaces. For more general issues regarding ultraprecision measurement instrumentation, the reader may refer to the recent book by Leach (see Section 5.8.8).

Morphology and chemistry are not independent at the nanoscale. Depending on how it is cut, the planar face of a crystal of a binary compound MX can vary dramatically from pure M to pure X. "Roughness" or texture at the nanoscale may actually be constituted from an intricate array of different crystal facets. The chemical effect of this morphology depends on the characteristic length scale of the phenomenon being investigated. Living cells, for example, are known to be highly sensitive to the crystallographic orientation of a substratum. This has been demonstrated by cell growth experiments on single crystals: epithelial cells attached themselves and spread only on the (011) faces of calcium carbonate tetrahydrate and not on the (101) faces, within tens of minutes following initial contact, but after

72 hours all cells on the (011) faces were dead, but well-spread and living on the (101) faces [117]. However, it should be noted that these two faces mainly differ in the surface distribution of lattice water molecules, to which the living cell may be especially sensitive. Note that cells actively secrete extracellular matrix (ECM) proteins when in contact with a substratum, which are then interposed to form a layer between the cell and the original substratum material; hence, the observed effects could have been due to the different conformations adopted by these ECM proteins due to the different chemistries and morphologies of the different crystal faces (cf. Section 4.1.4).

Dynamic measurements are likely to become of increasing importance as process as well as structure also falls under the scrutiny of the metrologist. *In situ* techniques are required here; rastering is preferably to be avoided because different parts of the surface are not imaged simultaneously; it is a disadvantage of the entire family of scanning nanoscopes.

Metrology of the nano/bio interface presents challenges of immense complexity, often requiring completely different approaches from those developed for wholly inanimate systems.

Although the functional characterization of nanodevices (device performance) could be legitimately included under nanometrology, it is really too vast and varied a field to be included in this book; selected aspects are considered in Chapters 7 and 10.

5.1 TOPOGRAPHY

Methods can be divided into "contact" and "noncontact". Microscopically, there is little ambiguity about the notion of contact but in the nanoscale one is aware that the Born repulsion prevents atoms from moving arbitrarily close to each other. Hence, although the AFM – Section 5.1.1 – is often considered to be a miniature, nanoscale stylus scanned over the sample, the repulsion between the atoms of the AFM tip and sample asperity is actually action at a distance; hence, the method could equally well be classified as noncontact. Scanning near-field optical microscopy (SNOM) is, however, included; it is usually grouped with AFM because the instrumentation is very similar; in essence it uses a stylus made of light, which can be considered truly noncontact.

5.1.1 CONTACT METHODS

Stylus-based profilers have long been used by engineers to determine the surface roughness of objects. A sharp-pointed stylus equipped with some means of determining its vertical displacement is simply dragged over the surface perpendicular to the surface plane. The vertical motions (deflexions from a mean) of the stylus are considered to more or less faithfully mimic the surface topography, which is recorded through a position transducer attached to the stylus. This device was being progressively miniaturized, and the ability to determine subnanometer vertical displacement was achieved for the first time by the Topografiner, invented by scientists at the US National Standards Institute [312]. This nanoscale vertical resolution was achieved using the electron tunneling effect, a quantum phenomenon [given the existence of two levels having the same energy, there is a finite probability for an electron occupying one of the energy levels to pass to the other one (if unoccupied), depending exponentially on the spacial distance separating the levels]. The current measured between an electrically conducting stylus and an electrically conducting sample can therefore be converted into sample topography.

An indispensable technological advance was the perfection of piezoelectric motion controllers in the (x, y) plane (i.e. that of the surface) and in the z direction (perpendicular to the surface). The stylus could then be raster-scanned very close to the surface. A feedback circuit can be arranged to appropriately adjust the z displacement required to keep the tunneling current constant. The perfected instrument is called the scanning tunneling microscope (STM) [25]. The principle of miniature styli moving over the sample surface and at each position returning some information about topography, or friction, or chemical nature, etc., has meanwhile been vastly extended to cover dozens of different SPMs, as the family is called. The most important (in the sense of being the most widely used) is called the AFM (Fig. 5.1).

It is a considerable advantage over electron microscopy (EM) that many of these probes can operate in air at atmospheric pressure, and even in liquids. On the other hand, the forces applied to the sample features are relatively large, and unless extreme care is taken may distort or even destroy the sample during imaging. Furthermore, the presence of liquid (even thin films of water coating the sample surface and scanning tip) gives rise to capillary forces, which may, for example, pull the tip towards the sample surface (Section 3.3; cf. Fig. 3.1). The technique is, however, being continuously improved. An important innovation has been the introduction of "tapping mode", in which the cantilever oscillates vertically, thereby minimizing contact of the tip with the sample, and permitting the use of lock-in amplification to reduce noise.

AFM resolution. The vertical resolution is limited only by the piezoelectric crystal that moves the sample relative to the tip, and the arrangement for detecting cantilever deflexion. An important consideration is that for the finest nanotexture, such as that mimicking a protein surface, the *lateral* features would be smaller than the 20 to 40 nm radius of the typical stylus of an SPM: current technology is able to routinely mass-produce silicon or silicon nitride tips with radius R equal to a few tens of nanometers. Hence, in the image generated by the microscope, the apparent lateral dimensions of features will be broadened (see, e.g. [265]). If r is the true feature radius, the apparent lateral dimension L of an object imaged by a tip of radius R is given by

$$L = 4(Rr)^{1/2}. \tag{5.1}$$

This problem can to some degree be overcome by independently measuring the precise shape of the tip [e.g. with a scanning electron microscope (SEM)] and then processing the deflexion profiles recorded using SPM in order to deconvolute the influence of tip shape – analogously to the modulation transfer function (MTF)

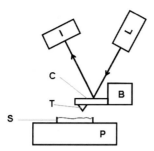

FIGURE 5.1

The AFM. The sample S is mounted on the platform P, in relation to which the block B can be moved in the x, y plane (parallel to the surface of S) and z direction (perpendicular to the surface of S). A flexible cantilever is mounted on the block, and on the end of the cantilever is a sharp tip. In order to record the vertical displacement of the tip as a result of the asperity of S, the beam from a diode laser L is reflected off the cantilever onto a split photodiode D. The tip is scanned across the sample (i.e. in the x, y plane) while remaining almost in contact with it; sample asperity displaces the tip vertically, and the corresponding deflexions of the cantilever are faithfully recorded as ratios of the signals from the two halves of the photodiode, from which sophisticated signal processing allows the three-dimensional topography of the surface to be extracted. A quadruply split photodiode enables sideways deflexions of the cantilever due to friction to be recorded as well.

approach used in optical image processing. Alternatively, finer styli (tips) can be used; for example, made by controlled etching of standard tips. Such tips enable subnanometer resolution to be obtained in principle. These ultrafine tips are, however, very fragile and easily broken during scanning. Furthermore, if imaging in liquid, impurities dissolved in the liquid may rapidly deposit on the tip. The effects on a conventional tip of $R \sim 30$ nm might be neglected, but such deposition is likely to significantly increase the radius of an ultrafine tip. CNTs (see Chapter 9), being extremely thin and rigid, offer a promising alternative to etched tips provided a convenient way of manipulating and attaching them to the microscope cantilever, and possibly replacing them *in situ* when broken or contaminated, can be found.

Contrast enhancement. A significant challenge in the field is the metrology of fabricated surfaces, especially the characterization of chemical variegation. If only morphology needs to be characterized, then the problem can be solved by either contact or noncontact techniques. The chemical contrast may, however, be insufficient to yield a clear distinction between the different components. A practical approach to overcome this problem is selective post-fabrication processing that only affects one of the components. This principle is applicable more widely than just to the imaging of complex surfaces. For example, iron is a catalyst for CNT growth using (plasma-enhanced) chemical vapor deposition, and hence if there is a heterogeneous deposit (islands) of Fe on the surface, each island will serve as an initiator of columnar CNT formation.

When the variegation resides in differing organic functionalities, the contrast between regions of differing functionalities may be too low. In this case, a useful technique may be to allow high contrast objects smaller than the smallest feature size to selectively bind to one functionality. Now that a plethora of very small nanoparticles is available commercially, and others can be chemically synthesized by well-established methods (see, e.g. [245]), this method has become very useful and practicable. Examples of the decoration of block copolymer films are given in references [215] and [184]. Using fairly extensive tabulations of single-substance surface energies, the adhesive force between two substances in the presence and absence of a fluid (in which the nanoparticles will usually be suspended) can be readily estimated (Section 3.2) and used to select a suitable nanoparticle.

Scanning ion current microscopy (SICM). A useful extension to the original SPM concept is SICM [103], in which the surface to be characterized is immersed in electrolyte containing a counterelectrode, and the scanning probe is a miniature working electrode inside a very fine glass capillary (Fig. 5.2). The closer the end of the capillary is to the surface, the smaller the ion current, which can therefore be used to generate a map of the topography. This method provides an innovative solution to the problem of excessive lateral force being applied to the sample and is particularly useful for mechanically imaging ultrafragile samples, such as a living cell, since the capillary never actually touches the surface, making this realization of the SPM concept truly noncontact.

Near-field optical microscopy. An important addition to the SPM family is the SNOM, also known as the near-field scanning optical microscope (NSOM). The scanning arrangements remain the same, but now an optical fiber brings light very close to the surface. Transmission, reflexion and fluorescence can all be measured. The principle is shown in Fig. 5.3. The obtainable resolution is below the diffraction limit applicable to far-field optics (Eq. (5.2)). The resolution depends on the fineness of the construction, especially the diameter of the optical fiber-based dielectric probe illuminating the sample. The relative motion, with subnanometer control, between

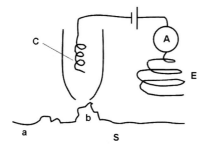

FIGURE 5.2

The SICM. A capillary C containing the work-ing electrode is moved relative to the sample S. The magnitude of the current (measured by the ammeter A) between the working electrode and the large counterelectrode E depends on the gap between the tip of the capillary and the sample surface. Sample and electrodes are bathed in an electrolyte. When C is above feature b, the gap is small, the resistance is high and the current is low. Above feature a, the gap is relatively large, the resistance low and the current high.

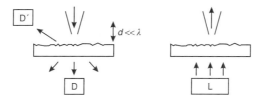

FIGURE 5.3

SNOM, also known as NSOM. On the left, SNOM in illumination mode: a dielectric probe (e.g. a tapered and surface-metallized optical fiber) positioned at a distance $d \ll \lambda$ from the surface illuminates the sample from above. Either the transmitted or the reflected light is collected in the far field (detectors D or D', respectively). On the right, SNOM in collection mode: the sample is illuminated from far below (source L). A dielectric probe in the near field collects the light transmitted through the sample.

sample and dielectric probe is accomplished using piezoelectric crystals as in the other SPMs. A related technique is thermal radiation scanning tunneling microscopy (TRSTM) [301].

5.1.2 NONCONTACT (OPTICAL) METHODS

Optical profilers are analogous to the mechanical stylus instruments but use focused beams to detect the location of the surface. They are therefore unlikely to have the resolution required to characterize nanotexture. The nanoscale equivalent is the SNOM.

Light scattering techniques. These are, in principle, more useful than the optical profilers, especially for characterizing statistical (ir)regularity. Conventional scattering techniques include specular reflexion, total integrated scattering and angle-resolved scattering; the newer speckle techniques (speckle contrast, speckle pattern illumination and angular- or wavelength-dependent speckle correlation [178]) are of particular interest. In the speckle pattern illumination method [179], based on doubly scattered coherent light, the (specularly reflecting) surface is illuminated with a monochromatic speckle pattern, whose phase distribution is then modulated by the rough surface. In polychromatic speckle autocorrelation [180], the (diffusely scattering) surface is illuminated with a collimated, partially coherent (i.e. polychromatic) light beam, either discrete (produced by a combination of laser diodes) or continuous (produced by superbright light-emitting diodes, for example). Figure 5.4 shows an example.

The distribution of collimated, typically partially coherent light scattered from a diffusely reflecting surface is suitable for determining its statistical roughness up to about a quarter of a wavelength (i.e. about 150 nm for typical visible light sources). If the surface is specularly reflecting, the illuminating light should itself be a speckle pattern, whose phase distribution is modulated by the asperity.

Evanescent optical wave-based techniques, especially its most recent variant, optical waveguide lightmode spectroscopy (OWLS) can yield structural data on ultrathin layers, including geometric thickness, refractive indices, molecular orientation (e.g. [250]), the refractive index profile perpendicular to the interface [197] and the lateral distribution of adsorbed objects (e.g. [9]). There are some limitations regarding substrata (Table 5.1). They are especially useful because they can be used to measure processes *in situ* with time resolution (Section 5.7) and the nano/bio interface (Section 5.8) and are hence described in more detail in those sections.

Imaging nanostructures. Ever since the invention of the microscope in the seventeenth century, science has been confronted with the challenge of exploring phenomena that are not directly visible to the human eye. The same extension of the senses applies to "colors" only visible using infrared or ultraviolet radiation, sounds of a pitch too low or too high to be audible and forces too slight to be sensed by the nerves in our fingers. Although artists sometimes maintain that there is a qualitative distinction between the visible and the invisible, scientists have not found this distinction to be particularly useful; for them, the problem of "visualizing" atoms is only technical, not conceptual.

Improvements in lenses, and other developments in microscope design, eventually enabled magnifications of about 2000-fold to be reached. With that, objects around 100 nm in size could just be visualized by a human observer peering through the

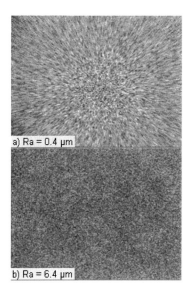

a) R_a = 0.4 µm

b) R_a = 6.4 µm

FIGURE 5.4

Polychromatic speckle patterns for two different roughnesses: (a) R_a = 400 nm and (b) R_a = 6.4 µm. Note the clearly elongated speckles in (a), and the completely decorrelated speckles in (b).

-*Reproduced with permission from J.J. Ramsden, et al., The design and manufacture of biomedical surfaces, Ann. CIRP 56(2) (2007) 687–711.*

Table 5.1 Summary of the Ranges of Applicability of Optical Techniques for Investigating Thin Films

Technique	Thin Film Material		
	Transparent dielectrics	*Opaque materials*	*Metals*
Scanning angle reflectometry (SAR)	√	√	√
Ellipsometry	√	√	√
Surface plasmon resonance (SPR)			√
Optical waveguide lightmode spectroscopy (OWLS)	√		

eyepiece of the microscope. The classical microscope runs into the fundamental limitation of spacial resolving power Δx, due to the wavelike nature of light (Abbe's limit)

$$\Delta x = \lambda/2(\text{N.A.}), \tag{5.2}$$

where λ is the wavelength of the illuminating light and N.A. is the numerical aperture of the microscope condenser. To address this problem, one can

- diminish the wavelength of the light
- operate in the near field, rather than the far field, as in SNOM (Fig. 5.3)
- renounce direct imaging
- use a totally different approach (profilers, Section 5.1.1).

Diminution of wavelength. Although shorter-wavelength varieties of radiation (ultraviolet, X-rays) are well known, as the wavelength diminishes it becomes very hard to construct the lenses needed for the microscope. However, one of the most important results emerging from quantum mechanics is the de Broglie relationship linking wave and particle properties,

$$\lambda = h/p, \tag{5.3}$$

where λ is the wavelength associated with a particle of momentum $p = mv$, where m and v are the mass and velocity, respectively, and h is Planck's constant, with a numerical value of 6.63×10^{-34} J s. Hence, knowing the mass and velocity of a particle, we can immediately calculate its wavelength.

The electron had been discovered not long before the formulation of the de Broglie relationship and was known to be a particle of a certain rest mass ($m_e = 9.11 \times 10^{-31}$ kg) and electrostatic charge e. We know that opposite charges attract; hence, the electron can be accelerated to a desired velocity simply by application of an electric field. In other words, the wavelength can be tuned as required! Furthermore, ingenious arrangements of magnetic fields can be used to focus electron beams. The transmission electron (ultra)microscope (TEM) was invented by Ernst Ruska and Max Knoll in the 1930s. Nowadays, high-resolution EM can indeed image matter down to atomic resolution. The space through which the electrons pass, including around the sample, must be evacuated, because gas molecules would themselves scatter and be ionized by fast-moving electrons, completely distorting the image of the sample. If the sample is very thin, the modulation (according to electron density) of electrons transmitted through the sample can be used to create an electron density map (TEM). Otherwise, a finely focused beam can be raster-scanned over the sample and the reflected electrons used to create a topographical image (SEM, first developed by Manfred von Ardenne, also in the 1930s, although not commercialized until the 1960s). In this case, if the sample is not electrically conducting, a thin layer of a metal, typically palladium, must be evaporated over its surface to prevent the accumulation of those electrons that are not reflected, which is liable to obscure the finest features.

Alternatively, if the sample is a semiconductor with a not-too-large band gap, it might be practicable to heat it in order to make it sufficiently conducting. Band gap illumination might also be useful, although it does not appear to have been attempted. Continuous incremental improvements in the technology of scanning electron microscopy now make it possible to obtain images in the presence of air at a pressure of a few thousandths of an atmosphere. This is called environmental scanning electron microscopy (ESEM). Some resolution is thereby sacrificed, but on the other hand it is not necessary to dehydrate the sample, nor is it necessary to coat it with a metal if it is nonconducting – the remaining air suffices to conduct excess electrons away. A combination of transmission and scanning microscopies (STEM, scanning transmission electron microscopy) was first realized by Manfred von Ardenne but again only commercialized several decades later). It has recently been demonstrated that a commercial instrument, the Nion UltraSTEM, has practically achieved the nanometrologist's goal of imaging and chemically identifying every atom in an arbitrary three-dimensional structure.

Instead of electrons, ions can also be used as imaging particles (FIB, fast ion bombardment). Originally developed as a way of controllably etching parts of samples in order to be able to more completely investigate their inner structures using electron microscopy, the instrumentation is now sufficiently advanced to enable it to be used as an ultramicroscopy technique in its own right.

5.2 MECHANICAL PROPERTIES – INDENTATION

Pressing a stylus into a surface constitutes indentation, and when the dimensions of the stylus are in the nanoscale, we have nanoindentation. It is an obvious application of the AFM (Fig. 5.1), although the AFM is not ideal for this purpose, because the tip cannot be truly perpendicular to the sample surface, resulting in ambiguity of interpretation, and the cantilevers appropriate for imaging are not stiff enough to apply an adequate indentation load. Hence, it is best to use a separate indenter, combining it with use of an AFM to control positioning and result [181]. The indenter yields a load (P)–displacement (h) curve and the AFM is used to obtain the area A of the indentation at the sample surface. The hardness H is defined as

$$H = P_{max}/A, \tag{5.4}$$

where P_{max} is the maximum load. The contact stiffness S is

$$S = \mathrm{d}P/\mathrm{d}h = 2\beta E_r \sqrt{A/\pi}, \tag{5.5}$$

the initial portion of the unloading curve (i.e. immediately after P_{max} has been reached) being used to obtain $\mathrm{d}P/\mathrm{d}h$; β is a geometrical constant, of order unity, of the indenter and E_r is a reduced elastic modulus

$$1/E_r = (1 - v)/E + (1 - v_i)/E_i \tag{5.6}$$

accounting for elastic deformation in the indenter; v is Poisson's ratio and E is elastic modulus, subscript i denoting indenter. Nanoindentation is particularly useful for measuring the mechanical properties of nano-objects resting on a hard substrate.

5.3 CHEMICAL SURFACE STRUCTURE (CHEMOGRAPHY)

Many of the available techniques are essentially nonimaging approaches because they lack lateral resolution in the nanoscale. Often nanoscale resolution is only achievable for regular structures. The best-known approach of this type is probably XRD (which could equally well be considered as a technique for determining structure). A beam of X-rays is made to impinge on the sample, making an angle θ with a plane of atoms within it, and the spacial distribution of the scattered X-rays is measured. Because the wavelength λ of X-rays is of the order of interatomic-plane distance d (tenths of a nanometer), crystalline material, or at least material with some order in its atomic arrangement, diffracts the beam. The key condition for constructive interference of the reflected beam is Bragg's law

$$d \sin \theta = n\lambda, \quad n = 1, 2, \ldots . \tag{5.7}$$

This metrology technique was developed soon after the discovery of X-rays by Röntgen in 1895, in other words long before the era of nanotechnology.

The main family of classical surface chemical analytical methods involves firing one kind of photon (or electron) at the sample and observing the energy of the photons (or electrons) whose emission is thereby triggered.

In Auger electron spectroscopy (AES), an incident electron beam ejects an electron from a core level; the resulting vacancy is unstable and is filled by an electron from a high level, releasing energy that is either transferred to another (Auger) electron from a yet higher level, or emitted as an X-ray photon. The measurement of the spectrum of the Auger electrons is called AES. In energy-dispersive X-ray spectroscopy (EDS, EDX), it is the X-ray photons whose spectrum is measured. Both these techniques are capable of good lateral resolution (within the nanoscale), because the incident electron beam can be finely focused. EDS is typically carried out within a SEM equipped with a suitable X-ray detector. It yields quantitative elemental abundances with an accuracy of around 1 atom%. AES, on the other hand, can additionally identify the chemical state of the element. All these techniques yield an average composition within a certain depth from the surface of the sample; that depth is a complicated function of the scattering of the incident and emergent radiations. Typically, AES samples only the first few nanometers from the surface, whereas EDS averages over a somewhat greater depth, which might be as much as 1 µm.

Techniques such as X-ray fluorescence (XRF) and X-ray photoelectron spectroscopy (XPS), in which the incident photon is an X-ray, have insufficient lateral resolution to be useful for mapping nanotexture.

A different kind of technique is secondary ion mass spectrometry (SIMS), in which a beam of energetic ions (typically gallium or oxygen) focused on the sample knocks out ions from it and are detected in a mass spectrometer. Recent advances in ion beam technology have resulted in the introduction of nanoSIMS, with a lateral resolution of a few tens of nanometers. The relationship between the detected ion abundances and the original sample composition strongly depends on the overall constitution of the sample, and hence quantification is a difficult challenge. One advantage of SIMS is that the incident ion beam can be used to systematically etch away the sample, allowing the depth profile of the chemical composition to be obtained.

Atom probe field ion microscopy (APFIM) requires an acicular sample (hence limiting its practical applications); upon applying a high electric field between the counterelectrode and the sample, some atoms evaporate from the sample as ions and move towards a detector, their position on which is directly related to their original position in the sample.

Continuing advances in instrumentation now make it feasible to nondestructively map the inner sections of three-dimensional objects, from which they can be completely reconstructed (tomography). Although initially demonstrated by Hounsfield with X-rays, tomography is now available for a variety of the techniques discussed above, including APFIM, TEM, STEM, AFM etc., with the capability of nanoscale resolution in all three coördinates (Section 5.5).

AFM (Fig. 5.1) can also be used to determine the chemical structure since the force–distance characteristics as the tip is made to approach and is then retracted from a particular patch on the surface are characteristic of the chemistry of that patch (Chapter 3). Careful measurement of those characteristics of each patch (Fig. 5.5), although laborious, allows a complete map of the chemical variegation of the surface to be obtained. Advances in nanoscale tip fabrication already offer considerable flexibility in choosing tips made from different materials in order to maximize the contrast between different patches. If a rather stiff cantilever is used, its motion will tend to follow the topography of the Born repulsion. A more flexible cantilever will be sensitive to the longer-range, but weaker, electron donor–acceptor and electrostatic interactions (Section 3.2), which depend upon the chemical composition of the sample at the point being measured.

Finally, we note that traditional vibrational spectroscopies (infrared and Raman–Mandelstam) are useful for identifying nano-objects, typically in reflexion mode. Sum frequency generation (SFG) spectroscopy is highly surface-specific. In this technique, a beam of visible light ($h\nu_1$) and a (tunable) beam of monochromatic infrared light ($h\nu_2$) are directed onto the surface. The intensity of the sum frequency $h(\nu_1 + \nu_2)$ is measured while varying ν_2; it is enhanced if ν_1 or ν_2 is near a resonance (e.g. a bond vibration). Summation, normally forbidden if there is an inversion center, is allowed at a surface since any inversion symmetry must be broken; hence, the technique is surface-specific. It is especially useful for buried interfaces, such as the surface of a solid in contact with water. Strong pulsed lasers are required because the summation efficiency is very low.

5.4 X-RAY DIFFRACTION AND RELATED TECHNIQUES

The main techniques are single-crystal diffraction and powder diffraction. The latter is generally applicable to nano-objects. When only very small amounts of

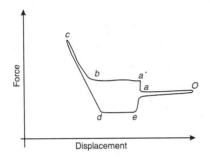

FIGURE 5.5

Prototypical force–displacement cycle in the AFM. At the selected point on the sample surface, the tip is slowly made to approach the sample (starting from position O, at which it is not in contact with the sample). At point a, the detector response moves from a to a'. This corresponds to movement of the monitoring beam reflected off the cantilever from an initial position near the perimeter of the four segment diode detector towards the center of the four segments, without the cantilever being deflected. During the displacement $a \rightarrow a'$ the beam is centered on the detector and a practically constant response, indicating a practically constant deflexion, that is a practically constant force, is observed. Moving the tip still further into the sample, from $b \rightarrow c$, causes a significant increase in detector response; the displacement in this region is the sum of cantilever deflexion and layer deformation. Hysteresis can be quantified as the force difference between the mean of $a' \rightarrow b$ and $d \rightarrow e$.

material are available, electron diffraction within the EM may be used instead. For nanocomposite materials, small angle neutron scattering (SANS) is a standard technique for obtaining morphology and size of the nano-objects embedded in a matrix. All these techniques are too well known to warrant detailed treatment in this book.

5.5 THREE-DIMENSIONAL MICROSCOPY

Although nanotechnology is often taken to be IF (surface) technology, and by far the greatest part of the nano-industry is the fabrication of VLSIs, which are, essentially, two-dimensional, there is growing interest in creating three-dimensional objects with nanoscale precision.

X-rays are attractive for ultramicroscopy because of their very short wavelength. The availability of synchrotrons, providing abundant sources of hard X-rays, further increase the attraction. They are, however, difficult to focus. Multilayer Laue lenses (MLLs) are able to achieve nanofocusing down to 10–20 nm [220].

X-rays are, however, highly penetrating. Variations in density with depth of an object are superimposed. One can only record the mean absorption of the materials through which the rays have passed.

This problem can be solved by measuring the attenuation of penetrating rays (such as γ-rays, X-rays or electrons) from a very large number of different angles and reconstructing the object in three dimensions from this information. This is the principle of computed tomography [131, 208]. The theoretical basis goes back to the Radon transform [241, 53].

A further refinement is to filter the energy of the acquisition, which allows a three-dimensional map of the chemical composition of the object under examination to be obtained [211].

5.6 THE REPRESENTATION OF TEXTURE

The ultimate representation of any surface is to list the height (relative to some reference plane) of each atom or cluster on the surface. This would amount to a topographical image and is, of course, often obtained, albeit with lower resolution. If, however, one wishes to correlate the texture of the surface with some functional parameter, the information needs to be interpreted (e.g. in order to determine the laws governing the effects of nanotexture on the response of – for example – living cells growing on it) and one needs to go beyond mere images. Hence, ways of reducing the data are sought. The simplest way of describing an uneven surface, reducing it to a single parameter, is to compute its roughness. At first sight roughness appears to be rather random and suitable for a statistical treatment. Although profilometry is capable of yielding a continuous description of the height $h(x)$ of a surface scanned in the x direction (parallel to the plane of the surface), in this digital age we should assume that it is measured discretely and has a value h_i for the ith measured patch.

5.6.1 ROUGHNESS

The quantification of roughness then amounts to a straightforward statistical analysis of the set of height data as summarized in Table 5.2. This list of parameters is not exhaustive. Others are in use, including some derived from the Abbott curve, a plot

Table 5.2 Table of Common Roughness Parameters

Symbol	Name	Definition	Synonyms
$\langle h \rangle$ or $\bar{h}$	Mean height	$(1/N)\sum_i^N h_i$	Zeroth moment; d.c. component
R_a	Arithmetical mean roughness	$(1/N)\sum_i^N \lvert h_i - \langle h \rangle \rvert$	First moment; centerline average (CLA)
R_q	Root mean square roughness	$\sqrt{\sum_i^N (h_i - \langle h \rangle)^2 / N}$	Second moment; interface width
R_t	Maximum peak to valley	$\max h_i - \min h_i$	Maximum roughness depth
R_z	Mean peak to valley height	R_t averaged over segments	
r	Roughness ratio	$\dfrac{\text{actual area}}{\text{apparent area}}$	

h_i is the height of the ith spot on the surface, with reference to some arbitrary fixed level; N is the number of measured spots. Additional parameters are the skew or skewedness (the third moment of the distribution of h) and the kurtosis (the fourth moment)

of the height of the surface as a function of the fraction of the surface exceeding that height. Most of these parameters are described in international standards (e.g. ISO 13565). A general and sophisticated surface characterization can be accomplished by determining the power spectral density (PSD) S of surface features, which encapsulates information about the lateral distribution of the individual roughness components,

$$S(f_x, f_y) = \lim \frac{1}{L^2} \left(\int_0^L \int_0^L h(x,y) e^{-i2\pi(f_x x + f_y y)} dx dy \right)^2, \qquad (5.8)$$

where the f are the spacial frequencies in the (x, y) plane of the two-dimensional surface profile $h(x, y)$ defined for a square of side L. The PSD is the Fourier transform of the autocorrelation function.

If there is some evidence of statistical self-similarity, the corresponding fractal dimension may be a compact way of representing the roughness. If the PSD is plotted (as its logarithm) against the logarithm of the spacial frequency, and if there is a straight line portion extending over two or three orders of magnitude of frequency, then a fractal dimension d_C can be derived from the slope. A perfectly flat surface has a (Euclidean) dimension of 2; a fractal surface would have a dimension >2. The formal definition of fractality relates to the capacity of the surface, and in particular how that capacity scales with the size of objects placed on the surface. One can determine it by measuring the number n of balls required to cover the surface, using a series of differently sized balls, and plotting $\log n$ against $\log r$, where r is the radius of a ball. For a perfectly flat surface, n will be proportional to $1/r^2$, but for the fractal surface one will be able to place proportionally more smaller balls, and the slope $-d_C$ of the log–log plot becomes less than 2. Experimentally, the fractal surface dimension could be determined by measuring the quantity of nanoparticles of a range of sizes ranging from a few nanometers to a few micrometers required to jam the surface.

A fractal dimension is actually the capacity, as originally defined by Kolmogorov [168],

$$d_C = \lim_{\varepsilon \to 0} \log N(\varepsilon)/\log(1/\varepsilon), \tag{5.9}$$

where $N(\varepsilon)$ is the minimum number of p-dimensional cubes of side ε needed to cover the set; p is the dimension of Euclidean space and the set in question is a bounded subset of that space.

The fractal concept highlights the importance of characteristic length as a parameter of surface roughness, even though very few real surfaces have self-similarity extending over a sufficiently wide range of lengths to warrant being described as fractals. The fact is that in many of the complex environments with respect to which biocompatibility is being determined, a single roughness parameter is quite inadequate to characterize surface topography. That is perhaps the main reason why the present state of knowledge regarding correlations between surface roughness and cell response is still in a fairly primitive state.

Lacunarity is an extension of the fractal concept that attempts to capture information about the spacial correlations of real surfaces (Section 5.6.4).

5.6.2 THE ACTUAL AREA OF A ROUGH SURFACE

A nanometric topographical measurement yields data (e.g. from the AFM) that consists of an $m \times m$ array, each element (x, y) of which gives the height z_{xy} of that point. The coordinates x and y designate the center of each pixel. The surface is then triangulated [152]: a central pixel is chosen and lines drawn to the centers of the eight adjacent pixels, which will normally have different heights z. The third side of each of the resulting eight triangles is the line joining the centers of each pair of neighboring pixels of the eight. The sides c_1, c_2, c_3 of these triangles are given by Pythagoras' theorem,

$$c = \sqrt{a^2 + b^2}, \tag{5.10}$$

where a is the planimetric (horizontal) distance and b is the difference in height. The area of each triangle is then given by

$$A = \sqrt{s(s - c_1)(s - c_2)(s - c_3)}, \tag{5.11}$$

with

$$s = (c_1 + c_2 + c_3)/2. \tag{5.12}$$

By this means (neglecting the borders of the chosen area) the entire real surface area $\mathfrak{A} = \sum A$ can be rapidly computed.

5.6.3 ONE-DIMENSIONAL TEXTURE

The techniques discussed in Sections 5.1 and 5.3 are aimed at mapping out the chemical and topographical variegation to as fine a resolution as that of any variegation that may be present. Raster techniques such as the SPMs can identify variegation pixel-by-pixel and produce a one-dimensional string of information *a priori* suitable for analysing according to the methods suggested in Section 5.6.3 (the to-and-fro motion of the scanning probe may impose a certain kind of correlation on the symbolic sequence if there are two-dimensional features present, i.e. thicker than a single scanning line). The main task is to determine the presence of compositional and/or topographical correlations within the plane of the surface.

   Preliminary to the discussion of nanotexture proper, which implies a two-dimensional arrangement of features important for molecular and cellular recognition, let us then consider the one-dimensional situation, which is of course much simpler. The purpose of this section is to look at some relatively straightforward parameters that might be used as means to quantify texture. A one-dimensional sequence of symbols (each symbol representing a topographical feature, such as a height, or a chemical element in a particular oxidation state) can be completely defined by the set of probability distributions: $W_1(yx)dy$, the probability of finding y, the value of the symbol, in the range $(y, y+dy)$ at position x (if one is moving along the sequence at a uniform rate, this might be equivalent to a time t); $W_2(y_1x_1, y_2x_2)dy_1dy_2$, the joint probability of finding y in the range (y_1, y_1+dy_1) at position x_1 and in the range (y_2, y_2+dy_2) at position x_2; and so on for triplets, quadruplets and higher multiplets of values of y. If there is an unchanging underlying mechanism generating the sequence, the probabilities are stationary and the distributions can be simplified as $W_1(y)dy$, the probability of finding y in the range $(y, y+dy)$; $W_2(y_1y_2x)dy_1dy_2$, the joint probability of finding y in the ranges (y_1, y_1+dy_1) and (y_2, y_2+dy_2) when separated by an interval $x = x_2 - x_1$; etc. If successive values of y are not correlated at all, i.e.

$$W_2(y_1x_1, y_2x_2) = W_1(y_1x_1)W_1(y_2x_2), \qquad (5.13)$$

all information about the process is completely contained in W_1, and the process is called purely random.

Markov chains

If, however, the next step of a process depends on its current state, i.e.

$$W_2(y_1y_2x) = W_1(y_1)P_2(y_2|y_1x), \qquad (5.14)$$

where P_2 denotes the conditional probability that y is in the range (y_2, y_2+dy_2) after having been at y_1 at an earlier position x, we have a Markov chain, defined as a sequence of "trials" (that is, events in which a symbol can be chosen) with possible outcomes **a** (possible states of the system), an initial probability distribution $\mathbf{a}^{(0)}$ and (stationary) transition probabilities defined by a stochastic matrix P. A random sequence, i.e. with a total absence of correlations between successive symbols, is a zeroth-order Markov chain. Without loss of generality, we can consider a binary string, that is a linear sequence of zeros and ones. Such sequences can be generated by choosing successive symbols with a fixed probability. The probability distribution for an r-step process is

$$\mathbf{a}^{(r)} = \mathbf{a}^{(0)} P^r. \tag{5.15}$$

The Markovian transition matrix, if it exists, is a compact way of representing texture.

If upon repeated application of P the distribution $\mathbf{a}$ tends to an unchanging limit (i.e. an equilibrial set of states) that does not depend on the initial state, the Markov chain is said to be ergodic, and

$$\lim_{r \to \infty} P^r = Q, \tag{5.16}$$

where Q is a matrix with identical rows. Now

$$PP^n = P^n P = P^{n+1} \tag{5.17}$$

and if Q exists it follows, by letting $n \to \infty$, that

$$PQ = QP = Q \tag{5.18}$$

from which Q (giving the stationary probabilities, i.e. the equilibrial distribution of $\mathbf{a}$) can be found. Note that if all the transitions of a Markov chain are equally probable, then there is a complete absence of constraint, in other words the process is purely random.

The entropy of a Markov process is the "average of an average" (i.e. the weighted variety of the transitions). For each row of the (stochastic) Markov transition matrix, an entropy $H = -\sum_i p_i \log_2 p_i$ (Shannon's formula) is computed. The (informational) entropy of the process as a whole is then the average of these entropies, weighted by the equilibrial distribution of the states.

Runs

Another approach to characterize texture is to make use of the concept of a run, defined as a succession of similar events preceded and succeeded by different events. Let there again be just two kinds of elements, 0 and 1, and let there be n_0 0s and n_1 1s, with $n_0 + n_1 = n$. r_{0i} will denote the number of runs of 0 of length i, with $\sum_i r_{0i} = r_0$, etc. It follows that $\sum i r_{0i} = n_0$, etc. Given a set of 0s and 1s, the numbers of different arrangements of the runs of 0 and 1 are given by multinomial coefficients and the total number of ways of obtaining the set r_{ji} ($j = 1, 2; i = 1, 2, \ldots, n_0$) is [212]

$$N(r_{ji}) = \begin{bmatrix} r_0 \\ r_{0i} \end{bmatrix} \begin{bmatrix} r_1 \\ r_{1i} \end{bmatrix} F(r_0, r_1), \tag{5.19}$$

where the terms denoted with square brackets are the multinomial coefficients, which give the number of ways in which n elements can be partitioned into k subpopulations, the first containing r_0 elements, the second r_1, etc.,

$$\begin{bmatrix} r \\ r_i \end{bmatrix} = \frac{n!}{r_0! r_1! \ldots r_k!}, \quad \text{with} \sum_i^{i=k} r_i = n, \tag{5.20}$$

Table 5.3   Values of the Special Function $F(r_0, r_1)$ used in Eqs. (5.19) and (5.21)

$\lvert r_0 - r_1 \rvert$	$F(r_0, r_1)$
>1	0
1	1
0	2

and the special function $F(r_0,r_1)$, given in Table 5.3, is the number of ways of arranging r_0 objects of one kind and r_1 objects of another so that no two adjacent objects are of the same kind. Since there are $\binom{n}{n_0}$ possible arrangements of the 0s and 1s, the distribution of the r_{ji} is

$$P(r_{ji}) = \frac{N(r_{ji})F(r_0,r_1)}{\binom{n}{n_0}}. \qquad (5.21)$$

The deviation of the actually observed distribution from Eq. (5.21) can be used as a statistical parameter of texture.

Genetic sequences

One very important physical instantiation of one-dimensional texture is the sequences of nucleic acids that encode proteins and also constitute the binding sites (promoters) for transcription factors – proteins that bind to DNA as a prerequisite for the transcription of the DNA into RNA that precedes the translation of the RNA into amino acid sequences (proteins). The nucleic acids have a variety of four (the bases A,U,C,G in natural RNA, and A,T,C,G in DNA). Many statistical investigations of intragenome DNA correlations group the four into purines (A and G) and pyrimidines (U or T and C) – rather like analysing texts in terms of vowels and consonants. Mere enumeration of the bases is of course inadequate to characterize texture.

Algorithmic information content (AIC)

This can be quantified by estimating the AIC, also called algorithmic or Kolmogorov complexity, which is essentially a formalization of the notion of estimating the complexity of an object from the length of a description of it. The first task is to encode the measured variegation in some standard form. The choice of rules used to accomplish the encoding will depend on the particular problem at hand. For the one-dimensional nucleic acid textures referred to in the preceding paragraph, encoding purines as 1 and pyrimidines as 0 may be sufficient, for example. The formal definition of the AIC of a symbolic string s (encoding the object being described) is the length of the smallest (shortest) programme P that will cause the standard universal computer (a Turing machine T) to print out the symbolic string and then halt. Symbolically (but only exact for infinite strings), denoting the AIC by K,

$$K(s) = \min\{|P| : s = C_T(P)\}, \tag{5.22}$$

where $|P|$ is the length of the program (in bits) and C_T is the result of running the program on a Turing machine. Any regularity present within the string will enable the description to be shortened. The determination of AIC is therefore essentially one of pattern recognition, which works by comparing the unknown object with known prototypes (there is no algorithm for discovering patterns *de novo*, although clusters may be established according to the distances between features). The maximum value of the AIC (the unconditional complexity) is equal to the length of the string in the absence of any internal correlations; that is, considering the string as random, viz.,

$$K_{\max} = |s|. \tag{5.23}$$

Any regularities, i.e. constraints in the choice of successive symbols, will diminish the value of K from $K_{\max}$.

Effective complexity (EC)

The concept of EC [95] was introduced in an effort to overcome the problem of AIC increasing monotonically with increasing randomness. EC is defined as the length of a concise description (which can be computed in the same way as the AIC) of the set of regularities of the description. A very regular symbolic sequence will have only a small number of different regularities, and therefore a short description of them; a random sequence will have no regularities, and therefore an even shorter description. There will be some intermediate descriptions with many different regularities, which will yield a large EC. In a certain sense, EC is actually a measure of our knowledge about the object being described, for it quantifies the extent to which the object is regular (nonrandom), and hence predictable. It presents the same technical difficulty as AIC: that of finding the regularities, both in compiling an initial list of them, and then in finding the regularities of the regularities.

5.6.4 TWO-DIMENSIONAL TEXTURE: LACUNARITY

Whereas AIC and EC can be straightforwardly applied to linear texture, it is not generally obvious how a two-dimensional pattern should be encoded. Of course it could be mapped in raster fashion (as is actually done in SPM and SEM), and patterns extending over many lines should appear as regularities.

Another approach to capture information about the spacial correlations of arbitrarily heterogeneous real surfaces is to extend the fractional dimension or fractal representation of roughness to encompass the quantification of voids in rough objects (the lacunarity Λ). Consider an image constructed from binary (black or white, corresponding to values of 0 and 1) pixels, and let the numbers of boxes of side r containing s white pixels have the distribution $n(s,r)$. $\Lambda(r)$ is defined as

$$\Lambda(r) = M_2/M_1^2, \tag{5.24}$$

where M_1 and M_2 are the first and second moments of the distribution,

$$M_1(r) = \sum_{s=1}^{r^2} s(r)n(s,r)/N(r) = \langle s(r) \rangle \qquad (5.25)$$

and

$$M_2(r) = \sum_{s=1}^{r^2} s^2(r)n(s,r)/N(r) = \sigma_s^2(r) + \langle s \rangle^2(r), \qquad (5.26)$$

where $\langle s \rangle$ and σ_s^2 are, respectively, the mean and variance of the distribution and the total number $N(r) = (M - r + 1)^2$ for a square pattern of size M of boxes of size r; i.e. a type of variance-to-mean ratio. The lacunarity can be thought of as the deviation of a fractal from translational invariance [94], or a scale-dependent measure of heterogeneity (i.e. "texture") of objects in general [4]. Its lowest possible value is 1, corresponding to a translationally invariant pattern (including the special case $\Lambda(M) = 1$). Practically, comparison of the lacunarity plots (i.e. log–log plots of the function $\Lambda(r)$) of artificially generated patterns with the experimental lacunarity may be used to analyse the texture of the sample.

5.7 THE METROLOGY OF SELF-ASSEMBLY

Hitherto we have been implicitly considering time-invariant situations. For self-assembly, the permanent fixing of the structure required by EM or the relatively slow rastering procedure required by the "contact" methods render them unsuitable.

A typical self-assembly scenario (see Section 8.2.1) is to disperse nano-objects randomly in a liquid and allow them to adsorb and assemble on a solid substratum (i.e. at the solid/liquid interface). In order to establish basic relations between the self-assembled structure and the characteristics of the nano-objects, the liquid and the substratum, some non-perturbing, *in situ* means of (a) counting the added particles, (b) determining their rate of addition and (c) determining the structure of the self-assembled film is required.

Optical methods [scanning angle reflectometry (SAR), ellipsometry] rely on monitoring changes in the reflectance of the solid/liquid interface due to the accumulation of the nano-objects. OWLS relies on perturbation of the evanescent field generated by waves guided along the substratum by the nano-objects, provided their polarizability is different from that of the liquid (Fig. 5.6).

If the substratum is a thin metal film, optically excited SPR can also be used to monitor the presence of the nano-objects, although it is less sensitive and less informative than OWLS. If an electrode-coated piezoelectric crystal can be used as the substratum, changes in the resonant vibration frequency of the crystal, and the dissipation of its oscillation, can also provide useful information (the quartz crystal microbalance, QCM and QCM-D).

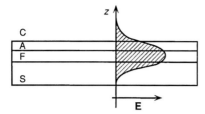

FIGURE 5.6

The electromagnetic field distribution of a zeroth-order guided wave in a four-layer slab waveguide [S, support; F, high refractive index film; A, self-assembled adlayer; C, cover medium (liquid in which nano-objects are suspended)]. Note the exponentially decaying evanescent fields in the zones S and A, C. The highest sensitivity of the phase velocity of the guided modes to adlayer structure is obtainable with thin waveguides whose thickness (i.e. that of the F-layer) is close to the cut-off limit of the zeroth modes, but in this case only two (orthogonal) modes can be excited in the waveguide.

5.8 METROLOGY OF THE NANO/BIO INTERFACE

Here, we come to the most difficult challenge for the nanometrologist. Whereas the techniques discussed in the preceding sections are closely related to well-established ones – indeed metrology was ably handing measurements in the nanoscale long before the term nanotechnology was introduced and even before Feynman's lecture [85] – when living organisms are introduced into the system the focus is moved into unfamiliar territory. Particular problems are associated with the very narrow range of ambient conditions in which life can be maintained (otherwise the life must be terminated and the sample fixed, but then one may not be measuring what one wishes to know); the fact that life is mainly built from soft, irregular structures that are constantly evolving on different timescales; and the fact that any living organism, even a single cell, is able to modify its environment (e.g. by secreting macromolecular substances).

One of the most important techniques yielding nanometer resolution is EM. In principle, however, this is ruled out for investigating the nano/bio interface because of the damage to delicate biostructures caused by the necessary sample preparation, typically including dehydration and impregnation with heavy metals [124]. Given that water is the dominant molecular species in nearly every biological system, the ability to operate in an aqueous environment is an essential prerequisite.

Some measurement problems associated with nanotechnology do not require nanometrology. For example, the toxicologist investigating the effect of nanoparticles on a living organism may merely administer the nanoparticles intravenously and observe the subsequent behavior of the animal.

5.8.1 DETERMINING THE NANOSTRUCTURE OF PROTEIN CORONAS

The protein corona is the collection of proteins that accumulate on an artificial surface introduced into a biofluid such as blood; that is, the process of opsonization. Since there are many hundreds of different proteins (far more if one includes all the IgG antibodies of different affinities as "different proteins") it is very likely that the surface becomes coated with a complex structure comprised of many different proteins and, moreover, constantly changing with time.

A major goal is to be able to design surfaces that can determine the nature of the corona. To achieve this, one must understand how the corona is assembled. Unlike other self-assembly problems, the corona is constantly changing its composition and structure. The main challenges are identifying the proteins and characterizing the structure. The first of these is very difficult. If one exposes the corona-decorated nano-objects to a series of different monoclonal antibodies, this would allow one to at least identify the outermost proteins, but the procedure becomes impracticable for more than a few proteins; apart from the expense of procuring the monoclonal antibodies, the initial bindings will block access by subsequent ones. It must, moreover, be repeated over a long succession of intervals. On the other hand, the reflectance techniques described in Section 5.7 are well applicable to measuring the nanostructural evolution of the corona *in situ*. OWLS in particular is well able to handle anisotropy [134] and refractive index profile determination [197], as well as overall thickness and mean refractive index of the corona.

5.8.2 MEASURING CELL ADHESION: THE INTERACTION OF AN EVANESCENT FIELD WITH A CELL

The evanescent field decays exponentially perpendicular to the plane of the waveguide (see, e.g. [288]), and the characteristic decay length $1/s$ (in which the field intensity decays to $1/e$ of its value at the waveguide surface), also called the penetration depth, depends on the refractive indices of the waveguide, the waveguide support S, the cover medium C (including the cells) and the film thickness d_F (see Fig. 5.6). For the transverse electric (TE) guided lightmodes, the inverse penetration depth is given by

$$s = k(N^2 - n_C^2)^{1/2} \tag{5.27}$$

where k is the wave vector of the light (numerically equal to $2\pi/\lambda$, λ being the vacuum wavelength), N is the effective refractive index of the guided mode and n_C is the refractive index of the composite cover medium (culture fluid and cells) as sensed by the guided lightmode. This is for decay into the cover medium; the inverse decay length into the substrate is

$$s = k(N^2 - n_S^2)^{1/2}. \tag{5.28}$$

In using these equations, one should bear in mind that N also depends on n_S as well as on all the other waveguide parameters.

To a first approximation, we can suppose that at any distance z from the waveguide surface, the cover medium refractive index is the arithmetic mean of that of the cells and that of the culture fluid,

$$n_C(z) = n_\kappa c A(z) + n_M(1 - c A(z)), \tag{5.29}$$

where n_κ is the refractive index of the cell material (which may itself depend on z), c is the number of cells per unit area of the waveguide, $A(z)$ is the cross-sectional area of the cell parallel to the waveguide at a distance z from the waveguide surface and n_M is the refractive index of the pure culture fluid. Hence, to determine the total contribution of the cells to n_C, we need to integrate the product $A(z)e^{-sz}$ from $z = 0$ to ∞; this is just the Laplace transform of $A(z)$, i.e.

$$v' = \int_0^\infty A(z)e^{-sz}dz = \mathcal{L}(A(z)); \tag{5.30}$$

v' is called the effective volume of the cell [255].

By taking the Laplace transform of the right-hand side of Eq. (5.29) and dividing by the total volume effectively sensed by the waveguide, equal to $\int_0^\infty e^{-sz}dz = 1/s$, we obtain the final expression for the cover refractive index

$$n_C = n_\kappa s c v' + n_M(1 - s c v'), \tag{5.31}$$

where we have assumed that n_κ is the refractive index throughout the cell. If this is considered to be inadmissible — for example, when taking into account the optically denser nucleus, the term $n_\kappa c A(z)$ in Eq. (5.29) can be replaced by $cA(z)[n_D D(z)/A(z) + n_\kappa(1 - D(z)/A(z))]$, where n_D is the refractive index of the nucleus, and $D(z)$ is its cross-sectional area, obviously assumed to be less than $A(z)$. Additional terms could, in principle, be added for the endoplasmic reticulum, Golgi body, etc., although probably only the nucleus warrants special treatment, its refractive index being significantly higher because of the relatively high concentration of phosphorus-containing nucleic acids.

The fundamental property that is measured with total internal reflexion is the phase shift Φ; the complex Fresnel reflexion coefficient can be written out as

$$\hat{R} = |\hat{R}|e^{i\Phi}. \tag{5.32}$$

The optical waveguide is constructed by sandwiching a slab of the high refractive index material F between lower refractive index materials, the mechanically strong support S and the medium C containing the cells. Both these materials are much thicker than $1/s$, and hence can be considered semi-infinite. The condition for the guided waves to propagate is for the different contributions to Φ to sum to zero or an integral multiple of 2π. This condition can be written as a set of mode equations linking the effective refractive index N of a particular mode with n_S, n_F, d_F and n_C. Only discrete values of N are allowed, corresponding to the mode numbers $m = 0, 1, 2, \ldots$, each value of m being represented by two orthogonal polarizations, transverse magnetic (TM) and TE. The sensitivities $\partial N/\partial n_C$ depend on the polarization, mode number and effective waveguide thickness: below a certain minimum *cutoff thickness* no propagation can take place and as the thickness increases the sensitivity rapidly reaches a maximum and then slowly diminishes; the sensitivities decrease for higher-order modes within conventional waveguides [288].

For reversed waveguides ($n_S < n_C$ – called "reverse (a)symmetry" waveguides in some earlier literature – see Section 5.8.5), the sensitivity at the cutoff thickness is 1 for all of the modes and slowly decreases above the cutoff thickness [288, 135]. If n_S, n_F and d_F are known, n_C can be determined from measurement of a single value of N. According to Eq. (5.31), any change in n_C can be interpreted as a change in cell concentration (number per unit area of substratum) c, or in cell size (characterized by the radius R), or in cell shape (if the shape is assumed to be a segment, then the contact area a will suffice as a characteristic parameter, if it is further assumed that the volume does not change upon spreading), or in cell refractive index (distribution).

In summary, then, the fundamental measurable quantity output by the optical measurement setup is one or more effective refractive indices N. Further interpretation of N depends on the model chosen, with which the distribution of polarizability within the evanescent field can be linked to changes in the cover medium refractive index n_C.

5.8.3 OPTICAL MEASUREMENT SCHEMES

The principal approaches to measure the effective refractive indices N with high precision are grating coupling and interferometry. In the former, a diffraction grating created in the waveguide is used to couple an external light beam incident on the grating with an angle α to the grating normal; measurement of α yields N according to

$$N = n_{\text{air}} \sin \alpha + 2\pi \ell / \Lambda, \tag{5.33}$$

where $\ell = 0, \pm 1, \pm 2, \ldots$ is the diffraction order and Λ is the grating constant. The diffraction grating can be created either by modulating the topography or the refractive index of the waveguide material. α can be determined by mounting the waveguide on a high-precision goniometer with the grating positioned at its axis of rotation. Photodiodes placed in contact with the ends of the waveguide enable the incoupling peaks to be determined while α is varied. It is customary to measure propagation in both possible directions and take the average, thereby correcting for any irregularity in the grating position and yielding the absolute incoupling angle. Alternatively, the grating can be used to couple light already propagating in the waveguide out of it. However, this requires a more elaborate arrangement for introducing the light into the waveguide, and it is more difficult to measure the absolute incoupling angle. Grating coupling is also called OWLS, because the spectrum of modes can be obtained by scanning α. As well as the high precision and ease with which absolute effective refractive indices can be determined of as many modes as the waveguide will support, OWLS also enables the incoupling peak shape to be determined; any broadening beyond the theoretical lower limit of width ($\approx \lambda/L_x$ in radians, where L_x is the illuminated grating length [288]) is highly informative regarding the distribution of objects on the waveguide surface [54].

Variants of the classical incoupling scheme described above include the use of chirped gratings (i.e. with a spacially varying Λ) and varying λ to determine N. Such schemes have been devised to eliminate moving parts, especially the expensive and relatively slow high-precision goniometer. However, they are generally less precise and accurate. Nevertheless, it might be worth sacrificing some precision in return for another benefit, such as high throughput. Hence, incoupling of broadband light and spectral monitoring of the outcoupled light by the same grating provides a very compact way of measuring relative effective refractive index changes (this scheme is also called the resonant waveguide grating (RWG) [73]).

The three main interferometric techniques that *might* be used (no measurements with living cells using interferometry have apparently been reported – often the waveguide geometry is unfavorable, e.g. too thin to underlie the entire cell, and excessive scattering engendered by the long waveguides used to increase sensitivity vitiates useful information being obtained) are:

1. allowing the TE and TM modes to interfere with one another; this allows a planar waveguide without any structuring to be used, but the measuring arrangement is more elaborate than in grating coupling;
2. the Mach–Zehnder interferometer, the integrated optical analogue of the Rayleigh interferometer – typically a ridge or channel waveguide is split into two channels, one of which being the cell substratum, and then recombined;
3. the "dual polarization interferometer" (DPI) in which a sandwich structure consisting of support, reference waveguide, intermediate layer (may be the same material as the support) and finally the waveguide acting as cell substratum is illuminated at one end, and interference between the two beams emerging at the other end is recorded in the far field.

The main advantage of an interferometer is that the length can be arbitrarily extended to increase the sensitivity (provided the waveguide can be made sufficiently uniform and the temperature maintained constant over the entire length). On the other hand, it is usually only practical to measure with one mode, and it is very difficult to measure absolute values of the effective refractive index. Moreover, in interferometry any lateral inhomogeneity information is averaged, in contrast to grating coupling, in which energy is injected continuously under the cells in the entire measuring zone, not just at the end of the waveguide. Therefore, the resonant peak shape carries information about the optical microstructure of the cover medium [54], information inaccessible to an interferometric measurement.

5.8.4 GRATING-COUPLED INTERFEROMETRY (GCI)

This technique combines waveguide interferometry with waveguide grating couplers [170]. In the experimental setup, a highly monochromatic and coherent light beam (e.g. from a He–Ne laser) is divided into two parallel beams. The waveguide is fabricated with two gratings separated by the sensing zone. The first beam impinges on the first grating, passes through the sensing zone (in which the evanescent field may interact with nano-objects adsorbed on, or in the vicinity of, the waveguide surface) and combines with the second beam that has impinged on the second grating, generating interference. Detection of the light takes place at the end of the waveguide.

To improve the signal/noise ratio, the first beam is phase-modulated in a temporally varying fashion. A further improvement is achieved by sampling both beams by a partially reflecting mirror placed before the beams impinge on the gratings. The sampled beams are combined and the interference measured with a separate detector, providing a reference signal.

5.8.5 REVERSED WAVEGUIDES

In the conventional asymmetric waveguide, the support refractive index (e.g. optical glass) is greater than that of the cover, and this continues to hold when the cover consists of an aqueous medium and living cells. Therefore, the penetration depth is limited to a distance of around 200 nm, implying that only a small part of a cell body is sensed by the evanescent field. The maximum probe depth is achieved at the cutoff point ($N = n_S$), at which the sensitivity of such a waveguide is zero [288, 135]; when the sensitivity is maximal, the probing depth is even shorter, around 100 nm. While this is completely sufficient for monitoring the spreading transition from sphere to segment (Fig. 4.3), because the change in distribution of polarizability overwhelmingly takes place in the immediate vicinity of the waveguide surface (and this is expected to hold even if the actual cell–substratum contact is mediated by filopodia), the monitored optical signal will be insensitive to changes of polarizability distribution occurring within the bulk of the cell body (which may arise through a reconfiguration of the cytoskeleton, or repositioning of the nucleus). If, however, $n_S < n_C$, the penetration depth into the cover medium can be readily increased and tuned without any upper limit [288, 135, 137] (note that the condition $max(n_S, n_C) < N < n_F$ must always be met). This is called the reversed waveguide configuration (Fig. 5.7). Typical practical values of waveguide parameters are constrained by $n_M \sim 1.34$ for culture medium, and $n_\kappa \sim 1.4$, making $n_C \sim 1.35$. This means that reversed waveguiding cannot be achieved with the usual waveguide support materials (optical glass or quartz, $n_S \sim 1.5$). Sufficiently mechanically strong supports with a very low refractive index can be fabricated by introducing a large volume fraction of air-filled nanoscale pores into the support material.

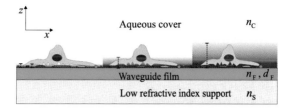

FIGURE 5.7

Cross-section of a cell on a reversed waveguide with (from left to right) successively increasing penetration depths.

5.8.6 THE INTERPRETATION OF EFFECTIVE REFRACTIVE INDEX CHANGES

Contact area and detachment force. The basic interpretation is encapsulated in Eq. (5.31). If only spherical cells are being investigated, then the number of cells c per unit area under a certain set of environmental conditions (e.g. a constant weak force of detachment) provides a basic quantification of adhesion. If the waveguide is positioned vertically, then the (gravitational) force acting to detach the cells is uniform; the evident complexity involved in placing an OWLS device in a centrifuge seems to have precluded any experimental attempt. If the waveguide is positioned horizontally, then the detaching force can be supplied by fluid flow through the cuvette. An approximate formula for the detachment force F imposed on a cell of (unspread) radius R by a wall shear stress γ is [140]

$$F = 32R^2\gamma. \tag{5.34}$$

This formula is insensitive to the exact shape. The actual situation is more complicated (see, e.g. Ref. [32]) because the cell can deform under the influence of shear stress before detachment.

Spreading kinetics. The dynamics of change of some parameter (such as contact area) may be characteristic of the cell. It may be indicative of adhesiveness (and its dependence upon the presence of calcium or other ions); the occurrence of spreading may distinguish between different cell types; it may reveal the importance of nutrient; it may reveal the presence of ECM molecules coating a mineral substratum; it may reveal the adverse response of cells to drugs or toxic chemicals.

Bulk versus surface effects. In order to separate bulk and surface effects, the differing depth sensitivity curves of the TE and TM modes can also be exploited to derive a weighted difference D_γ of sensitivities, and a *crossover depth* at which $D_\gamma = 0$ (below it is negative and above it is positive [137]). Writing the sensitivities as S, the sign of the linear combination

$$D_\gamma = \gamma S_{TM} - S_{TE} \tag{5.35}$$

depends on the magnitude of the (always positive) tuning parameter γ, selected according to the waveguide parameters and depth zone of interest. Thus, if $D_\gamma < 0$, then the predominant refractive index changes take place near the waveguide surface; if $D_\gamma > 0$, then the changes are predominantly in the bulk of the cell body.

5.8.7 THE INTERPRETATION OF COUPLING PEAK WIDTH CHANGES

The introduction of binary (i.e. cell–no cell) spacial inhomogeneity in the cover medium C broadens the incoupling peaks (Section 5.8.3). The greatest inhomogeneity must necessarily occur when exactly half the surface is covered with objects (i.e. cells) whose refractive index differs from that of the bathing medium. If the locally engendered effective refractive indices are sufficiently different two peaks may be observed. If the penetration depth is rather small, as the cell–substratum contact area a increases from zero (corresponding to the perfectly spherical initially attached state of the cell), some measure of peak width w (such as full width at half maximum) should first increase *pari passu* with effective refractive index N, up to the point at which the cells cover half the detected area, and then decline; the highest value attained by w depends on the optical contrast between the two materials (i.e. cells and culture medium) constituting the covering medium.

Microexudate. Very commonly, living eukaryotic cells respond to the solid part of their environment (e.g. their substratum) by excreting ECM molecules (mostly large glycoproteins), which adsorb onto the substratum, modifying it to render it more congenial. Even the most carefully prepared nanostructure may thereby rapidly lose its initial characteristics. The prokaryotic equivalent is the biofilm, which constitutes a kind of microbial fortress defending the population against external attack. These adsorbed secretions, whose refractive index is greater than that of the culture medium, cause the effective refractive index N to increase additionally to the increase caused by the cell shape changes (spreading). However, the coupling peak widths w will be negligibly affected because the discrete objects (glycoproteins) comprising the secreted microexudate are much smaller than the wavelength of light, and can in general be assumed to be adsorbed uniformly over the entire waveguide surface. Hence, by simultaneously comparing the evolutions of N and w, two processes contributing independently to the increase of N, namely the increase of a and the accumulation of ECM glycoproteins at the substratum surface, can be separately accounted for.

5.8.8 ELECTRICAL MEASUREMENT SCHEMES

If cells are cultured on a small electrode, the circuit being completed by a large counterelectrode placed elsewhere in the electrolyte solution (culture medium) bathing the cells, the impedance due to the cells dominates the overall impedance [97]. Both the resistance beneath the cell, impeding current flow parallel to the substratum, and the resistance between cells contribute to the impedance. Knowing the resistance of the bulk medium (measured independently) and the diameter of the cells (measured independently using an optical microscope) allows the height of the gap between the electrode and the underside of the cell to be determined. The best approach seems to be measuring the frequency-dependent impedance and fitting the data to the calculated impedance of an equivalent circuit consisting of a resistor and a condenser in series.

The electrical signals also display high frequency fluctuations. These appear to be due to micromotion of the cells, changing the two resistances mentioned above. They are clearly a measure of vitality since metabolism-depressing drugs cause the fluctuations to be strongly dampened [98].

SUMMARY

Metrology at the nanoscale imposes stringent requirements of accuracy and precision. A particular challenge is the fact that the size of the measuring instrument and the feature being measured become comparable. Should the nanoscale feature be absolutely small in the quantum sense, then we have the potential problem that its state is destroyed by the measurement. Given the identity of quantum objects, however, this does not usually pose a real practical difficulty – one merely needs to sample and destroy one of many. Regardless of the absolute size of the object being measured, if the measurement instrument has the same relative size as the object being measured, the measurement is subject to distortion. Previously, this could always be resolved by shrinking the relative features of the measurement instrument. If, however, the object being measured is already at the lower size limit of fabrication (i.e. in the nanoscale) no further shrinkage of the measurement instrument is possible. Nevertheless, this does not pose an insuperable difficulty. Indeed, it has already been encountered and largely overcome whenever light has been used to measure features of an object smaller than the wavelength of the light.

The ultimate goal of nanometrology is to provide a list of the coördinates and identities of all the constituent atoms of a nanoscale structure, and all practical metrology is aimed at that goal. Approaches can be categorized as imaging or non-imaging, each of which category contains methods that can be categorized as contact or noncontact. As one approaches the nanoscale, however, the distinction becomes harder to make. In another dimension, one can distinguish between topographical and chemical features; some techniques yield both. In particular, commercial STEM is now able to yield the atomic coördinates of arbitrary three-dimensional structures.

Techniques able to make time-resolved measurements *in situ* are very useful for monitoring actual processes.

The representation of structure by a list of the atomic coördinates and identities will usually lead to vast, unmanageable data sets. Therefore, there is considerable interest in capturing the salient features of topography and chemistry by identifying regularities, possibly statistically.

The nano/bio interface presents perhaps the greatest challenges to nanometrology, not least because the introduction of living components into the system under scrutiny moves the problem into territory unfamiliar to most metrologists. However, important advances are being made in this domain, which has already allowed it to assume a far more quantitative nature than hitherto.

The greatest current challenge is to devise instrumentation able to combine two or more techniques, enabling the simultaneous observation of multiple parameters *in situ* on the same sample.

FURTHER READING

1. D.A. Bonnell, et al., Imaging physical phenomena with local probes: From electrons to photons. Rev. Modern Phys. 84 (2012) 1343–1381.
2. P.W. Hawkes, From a fluorescent patch to picoscopy, one strand in the history of the electron. Nanotechnol. Percept. 7 (2011) 3–20.
3. R. Leach, Fundamental Principles of Engineering Nanometrology, Elsevier, Amsterdam, 2009.
4. R.E. Kunz, Miniature integrated optical modules for chemical and biochemical sensing. Sensors Actuators B 38–39 (1997) 13–28.
5. J.J. Ramsden, High resolution molecular microscopy, in: Ph. Dejardin (Ed.) Proteins at Solid-Liquid Interfaces, Springer-Verlag, Heidelberg, 2006, pp. 23–49.
6. J.J. Ramsden, OWLS—a versatile technique for drug discovery. Front. Drug Design Discover. 2 (2006) 211–223.

Noncarbon nanomaterials and their production

6

CHAPTER CONTENTS

INTRODUCTION

A nanomaterial is a material having one or more external dimensions in the nanoscale or having internal or surface structure in the nanoscale (Fig. 6.1). If the description is to be kept deliberately vague, one could use the word nanosubstance merely to signify that some nanoscale features are involved. Nano-objects are categorized according to the number of their external dimensions in the nanoscale (Fig. 6.2). The concept

Nanotechnology: An Introduction. DOI: 10.1016/B978-0-323-39311-9.00012-1
Copyright © 2016 Elsevier Inc. All rights reserved.

FIGURE 6.1

Fragment of a concept system for nanotechnology (cf. Fig. 1.1).

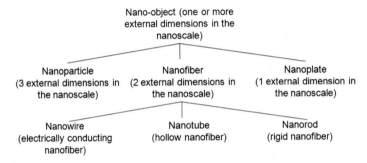

FIGURE 6.2

Concept system for nano-objects (cf. Figs 1.1 and 6.1). See text for further explanation.

of nano-object includes additional entities not shown in the figure but sometimes referred to, such as nanohorns, whose external dimension in the nanoscale tapers from large to small – that is, a hollow nanocone, except that the principal axis may not be straight. The concept of nanoplate includes all ultrathin coatings, even if their substrate is not planar, as long as the substrate's radius of curvature is much greater than the thickness of the coating.

A nanostructured material is defined as possessing internal or surface structure in the nanoscale. This structure may arise due to the presence of contiguous elements with one or more dimensions in the nanoscale. Primary atomic or molecular (in the chemical sense) structure is excluded (otherwise all materials would necessarily be categorizable as nanostructured). Ultimately nanostructured materials should be produced using bottom-to-bottom technology (i.e. mechanosynthesis, see Section 8.3), but currently this is only available for producing minute quantities of matter. At present, the largest exemplar of nanostructured materials is composites comprising nano-objects embedded in a matrix. In order to distinguish them from materials in which the embedded objects are larger than nanoscale, nanostructured composites may be called nanocomposites. In such materials, the overall structure is of a statistical nature. It would not, however, be appropriate to call such materials "randomly nanostructured", since almost certainly spacial correlations exist.

Since the above definition refers to "contiguous elements", it would appear that a heap of nano-objects would also fall into the category of nanostructured material.

Table 6.1 The Ontology of Nanodispersions

Matrix	State of the Nano-objects		
	Solid	**Liquid**	**Gaseous**
Solid	Nanocomposite[a]	–	Nanofoam[b]
Liquid	Nanosuspension[c]	Nanoemulsion	Nanofoam[b]
Gaseous	Aerosol	Aerosol	–

[a] Other materials falling into this category are nano-alloys, metal–matrix composites, etc.
[b] These materials may be described with the adjective "nanoporous" (or, in French, using the noun "nanoporeux").
[c] Nanofluids are included here (Section 6.6.4).

If the attractive energy of interaction between the objects is insignificant compared with thermal energy (i.e. $mgd > k_BT$, where m and d are, respectively, the mass and diameter of the object), one may refer to a powder (or an ultrafine powder or a nanopowder if the smallness of the objects needs to be explicitly stated, or indeed "nanoscale granular material"). If the objects attract each other weakly, one may refer to an agglomerate; typically, the external surface area of an agglomerate is similar to the sum of the surface areas of the individual objects comprising it. If the objects are strongly bonded to one another one may refer to an aggregate (typically the external surface area of an aggregate is significantly smaller than the sum of the surface areas of the individual objects prior to aggregation). In terms of interaction energies (Fig. 4.4), the particles in an agglomerate are separated by a distance ℓ_1, and those in an aggregate by a distance ℓ_0.

If a material comprises two or more distinct phases, if one or more dimensions of the zones of the phases are in the nanoscale one may refer to a nanostructured material. Certain alloys would fall into this category but not, presumably, solid solutions.

If nano-objects are dispersed into a continuous medium (matrix), one may describe the result as a nanodispersion. Depending on the states of the objects and the matrix, different terms are used to describe them (see Table 6.1).

The remainder of this chapter first covers the fabrication of the various kinds of nano-objects mentioned above, except for carbon-based materials, which are dealt with separately (Chapter 9) because of their unique importance and versatility. Second, the fabrication of composites is described.

6.1 NANOPARTICLES

Currently one can use either a top–down (comminution and dispersion, see Section 6.1.1) or a bottom–up (nucleation and growth, see Section 6.1.3) approach. The decision which to adopt depends, of course, on which can deliver the specified properties, and on cost. A third approach is bottom-to-bottom mechanosynthesis;

that is, "true" nanofacture, according to which every atom of the material is placed in a predefined position. At present this has no commercial importance because only minute quantities of material can be prepared. Nevertheless, it is the only approach that would solve the imperfections associated with top–down and bottom–up processes (the presence of defects, polydispersity, etc.).

Despite being the largest current commercial segment of nanotechnology, nanoparticles have as yet relatively few direct large-scale commercial uses (excluding photographic emulsions and carbon black, both of which long precede the nano-era); mostly their applications are in composites (i.e. a mixture of component A added to a matrix of component B, the latter usually being the majority component) – a nanocomposite differs from a conventional composite only insofar as the additive is nanosized and better dispersed in the matrix. Applications such as reagents for the remediation of contaminated soils (e.g. iron nanoparticles for decomposing chlorinated hydrocarbons) are being investigated, although the present very imperfect knowledge about the effects of such nanoparticles on soil ecosystems is preventing rapid exploitation in this area.

6.1.1 COMMINUTION AND DISPERSION

This top–down approach involves taking bulk material and fragmenting it. Crushing and grinding have typically been treated as low-technology operations; theoretical scientists seeking to formalize the process beginning with the formulation of mechanistic phenomenological rules (e.g. the random sequential fragmentation formalism) have hitherto had little industrial impact. The main advantages are universality (i.e. applicability to virtually any material) and low cost. Even soft organic matter (e.g. leaves of grass) can be ground by first freezing it in liquid nitrogen to make it brittle.

The main disadvantages are polydispersity of the final particles, the introduction of many defects (including contamination by the material used to make the grinding machinery – the smaller the particles the worse the contamination because it is introduced at the surface of the particles) and the impossibility of achieving nanoscale comminution, depending on the material: as a compressed brittle body is made smaller, its fracture strength increases until at a certain critical size crack propagation becomes impossible [156], cf. Section 2.7. This explains the well-known size limit to the crushing of materials – typically above the nanoscale (e.g. around 0.8 μm for calcium carbonate).

Crushing and grinding are venerable industrial processes in the form of hammering, rolling or ball-milling, but the advent of nanotechnology has given rise to novel, very well-controlled methods of achieving monodisperse nanoparticle generation by comminution and dispersion. One such process is electroerosion dispersion (EED) [213], in which granulated metal is ground into a fine powder by electrical discharges – typically a few hundred volts are discharged in a microsecond. The plasma temperature in the discharge filament is 10,000–15,000 K, sufficient to melt any metal. Fig. 6.3 shows a typical experimental installation.

FIGURE 6.3

Installation for the EED of metals. 1, reactor; 2, sedimentation chamber; 3, control cabin.

-Reproduced with permission from M.K. Monastyrov et al., Electroerosion dispersion-prepared nano- and submicrometer-sized aluminium and alumina powders as power-accumulating substances, Nanotechnol. Percept. 4 (2008) 179–187.

Even traditional ball milling has been greatly improved by advanced materials. It has been found that if ultrahard, ultradense small spheroids are used as the milling medium, with a surface nanotexture in the form of striations, the material to be comminuted is not crushed by mechanical impact between the balls, but instead subjected to shearing along its crystal planes due to the very high hydrodynamic shear gradient in the vicinity of the spheroids, without actually contacting them [61]. This results in far more monodisperse nano-objects, without surface defects, and whose shape may be controlled by adjusting the milling parameters.

6.1.2 "MECHANOSYNTHESIS" *QUA* GRINDING

Although we prefer to use the terms "mechanosynthesis" and "mechanochemistry" as synonyms for bottom-to-bottom methods (Section 8.3), it is also used to describe synthesis by means of grinding precursors [277], typically in a ball mill. For example, Ref. [244] describes making gold nanoparticles by milling a mixture of $HAuCl_4$ and aliphatic amines as stabilizers with steel balls; it appears that the steel galvanically reduces the gold. Mechanistically, the process appears to be exceedingly complex, defying detailed analysis.

6.1.3 NUCLEATION AND GROWTH

This process involves a first-order phase transition from an atomically dispersed phase to a solid condensed phase. During the first stage of the transition, fluctuations

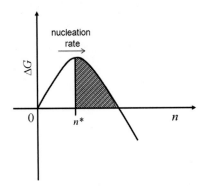

FIGURE 6.4

Sketch of the variation of free energy of a cluster containing n atoms (cf. Fig. 2.2). The maximum corresponds to the critical nucleus size. Clusters that have managed through fluctuations to climb up the free energy slope to reach the critical nucleus size have an equal probability to shrink back and vanish, or to grow up to microscopic size.

in the homogeneous, metastable parent phase result in the appearance of small quantities of the new phase. The unfavorable process of creating an interface opposes the gain in energy through the reduction in supersaturation of the parent phase, leading to a critical size of nucleus, n^*, above which the nuclei develop rapidly and irreversibly into the new "bulk" (albeit nanoscale) phase. Many syntheses were reported in the nineteenth century (e.g. [74, 104]).

When atoms cluster together to form the new phase, they begin to create an interface between themselves and their surrounding medium, which costs energy. Denoting the IF tension by γ, and using subscripts 1 and 2 to denote the new phase and surrounding medium, respectively (see Section 3.2), the energy cost is $A\gamma_{12}$, where A is the area of the cluster's surface, equal to $(4\pi)^{1/3}(3nv)^{2/3}$, where n is the number of atoms in the cluster, and v the volume of one atom. At the same time, each atom contributes to the cohesive energy of the new phase. Summing these two contributions, at first the energy will increase with increasing n, but ultimately the (negative) cohesive energy of the bulk will win (Fig. 6.4).

In order to synthesize nanoparticles via nucleation and growth, first the atoms are dispersed (dissolved) in a medium under conditions such that the dispersion is stable. Then, one or more of the external parameters is changed such that the bulk phase of the material now dispersed is stable. This could be accomplished, for example, by cooling the vapor of the material. The formation of the new bulk phase is a

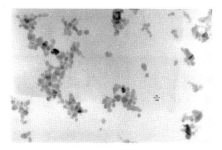

FIGURE 6.5

Transmission electron micrograph of CdS nanoparticles prepared in the author's laboratory by nucleation and growth in aqueous solution [245]. The apparent agglomeration is an artifact of the sample preparation, in which a drop of the particle suspension was placed on a carbon-coated copper grid and the water allowed to evaporate. This final stage of the evaporation process can be observed *in situ* in the electron ultramicroscope: the water forms a receding thin film and capillary forces (Section 3.3) at its edge drag the particles together.

first-order phase transition involving nucleation. Chance fluctuations will generate critical nuclei (see Fig. 6.4).

Compound particles can be synthesized by chemical reaction [245]. Suppose the formula of the desired substance is MX, where M represents a metal such as silver or cadmium, and X a metalloid such as sulfur or selenium. One then prepares two solutions of soluble compounds of M and X (for example, silver nitrate and sodium sulfide), which are then mixed together. Figs 6.5 and 6.6 show some examples.

Two key challenges in this process are (i) to obtain particles that are as uniform (monodisperse) as possible and (ii) to be able to control the mean size. In the case of synthesis by chemical reaction, the key parameter is the rate of mixing. Two extreme situations yield the desired quasimonodispersity: ultrarapid mixing of very concentrated solutions, and ultraslow mixing of very dilute solutions. In the former case, a very large number of critical nuclei are formed almost simultaneously (the rate of creation of critical nuclei is proportional to the supersaturation, that is, the ratio of the actual concentration to the solubility product of MX); growth of material onto the initially formed nuclei is too slow to compete with fresh nucleation in sinking the added mass. Conditions should be chosen such that the nuclei are just able to grow

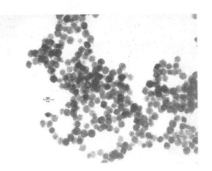

FIGURE 6.6

Transmission electron micrograph of hematite nanoparticles prepared in the author's laboratory by nucleation and growth in aqueous solution (cf. Fig. 6.5).

sufficiently large to be effectively irreversibly stable before all the free M and X ions have been scavenged by the formation of nuclei. Further growth to any desired size can then be achieved in a separate, subsequent, stage by adding fresh material at a rate just sufficient to allow all the nuclei to grow without creating any new ones.

In the latter case, nuclei are formed extremely rarely and are unable to grow beyond the size of minimum stability because of the lack of material; diffusion of fresh material to the few nuclei formed initially is too slow to prevent new nuclei being formed in order to sink the added reagents. Once a sufficient number of nuclei have been synthesized, they can be grown up to the desired size as in the previous case. This approach is very effective for synthesizing monodisperse noble metal particles (e.g. gold) by very slowly reducing the solution of a salt of the metal [74].

Because of the Kelvin relation (Eq. (2.7)), larger particles will have a slightly lower solubility than smaller ones. Therefore, there will be a slight tendency for the smaller ones to dissolve, and for their material to be deposited onto the bigger ones. This process is known as Ostwald ripening and under certain conditions may permit the size distribution of a collection of particles to be narrowed, albeit at the price of increasing the mean size.

Once a collection of nuclei has been synthesized, it is very easy to grow shells of different materials around them; one simply needs to ensure that the new material is added at a sufficient rate to allow all the particles to grow uniformly, and not so rapidly that fresh nuclei are formed.

The IF free energy for aggregation of particles made from material 1 in the presence of medium 2 is given by (see Section 3.2.1)

$$\Delta G_{121} = \Delta G_{11} + \Delta G_{22} - 2\Delta G_{12}, \tag{6.1}$$

where ΔG_{11} and ΔG_{22} are the cohesive energies of materials 1 and 2, and ΔG_{12} is the solvation energy. Note that water has a very large cohesive energy. Therefore, particles of almost any insoluble material synthesized in water are likely to aggregate, unless appropriate measures to ensure their hydration are taken. A useful strategy is to synthesize the particles in the presence of a very hydrophilic material such as polyethylene glycol or a polyion such as hexametaphosphate, which is able to adsorb on the surface of the particles and effectively hydrate them. Michael Faraday's famous synthesis of gold nanoparticles used citrate ions to hydrate their surface [74]. Crystals of silver chloride, silver bromide and silver iodide ranging in size from tens of nanometers to micrometers, which form the basis of conventional silver halide-based photography, are stabilized in the emulsions used to coat glass or polymer substrates by the natural biopolymer gelatin.

Micelles and superspheres are dealt with in Section 8.2.9.

6.1.4 COMPLEX NANOPARTICLES

For many technical applications nanoparticles more complex than those made from a single material are required. Two very common modifications are core–shell nanoparticles and porous, usually called mesoporous, nanoparticles.

The term "core–shell" signifies a nanoparticle of one material around which is a shell of another material. They have long been used in the silver halide photographic industry in order to confer particular properties on an emulsion [127]. Another example of the use of a shell to modify the electronic properties of nanoparticle is to coat a photoluminescent semiconductor particle with a wider-gap semiconductor (e.g. a CdSe core coated with a ZnS shell). This procedure tends to eliminate nonradiative recombination sites on the cadmium selenide surface, thereby augmenting the photoluminescent quantum yield. If these chalcogenides are made by nucleation and growth (Section 6.1.3), then the initial reaction between the cadmium and selenium compounds can be followed by reaction between a zinc and a sulfur compound in a continuous process: by appropriately selecting the rate of zinc and sulfur addition, the pre-existing cadmium selenide particles act as nucleation sites for the zinc sulfide, which crystallizes exclusively as a shell around the CdSe.

Another motivation for creating a core–shell particle is to confer environmental compatibility. For example, magnetic metallic or intermetallic nanoparticles have been proposed, and are being tested, as therapeutic agents for anticancer treatment. Using an external magnetic field, they can be steered to the site of a tumor and then heated using radio frequency electromagnetic excitation in order to kill the neoplasm (cf. Section 4.2). These metallic particles are, however, usually opsonized and eliminated from the body by the immune system before they can accomplish their intended action. If, however, they are coated with a thin shell of a biocompatible material, such as titanium dioxide, perhaps augmented with an organic material such as polyethylene oxide, they become stealthy with respect to the immune system ("stealth" nanoparticles) and may evade elimination.

A further common motivation for creating a shell around a core particle is to prevent agglomeration or aggregation of the particles when they are dispersed

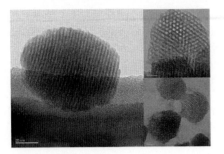

FIGURE 6.7

Transmission electron micrograph of mesoporous silica nanoparticles showing their detailed structure. Pores with cavities of diameter 2–3 nm can be clearly seen throughout
the particle.

*-Reproduced with permission from J. Miao et al., Nano-
materials applications in "green" functional coatings,
Nanotechnol. Percept. 8 (2012) 181–189.*

in a liquid. Bipolar materials will naturally tend to aggregate (Section 3.2.1), but
this can be prevented by coating them with monopolar substance. Sometimes the
shell can be very thin, such as the adsorbed monolayer of citrate around a sol
of gold nanoparticles, or hexametaphosphate or nitrilotriacetate around cadmium
chalcogenide nanoparticles [245]. These oligoanions confer strong electrostatic
repulsion between the nanoparticles when they are dispersed in an aqueous solution.

Some of the most elaborate core–shell type structures are constructed for the sake
of enhancing catalytic properties [313]. An interesting example of the use of the
core–shell concept during processing to create particularly perfectly formed spheres
is the doping of ceria with titanium [76]. During nanoparticle formation, the ceria
core is completely encapsulated by a thin shell of titania. Minimization of the IF
energy ensures perfect sphericity. This, and size monodispersity, is very important
for ceria nanoparticles used to abrade very large-scale integrated circuits.

Mesoporous nanoparticles – the pore size of which is much smaller than the
nanoparticle diameter (Fig. 6.7) – are useful for a variety of purposes, including
"environmental paint" for conferring superior fire resistance on wood [207].

6.2 NANOFIBERS

"Nanofiber" is the generic term describing nano-objects with two external dimensions
in the nanoscale. A nanorod is a rigid nanofiber, a nanotube is a hollow nanofiber
and a nanowire is an electrically conducting nanofiber (Fig. 6.2).

Two approaches are presently mainly used to synthesize nanofibers. For some substances, under certain conditions, the natural growth habit is acicular. Therefore, the methods described in Section 6.1.3 can be used to generate nuclei, followed by a growth stage to elongate them. Heterogeneous nucleation can be induced at the solid/gas interface by predepositing small catalytic clusters. Upon addition of vapor, condensation on the clusters and growth perpendicular to the solid substrate takes place. This is used as an efficient way of synthesizing CNTs (see Section 9.4). If uniform nanopores can be formed in a membrane (e.g. by laser drilling or by self-assembly) they can be used as templates for nanofiber formation. The material for the fiber should be deposited as a shell on the inner surface of the pores (if the goal is to make nanotubes), or else should completely fill the pores (for nanorods). Nanofibers, especially nanorods, formed by either of the two previous methods can also be used as templates for making nanotubes of a different material. Lieber has reported the general synthesis of semiconductor nanowires with control of diameter [63, 111].

Organic nanowires with extraordinary morphological and optoelectronic properties can be grown epitaxially on surfaces using organic molecular beams [163].

6.3 NANOPLATES AND ULTRATHIN COATINGS

Many of the traditional engineering methods of fabricating thin coatings on a substrate have not produced objects in the nanoscale because typically they have been more than 100 nm thick. Nevertheless, the trend is to develop thinner functional surfaces by coating or otherwise modifying bulk material, and insofar as the coating or modification is engineered with atomic precision, it belongs to nanotechnology, and if it is nanostructured either laterally or perpendicular to the substrate, it will rank as a nanomaterial even if its overall thickness exceeds 100 nm.

The surface treatment of bulk material, especially metals, is an ancient technology. In the cases where nanoscale structural features were apparently necessary to ensure having the required attributes (e.g. in Damascus swords [257]), although the structures were created deliberately the nanoscale aspect might be considered as essentially inadvertent since the technologist is unlikely to have explicitly envisioned the nanoscale structuring. This is in contrast to, for example, the medicinal use of nanoparticulate gold (introduced by Paracelsus in the sixteenth century), when it was realized that a metal could only be assimilated by a living human organism if it were sufficiently finely divided.

Completely in the spirit of nanotechnology are the monomolecular layers now called Langmuir films, transferred to solid substrata using the Langmuir–Blodgett (LB) and Langmuir–Schaefer (LS) techniques; these films might only be a few nanometers thick. Their preparation (Section 6.3.2) belongs to the "top–down" category.

Many physical vapor deposition (PVD) techniques (such as an evaporation and magnetron sputtering) create films thicker than the nanoscale and hence are out of the scope of this book. Molecular beam epitaxy (MBE), a technique of great importance in the semiconductor processing industry, is briefly covered in Section 6.3.1 (see also Section 8.1.1).

6.3.1 MOLECULAR BEAM EPITAXY (MBE)

MBE can be considered as a precise form of PVD. Solid source materials are placed in evaporation cells around a centrally placed, heated, substrate. Pressures less than 10^{-5} torr ensure that the mean free path of the vapor exceeds practical vacuum chamber dimensions (~1 m), the molecular beam condition. Ultrahigh vacuum (UHV) conditions are needed to ensure the absence of contamination from residual gases (from the chamber walls, etc.). A few seconds are typically required to grow one monolayer. The technique has been developed very successfully using a practical, empirical approach – thermodynamic analysis is difficult because the various parts of the system (sources, substrate, chamber wall) are at different temperatures.

6.3.2 LANGMUIR FILMS

Langmuir films, first reported by Pockels at the end of the nineteenth century, consist of a monomolecular layer of amphiphiles (molecules consisting of an apolar nanoblock conjoined to a polar block of roughly the same size) floating on water. The polar "heads" dip into the water and the apolar "tails" stick up into the air. In other words, the film precursors are molecules of general formula XP, where X is (typically) an apolar chain (e.g. an alkyl chain), called the "tail", and P is a polar "head" group such as oligoethylene oxide, or phosphatidylcholine. When spread on water they mostly remain at the water/air interface, where they can be compressed (e.g. using movable barriers) until the molecules are closely packed to form two-dimensional liquid-like and solid-like arrays. Slowly withdrawing a hydrophilic plate perpendicular to and through the floating monolayer from below will coat the plate with a packed monolayer (Fig. 6.8(a)), as was extensively investigated by Blodgett; these supported layers are called LB films. The LB technique refers to the transfer of the floating monomolecular films to solid substrata by vertically dipping them into and out of the bath. In the LS technique, the substratum is pushed horizontally through the floating monolayer (Fig. 6.9). Very stable multilayer films can be assembled by making P a chelator for multivalent metal ions, which bridge lateral neighbors and/or successive layers (assembled head–head and tail–tail). Lateral stability can be increased by UV-irradiation of films with an unsaturated alkyl chain (photopolymerization). The process of perpendicular traversal of the floating monolayer can be repeated many times to build up multilayer films, provided the close packing of the monolayer is maintained, e.g. by adding fresh material *pari passu* in an appropriate fashion. Exceptionally laterally cohesive and rigid Langmuir films can be manipulated as free-standing objects.

   According to the nanocriterion of "absence of bulk" (Section 2.2), a few layers (Fig. 6.8(b)) would be nanoscale. But even though the multilayer (Fig. 6.8(c)) could, in principle, be of microscopic or even macroscopic thickness, it has no bulk equivalent. Simply throwing amphiphilic molecules into a large vessel and stirring

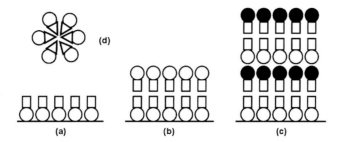

FIGURE 6.8

LB films. (a) A monolayer; (b) a bilayer; (c) a Y-type multilayer. The circles represent the polar heads and the squares the apolar heads of the amphiphilic molecule. (d) A micelle, which can form spontaneously upon dispersal in water if the amphiphilic molecules have a tail smaller than the head (see Section 8.2.9).

would lead to a highly disordered material and it would take an inordinately long time to thermally anneal it into any kind of ordered structure. Consider a LB film made with two types of fluorescent polar head, let us call them A and B (suppose that B is represented by the black circles in Fig. 6.8), with differing optical absorption and emission spectra, such that A's emission can be absorbed by B (the process is called Förster energy transfer), which in turn would emit light at wavelength $\tilde{\lambda}_B$. Then, exciting A (with light of wavelength λ_A) would result in emission of wavelength $\tilde{\lambda}_B$. On the other hand, if A and B were simply mixed at random, irradiating the mixture with λ_A would mostly result in emission of wavelength $\tilde{\lambda}_A$, since A and B have to be very close to each other in order for Förster energy transfer to occur: *nanoscale structure* results in qualitatively different behavior. Amphiphilic molecules can also form a bicontinuous cubic phase, which although it can be of indefinite (i.e. macroscopic) extent, it is pervasively nanostructured.

If two different immiscible amphiphiles are mixed on the Langmuir trough, the resulting variegated structures can be imaged using SPM after transferring the Langmuir film to a solid support using the LB technique (Fig. 6.10). These experiments have shown (a) how difficult it is to find convenient mixtures and (b) how difficult it is to predict theoretically what patterns result. There are moreover experimental difficulties in imaging texture (see Chapter 5) at scales below a few tens of nanometers. The lack of metrology techniques suited for this purpose is in fact one of the main current hindrances to progress in the field.

Although there has been recurrent industrial interest in LB films for numerous applications in microelectronics and integrated optics, the cumbersome nature of the fabrication methodology, well suited to laboratory research but difficult to automate, and the difficulty of creating defect-free films, have militated against its adoption. The problem of defects [e.g. pinholes in the insulating layer of a field effect transistor (FET), Fig. 7.6] can, however, be overcome by reducing the component

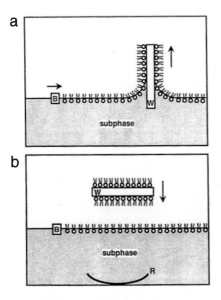

FIGURE 6.9

The LS technique. (a) The polar substrate is slowly drawn upwards through the Langmuir film to deposit a monolayer. (b) The coated substrate, now apolar, is pushed horizontally through the Langmuir film to deposit a second monolayer (and is caught by the receptacle R). This operation is equivalent to very rapid vertical descent. The rapidity is needed to ensure that the mensicus curves downwards during deposition.

area (Section 10.5), suggesting that nanoscale electronics might see a revival of interest. One attractive feature is the versatile chemistry allowing environmentally responsive films to be synthesized; bio-inspired films in a chemFET configuration can be responsive to both total captured particle number and the state of aggregation, for example, [249]. Bio-inspired LB films are typically in the liquid-crystalline phase, in which defects are self-annealed. Moreover, for electronics applications only very thin films are needed, whereas an optical waveguide requires ~100 monolayers, the defect-free assembly of which is cumbersome.

Post-processing LB films expands the fabrication possibilities. Blodgett herself explored "skeleton" films prepared by selectively dissolving away one component of a mixed film. Any multivalent metal or metal compound nanoplate can be prepared

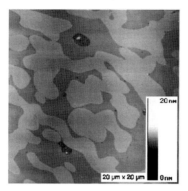

FIGURE 6.10

Scanning force micrograph (contact mode) of a mixed behenic acid/pentadecanoic acid film transferred horizontally onto muscovite mica at a surface pressure of 15 mN/m. Image size: 20×20 μm.

-*Reproduced with permission from S. Alexandre, C. Lafontaine, J.-M. Valleton, Local surface pressure gradients observed during the transfer of mixed behenic acid/pentadecanoic acid Langmuir films, J. Biol. Phys. Chem. 1 (2001) 21–23.*

from a metal–fatty acid LB film (e.g. dysprosium behenate) which is then pyrolyzed to remove the organic moiety.

Under certain conditions, regular striated patterns can be formed during LB deposition [274]. The cause is dynamic instability of substrate wetting.

6.3.3 SELF-ASSEMBLED MONOLAYERS (SAMS)

If particles randomly and sequentially added to the solid/liquid interface are asymmetrical, i.e. elongated, and having affinity for the solid substratum at only one end (but the ability to move laterally at the interface), and with the "tail" (the rest) poorly solvated by the liquid, they will tend to adsorb in a compact fashion, by strong lateral interaction between the tails (Fig. 6.11). This is a practical procedure of some importance for modifying the surfaces of objects fabricated by other means. The precursors are molecules of general formula XL, where X is (typically) an apolar chain (e.g. alkyl), and L is a ligand capable of binding to the substratum. Addition of XL to the metal surface results in a closely packed array of XL. The film is stabilized by hydrogen or chemical bonds to the substrate, and lateral LW forces between the X.

SAMs were discovered in the 1940s by Bigelow et al. [22]. Currently the two main types of L are –SH (thiol or mercaptan), which binds strongly to Au, Ag, Pt,

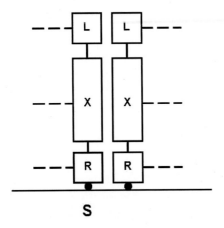

FIGURE 6.11

A (fragment of a) SAM. The component molecules have the general formula LXR, where X is an apolar chain (e.g. alkyl), and R is a reactive group capable of binding to the substratum S. X can be functionalized at the end opposite from R with a group L to form molecules L–XR; the nature of L can profoundly change the wetting properties of the SAM.

Cu, Hg, etc., and organosilanes, which bind strongly (bond covalently) to silica. These chemical requirements are the main constraints limiting the versatility of the technology. The original example was eicosyl alcohol ($C_{20}H_{41}OH$) dissolved in hexadecane ($C_{16}H_{34}$) adsorbing on silica glass. Later, molecules with R = –SH (thiol or mercaptan), which binds strongly to metals such as Au, Ag, Cu, Hg, etc., were investigated, and organosilanes, which bind strongly (bond covalently) to silica. If the tail moiety X is poorly solvated by the liquid, its flexibility may enable it to be compactified while the molecule is in the bulk solution, tending to prevent self-aggregation, and only unfurling itself after R is attached to the solid surface — a rudimentary form of programming. SAMs provide a very convenient way to change the wettability of a solid surface and hence may be useful as a precursor for other deposition processes. The monolayers of Bigelow et al. were both hydrophobic and oleophobic. An octadecanethiol (L = –H) film adsorbed on gold would be both oil- and water-repellent; if L = –OH it will be hydrophilic (see Chapter 3).

X can be functionalized at the end opposite from L with reactive groups to form molecules RXL. These can profoundly change the wetting properties of the assembled monolayer. For example, whereas octadecanethiol (R = –H) films are both oil- and water-repellent, if R = –OH then oil and water will spread. If they are bulky, the functionalized molecules should be mixed with unfunctionalized ones to avoid packing defects. Mixtures with different chain lengths (e.g. X = C_{12} and C_{22}) give liquid-like SAMs. The biologically ubiquitous lipid bilayer membrane could be considered to belong to this category. The component molecules are of type XR, where X is an alkyl chain as before, and R is a rather polar "head group", the volume of which is typically roughly equal to that of X. Placing XR molecules in water and gently agitating the mixture will spontaneously lead to the formation of spherical bilayer shells called vesicles. The vesicles will coat a planar hydrophilic substratum with a lipid bilayer when brought into contact with it [55].

SAMs can be patterned using photolithography, or "stamping" (microletterpress), to create patterns on substrata (e.g. gold and/or silica) to which the SAM precursor molecules will bind, leaving other zones free. In this procedure, the required pattern is the first created in relief on a silicon substrate, and which is used as a mold for the elastomeric polymer PDMS (polydimethylsiloxane). The SAM molecules can be used directly as ink to coat the projecting parts of the relief pattern, which is then stamped onto the substratum, or else the ink is some substance that passivates the substratum with respect to the SAM molecules, which then selectively bind as required.

6.3.4 NANOSTRUCTURED SURFACES

Very high power femtosecond pulsed lasers can be advantageously used to structure solid surfaces in the nanoscale. In the case of metals, highly localized melting appears to be the underlying mechanism. Structure at multiple scales can be generated (Figs 6.12 and 6.13). Pulse repetition rate and power are key machining parameters [109]. As seen in the figure, once the mold has been created in a hard material, low-cost replicas (in this case using polypropylene) can be fabricated *en masse*. In this example, the purpose was to create an ultrahydrophobic surface (see Section 3.2.1 for the definition of hydrophobicity) from a material such as polypropylene that is already fairly hydrophobic. The study of some natural plant and insect surfaces (e.g. Fig. 3.8) has revealed that they exploit the Wentzel and Cassie–Baxter laws to augment the intrinsic hydrophobicity by roughness and heterogeneity (Section 3.4.1). Intriguingly, nature uses multiple characteristic length scales for the roughness and heterogeneity, presumably in order to minimize material or fabrication cost. It was a serendipitous discovery that the femtosecond pulsed laser structuring achieves a similar result by judicious design of the pulse sequence.

Another application of femtosecond pulsed lasers is direct laser writing (DLW) via multiphoton polymerization (MPP) of an organic precursor. Multiphoton adsorption was predicted to occur by Göppert-Mayer [99]; it can be practically achieved using very high power, such as with femtosecond pulsed lasers. When the photon density is comparable to or higher than the atomic density of the irradiated material, new phenomena emerge, notably an avalanche of physical (dielectric breakdown) and chemical (following the direct photocleavage of interatomic bonds, resulting in free

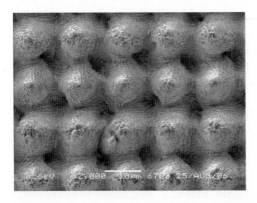

FIGURE 6.12

Scanning electron micrograph of a steel mold surface-nanostructured with a femtosecond pulsed laser.

-Reproduced with permission from J.J. Ramsden et al., The design and manufacture of biomedical surfaces, Ann. CIRP 56/2 (2007) 687–711.

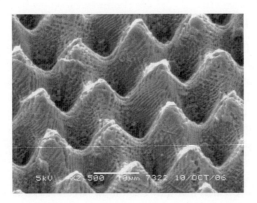

FIGURE 6.13

Scanning electron micrograph of the texture of the steel mold shown in Fig. 6.12 replicated in polypropylene.

-Reproduced with permission from J.J. Ramsden et al., The design and manufacture of biomedical surfaces, Ann. CIRP 56/2 (2007) 687–711.

radical generation) effects [194]. Hence, many polymers do not even need a photoinitiator. As with the nanostructuring of metals, the course of photopolymerization depends, often in a complex fashion, on pulse duration and repetition rate.

6.4 CRYSTALLIZATION AND SUPRAMOLECULAR CHEMISTRY

It is a remarkable fact that many or even most organic molecules, despite their usually very complicated shapes, are able to spontaneously form close-packed crystals. Perhaps only familiarity with the process prevents it from occupying a more prominent place in the world of self-assembly. In seeking to understand this remarkable phenomenon better, Kitaigorodskii formulated an Aufbau principle [162], according to which the self-assembly (cf. Section 8.2.1) of complicated structures takes place in a hierarchical fashion in the following sequence:

Stage 0 a single molecule (or a finite number of independent molecules)

Stage 1 single molecules join up to form linear chains

Stage 2 the linear chains are bundled to form two-dimensional monolayers

Stage 3 the two-dimensional monolayers are stacked to form the final three-dimensional crystal.

Many natural structures are evidently hierarchically assembled. Wood, for example, derives its structural strength from glucose polymerized to form cellulose chains, which are bundled to form fibrils, which are in turn glued together using lignin to form robust fibers. Note that the interactions between the molecular components of the crystal may be significantly weaker than the covalent chemical bonds holding the atomic constituents of the molecules together. Such weakness enables defects to be annealed by locally melting the partially formed structure. There is an obvious analogy between "crystallization" in two dimensions and tiling a plane. Since tiling is connected to computation, self-assembly, which can perhaps be regarded as a kind of generalized crystallization, has in turn been linked to computation. Just as quantum dots containing several hundred atoms can in some sense (e.g. with regard to their discrete energy levels) be regarded as "superatoms" (Section 2.5), so can supramolecular assemblies (typically made up from very large and elaborate molecules) be considered as "supermolecules". The obtainable hierarchically assembled structures – an enormous literature has accumulated – provide powerful demonstrations of the Kitaigorodskii Aufbau principle. The use of metal ions as organizing centers in these assemblies has been a particularly significant practical development.

6.5 METAL–ORGANIC FRAMEWORKS (MOFS)

These hybrid materials are diversely called soft porous crystals, porous hybrid matter, porous coordination polymers (PCPs), MOFs, modular porous solids and so forth [161, 79, 80, 133, 121]. We shall call them MOFs for convenience, without prejudice against the other terms. They consist of an array of metal coördinating centers (nodes or vertices in graph-theoretic language), which are usually ordered clusters of metal ions rather than single metal atoms, linked by organic groups (edges), which must fulfill two rôles, they must be ligands for the metals and they must have some structural attribute that constitutes the framework. MOFs seem to have emerged in the late 1980s/early 1990s. Their most obvious congeners are zeolites, but these materials are natural (as well as synthetic), rigid and wholly inorganic. In contrast, polymers of intrinsic microporosity (PIMs) are flexible and wholly organic [34]. In PIMs, the randomly contorted shapes of the polymers prevent their efficient packing, hence ensuring the existence of pores. In molecularly imprinted polymers (MIPs), polymerization of monomer precursors takes place around guest molecules, which serve as templates for the pores; these guests can be reversibly removed and reintroduced (that is, the pores retain their integrity even in the absence of the guest). Unlike MOFs, however, PIMs and MIPs have random not regular structures.

The existence of pores implies the ability to accept guests, small molecules that can enter and reside in the framework. The unit cells of MOFs can be very large, in some cases containing tens of thousands of atoms, making modeling difficult.

Kitagawa has attempted some classification of these materials. Temporally there has been a succession of generations: first generation materials collapse upon guest removal; second generation materials have robust and rigid frameworks that retain their crystallinity when the guest is absent; third generation materials can be structurally transformed by guest removal, and other external stimuli such as temperature, pressure and electromagnetic fields.

Kitagawa has also proposed a structural classification: Class I comprises one-dimensional chains; voids in the packing can be occupied by guests, which engender geometrical changes in the framework. Class II consists of stacks of two-dimensional layers. In class III, the layers are interdigitated to form one-dimensional channels. Class IV is considered to be truly three-dimensional, with pillared layers. Class V comprises three-dimensional frameworks that can expand or shrink with guest insertion or removal. In class VI, the grids interpenetrate.

Terminology (nomenclature) is presently unsystematic. MOFs are most typically designated by three capital letters denoting their laboratory of origin (for example, MIL stands for "Matériaux de l'Institut Lavoisier", perhaps the most prolific synthesizer) followed by a hyphen and a number, which usually merely indicates the sequence in which the material was synthesized, although similar materials may be given adjacent numbers regardless of temporal sequence. Sometimes the metal element is indicated in parentheses, which is useful when MOFs differ only in the identity of the metal and are otherwise structurally identical. Another common acronym is ZIF (zeolitic imidazole framework).

Despite their complicated structures and recent emergence, many MOFs are remarkably easy to synthesize. For example, to make ZIF-8 it suffices to mix solutions of zinc nitrate (zinc forms the vertices) and 2-methyl imidazole (which forms the edges). The MOF spontaneously assembles on any substrate in the solution. Sometimes these surface-assembled structures are called SURMOFs. Another way of generating surface-assembled MOFs is via APED (Section 6.6.6). Many thousands of different MOFs have already been synthesized, and the comprehensive investigation of their properties lags far behind the reporting of their synthesis and structure.

Potential applications of MOFs are legion. The most obvious are derived from their porosity – as molecular sieves for separation, and for gas capture and storage. Typically MOFs have immense BET surface areas, of several thousand m^2/g, exceeding those of nanoparticles by one or two orders of magnitude. Many MOFs have been found to have interesting catalytic functionality. Indeed, with their flexibility they bear a certain resemblance to proteins (Section 11.5). Most multivalent metals appear to be amenable to serve as the nodes, and the potential variety of linkers is almost infinite. It would be extremely desirable if it were possible to design MOFs to have a given functionality but this is presently far from possible; at best the structure of a proposed design might be predictable, but rather on the basis of the very large knowledge base that already exists.

6.6 COMPOSITES

Most of the materials around us are composites. As mentioned in Section 6.4, natural materials such as wood are highly structured and built upon sophisticated principles. The basic structural unit is cellulose, which is a polymer of the sugar glucose, but cellulose on its own makes a floppy fabric (think of cotton or rayon), hence to give it strength and rigidity it must be glued together into a rigid matrix. This is accomplished by the complex multiring aromatic molecule lignin. Another example is the shell of marine molluscs such as the abalone. They are composed of about 95% by volume of thin platelets, small enough to rank as nano-objects, of aragonite, a form of calcium carbonate and about 5% of protein. The toughness of the composite seashell exceeds by more than tenfold the toughness of pure calcium carbonate.

A striking thing about these natural composites is the enormous increase in desirable properties, such as stiffness or toughness, through small (by volume or mass) inclusions of another material that would, by its nature, normally rank as the matrix in artificial composites. In contrast, the matrix in artificial composites is usually the majority component (in mass or volume or both). The principle of combining two or more pure substances with distinctly different properties (which might be mechanical, electrical, magnetic, optical, thermal, chemical and so forth) in order to create a composite material that combines the desirable properties of each to create a multifunctional substance has been refined by humans over millennia, presumably mostly by trial and error. Typically, the results are, at least to a first approximation, merely additive. Thus, we might write a sum of materials and their properties like

	cellulose	high tensile strength	self-repellent
+	lignin	weak	sticky
=	wood	strong	cohesive

Empirical knowledge is used to choose useful combinations, in which the desirable properties dominate – in principle one might have ended up with a weak and repellent material. The vast and always growing accumulation of empirical knowledge usually allows appropriate combinations to be chosen. Indeed, the *motif* of strong fibers embedded in a sticky matrix is very widely exploited, other examples being glass fiber- and carbon fiber-reinforced polymers.

There are two principal motivations for creating composites. The first is to increase the toughness of the majority component (as in the seashell). This occurs by dividing up the monolithic material into nano-objects and gluing the objects together. The increase in toughness comes partly through the increased ductility of the comminuted objects (Sections 2.7 and 6.1.1) and partly through the fact that any fracture that does occur can only extend to the object boundary or, if it occurs in the matrix, to the nearest object lying in its path, a distance which is very likely to be shorter than the object size. Particularly in this kind of composite of the wetting of the principle phase by the glue is of extreme importance. Empirical knowledge is now backed up and extended by fundamental knowledge of the molecular-scale forces involved, but natural materials still far surpass artificial ones in this respect. Proteins, in particular, are very versatile glues because of the variety of chemical functionalities possessed by amino acids. The second motivation is the combination of properties – these composites might well be called combosites to distinguish them from the other kind; for example, ultrahard nanoparticles might be added to a relatively soft polymer matrix to create a hard, plastic material.

6.6.1 POLYMER–NANO-OBJECT BLENDS

In fact, most of the recognized successes in nanomaterials so far have been not in the creation of totally new materials through mechanosynthesis (see Section 8.3), which is still an unrealized goal, but in the more prosaic world of blending, which is the simplest form of combination. For example, one adds hard particles to a soft polymer matrix to create a hard, abrasion-resistant coating. As with atomically based mechanosynthesis, the results are, to a first approximation, additive. Thus, we might again write a sum like

	polypropylene	flexible	transparent
+	titanium dioxide	rigid	opaque
=	thin film coating (paint)	flexible	opaque

This is not actually very new. Paint, a blend of pigment particles in a matrix (the binder), has been manufactured for millennia. What is new is the detailed attention paid to the nanoparticulate additive. Its properties can now be carefully tailored for the desired result. If one of components is a recognized nanosubstance – a nanoparticle or nanofiber, for example – it is acceptable to describe the blend as a nanostructured material. Other applications of this kind include ultralow permeability materials for food packaging, for which it is often desirable to restrict the ingress of oxygen. The main principle here is to blend particles of a very high aspect ratio (fragments of nanoplates rather than nanoparticles) into the polymer such that the principal plane of the objects is perpendicular to the route of ingress. The tortuosity of gas diffusion is vastly increased in such materials. Very often it is sufficient to apply them as an ultrathin coating onto a conventional flexible polymer substrate to achieve the desired diminution of gas permeability. The design of friction damping composites requires careful consideration of the viscoelastic properties of the material. "Self-cleaning" and anti-graffiti coatings rely on the active, typically photo-induced, chemistry taking place at the surface of the nanoparticles included in the matrix.

The paradigm of paint. The biggest range of applications for such nanocomposites is in thin film coatings – in other words paint. This is a much older composite than most other nanomaterials. Paint consists of a pigment (quite possibly made of nanoparticles) dispersed in a matrix of varnish. Paint can be said to combine the opacity of the pigment with the film-forming capability of the varnish.

Traditional pigments may comprise granules in the micrometer size range; grinding them a little bit more finely turns them into nano-objects. Compared with transparent varnish, paint then combines the attribute of protection from the environment with the attribute of color. The principle can obviously be (and has been) extended practically *ad libitum*: by adding very hard particles to confer abrasion resistance; metallic particles to confer electrical conductivity; tabular particles to confer low gas permeability and so on.

The purpose of adding materials to polymer matrix is, then, to enhance properties such as stiffness, heat resistance, fire resistance, electrical conductivity, gas permeability, and so forth; the object of any composite is to achieve an advantageous combination of properties. If the matrix is a metal, then we have a metal–matrix composite (MMC, Section 6.6.2). A landmark was Toyota's demonstration in 1991 that the incorporation of a few weight percent of a nanosized clay into a polyamide matrix greatly improved the thermal, mechanical (e.g. doubled the tensile modulus) and gas permeability (barrier) properties of the polymer compared with the pure polymer or the conventional composite made with micrometer-sized additive. This was the first commercial nanocomposite.

Another mineral–polymer composite is the material from which many natural seashells are constructed: platelets of aragonite dispersed in a protein matrix (see Section 6.6.6). In this case, however, the "matrix" only constitutes a few percent of the volume of the composite.

There is no general theory suggesting that the advantage scales inversely with additive size; whether a nanocomposite is commercially viable depends on all the parameters involved. There is such a huge variety of materials that it is perhaps futile to attempt a generalization. However, the very small size of individual nanoparticles would make it feasible to incorporate a greater variety of materials within the matrix for a given additive weight percent. Very often, ensuring wetting of the particle by the matrix presents a significant technological hurdle. Most successful composites require the additive to be completely wetted by the matrix. Wetting behavior can be predicted using the Young–Dupré approach (see Section 3.2); if, however, the particle becomes very small, the surface tension will exhibit a curvature-dependent deviation from the value for a planar interface.

The three main fabrication routes for polymeric nanocomposites are as follows:

1. Blending preformed nanoparticles with the matrix, which is typically in the molten state.
2. Dispersing preformed nanoparticles in monomer and polymerizing the matrix around the nanoparticles (the Toyota composite mentioned above followed this route: modified clay was swollen in the presence of caprolactam followed by polymerization).
3. Synthesizing the nanoparticles *in situ* within the matrix.

In each case, the goal is to disperse the composites uniformly in the matrix. Hence, the chemistry of the particle–composite interface is very important. A useful way of *in situ* monitoring of dispersion is dielectric analysis (dielectrometry) [231].

Once good dispersion has been achieved, a very effective way of introducing the nanocomposite into a production line is by preparing a masterbatch, in which the nano-additive may be an order of magnitude more concentrated than in the final composite. This masterbatch can simply be mixed with the pure polymer at the point of use, essentially obviating the need for any modification to the production line. Typically polymers are handled as pellets and the masterbatch can be prepared in the same form.

Nanocomposites can, in principle, be substituted for pure polymer in most applications. Nevertheless, there is persistent reluctance to use them in structurally vital components since extensive data on their long term performance are still lacking. A general difficulty is that the improved strength, stiffness etc. of the composites inevitably result in their being subjected to increased designed stress – otherwise there would be no advantage in using them – and hence diminished tolerance to damage.

Polymers with added electroactive materials. This is a typical functionality-enhancing technology. Antimony tin oxide (ATO) has become a popular additive for applications such as protection against electromagnetic interference (EMI) and electrostatic discharge. Antistatic coatings incorporate additives with both hydrophobic and hydrophilic radicals, ideally concentrated in the IF zone between the polymer and the air: the hydrophobic radicals are oriented towards the polymer and the hydrophilic radicals towards the air, whence they attract moisture, whose conductivity allows static charge accumulation to be dissipated.

Polymers made conductive by the addition of electroactive materials are of growing importance in the automotive industry. If used for body panels, for example, electrostatic spray painting is possible, thereby reducing the need for primers, and which in turn improves painting reliability, reduces costs and lessens the impact on the environment by using less or no volatile organic solvents. Conductive plastics have also been used to make fuel pipes for automobiles, eliminating the buildup of static electricity and the danger of inadvertent ignition from a spark.

Much effort is currently being devoted to creating photovoltaic solar cells from semiconductor nanoparticles embedded in a matrix (the dielectric constant of which must be less than that of the particles if the electrons are to be properly routed)

[136]. The main advantage is that if the nanoparticles are small enough to show size-dependent optical absorption (see Section 2.5), particles of different sizes can be blended to optimally cover the solar spectrum. In addition, the fabrication of a nanoparticle–polymer composite should be significantly cheaper than the PVD of a thin semiconductor film on a substrate, especially if the film should be monocrystalline.

6.6.2 METAL–MATRIX COMPOSITES (MMCS)

Conventional MMCs use two types of reinforcement: microparticles – usually silicon carbide – dispersed in an aluminium or magnesium matrix to about 10–15 vol%; or continuous fibers – the most studied system is titanium with SiC reinforcement. The existing materials show enhancements of a few % in stiffness, about 20% in strength and in superior wear resistance and fatigue initiation behavior.

Pure magnesium reinforced with 30 nm alumina particles at up to 1 vol% gives Young's modulus improvements of about 20%, yield strength increases of 45% and some ductility improvement as well. The wear and creep performance is also improved significantly in both aluminium and magnesium alloys. Silicon carbide nanoparticles used to reinforce magnesium result in a 100% increase in microhardness for 5 vol% SiC nanoparticles. SiC nanowire reinforcement of ceramics improves toughness and strength by up to 100% for a few vol%.

Although extensive empirical work has been conducted, the parameter space is so vast, especially if one considers that distributions of sizes and shapes, such as mixing particles and rods of different lengths, may offer better toughness and ductility, that our current knowledge must be considered as being very limited. One general principle that has emerged is that a much smaller volume fraction of nanoparticles is required compared with microparticles of the same chemical substance to achieve equivalent improvements in strength and stiffness. Thus, the degradation of toughness and ductility will not be as extreme and these nanomaterials may be more damage-tolerant.

It is important that the particles are well-dispersed and well-bonded to the matrix in order to create good reinforcement. Ultrasonic waves are used for dispersion of nanoparticles in the melt. There are interactive influences of the intrinsic properties of the reinforcement and matrix and the size, shape, orientation, volume fraction and distribution of the reinforcement. Usually the reinforcement causes diminution of the ductility and toughness compared with the matrix alone. In the case of high-strength magnesium and aluminium alloys ductility is already limited; consequently attempts have been made to use ductile pure metals as the matrix.

Principle process routes are as follows:

Stir casting: the nano-objects are distributed and suspended in the molten metal, e.g. via high energy mixing. Melt sizes as large as 7 tons are practicable. The slurry is then cast as billets or rolling bloom. Product with volume fraction of reinforcement ranging from 10% to 40% is available. The microstructure of the stir cast has to be finely controlled to develop uniform distributions of the particles. Spray casting followed by hot extrusion is also used.

Liquid metal infiltration: this method requires a preform consisting of a mat or other assembly of the ceramic fibers and particles to be infiltrated. A volume fraction of between 40% and 70% is required to provide sufficient mechanical stability to withstand the infiltration forces without crushing. Liquid metal infiltration is extensively used by the automotive industry and is now a preferred process for the thermal management industry as it has the ability to produce near net shape components. Because relatively high volume fraction MMCs are produced by this process, applications tend to be the ones requiring wear resistance and high stiffness, rather than the ones requiring good toughness and ductility. The majority of applications use aluminium as the matrix material.

Powder metallurgical routes are used to produce continuous and discontinuous MMC in aluminium and titanium matrices. Sintered preforms can be hot-extruded to produce rods suitable for testing.

Mechanical milling: this method has been used for creating CNT-reinforced aluminium [283]. Multiwalled CNTs were ball-milled with aluminium powder in an inert atmosphere. After milling, the material was uniaxially cold compressed and then sintered. The smaller the metal particles, the more evenly distributed the nanotubes.

Current applications in the ground transportation industry account for about 60% of the MMC market by mass. Aerospace applications are 5% by mass, and general industrial markets are about the same [including cemented carbide and ceramic–metal composite (cermet) materials for general tooling].

6.6.3 SELF-REPAIRING COMPOSITES

The basic principle is the incorporation of hollow fibers and microspheres into conventional polymer matrix composites. These are presently typically microsized (20–50 μm in diameter) rather than nanosized. Fibers and microspheres contain fluorescent dyes and polymer adhesives. If the composite is damaged by fiber or resin cracking, the cracks intersect the hollow fibers and microspheres, which liberates the dyes and adhesives, providing more visible indications of damage than the cracks themselves, and self-repairs the matrix cracks. Catalysts for the cure of the adhesive can be placed in the resin separated from the fibers and capsule contents until the moment of release by damage. The concept has also been applied to self-repair concrete structures.

Manufacture of such materials is relatively easily accomplished on a laboratory scale. Hollow glass fibers of 20–50 μm diameter can be easily made, and hollow carbon fibers may also be made. Microcapsules containing adhesive can be created using *in situ* polymerization, followed by incorporation into a conventional structural epoxy matrix material (Section 6.6.1). The required infiltration of the repair adhesive into the hollow fibers is often a difficult manufacturing step.

The self-repair action repairs only cracks in the polymer matrix; there are no suggested routes to the repair of continuous fibers. It is important that the repair material has sufficient time to flow from the containment vessels into the damaged region and has time, very possibly long, of the order of one day, to cure. In the case of microspheres, it is important that the crack intersects and fractures them rather than propagating between them; whether this happens depends on local elastic conditions at the crack tip.

If successful repair can be demonstrated to have high structural integrity, it is envisaged that these composites will be used in aircraft structures, and in other applications where high structural integrity in safety-critical structures is required. It is difficult to envisage how these concepts can be extended into metallic structures, but they could be used in civil concrete structures. Presently, the technology is in its infancy. The approaches used so far are very primitive, and there needs to be substantial further development work before they could enter service. In particular, nondestructive techniques for evaluating the strength of the repair are required, together with engineering data on the delamination resistance, the fatigue resistance and the environmental resistance of the repair.

The concept of dye bleeding for damage detection works reasonably well, but an obvious disadvantage is the lack of capability to distinguish damage that needs attention from damage that can be left in service. Current synthetic realizations are a very long way from the type of self-repair schemes used in biological structures, which show vastly better flaw tolerance (Eq. (2.28)).

6.6.4 NANOFLUIDS FOR THERMAL TRANSPORT

Nanofluids are suspensions of nanoparticles or nanofibers in a liquid medium. The addition of nanoparticles should substantially enhance the heat capacity of a fluid, and hence nanofluids have been proposed for heat transfer fluids. The reported enhancements of critical heat flux are especially interesting in boiling heat transfer. Applications are found wherever cooling is important: in microelectronics, transportation, solid-state lighting and manufacturing.

Some reports of the degree of thermal conductivity enhancement are at variance with established theoretical predictions; however, both the experiments and the theory need to be scrutinized carefully. Current theory cannot explain the strong measured temperature dependence of the thermal conductivity of the nanofluids [302].

The nanoparticles may be fabricated in dry form and then dispersed in the fluid, or they may be produced *in situ* in the liquid. The latter route was established in the Soviet Union some decades ago.

About 1% by volume of copper nanoparticles or CNTs dispersed in ethylene glycol or oil increases the thermal conductivity by 40% (copper) and 150% (CNTs). To obtain similar enhancements using ordinary particles would require greater than

10% by volume. The nanofluids are also much more stable than suspensions of conventional particles.

6.6.5 NANOLUBRICANTS

Nanolubrication can mean at least three things: the lubrication of nanodevices; the use of nano-objects to enhance lubrication and scrutiny at the nanoscale in order to better understand lubrication phenomena.

The lubrication of nanodevices has been dominated by the need to lubricate computer hard disk drives. This kind of lubrication is achieved by tenacious, long-lasting surface films [138]. Current practice is to use a thin, 1–2 nm, film of a perfluoropolyether.

The addition of nanoparticles to bulk lubricants has become popular in recent years [198]. Different kinds of nano-objects have been explored, including micelles of amphiphilic molecules (e.g. fatty acids, often conjugated with a metal ion), lamellar nanoplatelets (e.g. molybdenum sulfide, graphene) and deformable metal nanoparticles (e.g. copper). The detailed mechanism is different in each case. The amphiphiles build up a surface film; zinc dialkyldithiophosphate (ZDDP) has been much investigated [198]. The nanoplatelets readily delaminate; the supposed advantage of their small size is efficient feeding into the sliding interface. The copper nanoparticles are supposed to become momentarily trapped between impinging asperity whereupon they instantaneously deform.

Nanometrology tools are increasingly exploited in order to understand friction and lubrication in the nanoscale [292]. This is further discussed in Section 7.9.

6.6.6 ALTERNATING POLYELECTROLYTE DEPOSITION (APED)

If the substratum is electrified (via the Gouy–Chapman mechanism) and the dissolved molecule is a polyion with an electrostatic charge of opposite sign, then it will adsorb on the surface and invert the charge; the strong correlations within the polymeric ion render the Gouy–Chapman mechanism invalid [110]. The polyion-coated substratum can then be exposed to a different polyion of opposite sign, which will in turn be adsorbed and again invert the charge; the process can be repeated *ad libitum* to assemble thick films [254].

The method of APED, invented by Iler [145], appears to have immense potential as a simple, robust method of surface modification. It requires the substrate to be electrostatically charged when immersed in water. It is then dipped into an aqueous solution of a polyelectrolyte of opposite charge, with which it rapidly becomes coated. Any excess is then washed off, and the coated substrate is dipped into a polyelectrolyte of the opposite charge, with which it now becomes coated, and whose excess is again washed off, and so on (Fig. 6.14).

There are few restrictions on the choices of polyelectrolytes. Much early work was done with polyallylamine as the polycation, and polystyrene sulfonate as the polyanion. The essential feature of the technique is that at each dipping stage

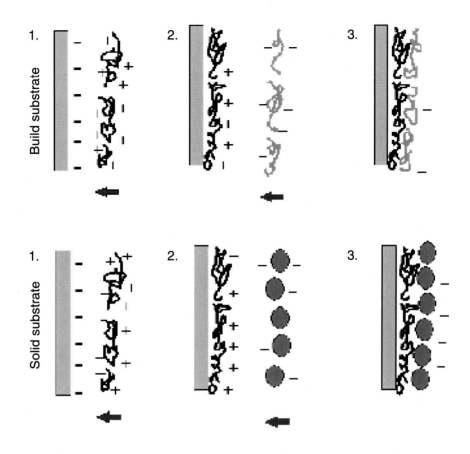

FIGURE 6.14

Upper panel: deposition of a polycation onto a negatively charged substrate followed by a polyanion. Lower panel: deposition of a polycation followed by a negatively charged nanoparticle onto a negatively charged substrate [188].

the substrate charge is not only neutralized but reversed ("overcharging"), hence allowing the deposition to be repeated indefinitely. This phenomenon contradicts the predictions of the mean-field theories, Gouy–Chapman and Debye–Hückel, of the distribution of ions in the vicinity of charged surfaces ("electrified interfaces"). The discrepancy arises because the charges of the polyions are correlated. Imagine a polyion approaching a surface already covered with its congeners. The new arrival will repel the already-adsorbed ones, creating a correlation hole (i.e. a negative image) permitting attraction (Fig. 6.15) [110]. Since the gaps between already-adsorbed polyions may be smaller than the size of one polyion, adsorption results in charged tails, ensuring overcharging (Fig. 6.16).

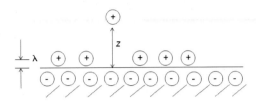

FIGURE 6.15

A polyion approaching a surface already covered with its congeners (see text).

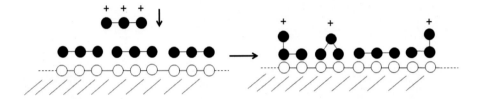

FIGURE 6.16

Overcharging resulting from adsorbed polyion loops and tails (see text).

(Monovalent) counterions screen the polyions in the usual Debye–Hückel fashion, diminishing the charging energy of the polyion more than its correlation energy, enhancing the charge inversion. (If the monovalent counterion concentration is very high the correlation disappears and APED is no longer possible.) Multivalent counterions are more difficult to treat theoretically and APED in their presence would appear to be a fruitful area of investigation. The hydrogen ion may play a special role, for example, it has been found that the porosity of built layers can be reversibly controlled by varying ion concentrations (e.g. pH).

Instead of polymeric polyions, nanoparticles composed of materials with ioniz-able surface groups can be used. In this case, although the electrostatic charge of the surface of the coating is always reversed, typically not all the inner charges are compensated because of steric hindrances, hence electrostatic charges build up, and the long-range electrostatic force ultimately prevents further particles from being deposited [201]. This is an example of a two-dimensional supersphere (Section 8.2.9).

If polymeric polyions are used as the polyelectrolyte of one sign, and ionizable particles as the polyelectrolyte of the opposite sign, the particles act as stress concentrators, thus greatly increasing the toughness of the built material. Large aspect ratio nanoparticles are very useful for diminishing the deleterious effects of defects (pinholes) in multilayer films. In this way, sophisticated coatings can be built up. It has been established that the shells of many marine organisms, such as the abalone, are assembled using this principle, producing materials that are both robust (cf. Eq. (2.28)) and beautiful: anisotropic nanoparticles are dispersed in a biopolymer matrix, which only occupies a few percent of the total mass. Natural biopolymers, which are nearly all heteropolymers, primarily based on amino acids as monomers, but also possibly incorporating polysaccharides and nucleic acids, can incorporate enormous functional variety, in ways that we can only dream about at present in synthetic systems. Nevertheless, developments are moving in this direction, as exemplified by the APED variant known as the surface sol–gel process [143].

SUMMARY

The terminological framework for nanomaterials is given. The two main divisions are nano-objects and nanostructured materials. The various kinds of nano-objects (particles, fibers and plates) are described, along with their methods of manufacture. The most important nanostructured materials are currently nanocomposites. Fabrication and applications are discussed. It must be emphasized that there is a vast body of detailed knowledge concerning this area, part of which is published but unpublished work undertaken in commercial confidence probably preponderates. Wetting of the embedded nano-objects by their matrix is of crucial importance, and here present synthetic capabilities reveal themselves as far inferior to what nature can do. Composites with enhanced mechanical properties require careful consideration of how load is transferred between matrix and particle or fiber, and how many cycles of repeated loading and unloading fatigue the material.

FURTHER READING

1. J.D. Swalen et al., Molecular monolayers and films, Langmuir 3 (1987) 932–950.
2. J.J. Ramsden, Langmuir–Blodgett films, in: P.A. Gale, J.W. Steed (Eds.), Supramolecular Chemistry: From Molecules to Nanomaterials, Wiley, Chichester, 2012, pp. 529–542.

Nanodevices

CHAPTER CONTENTS

Nanotechnology: An Introduction. DOI: 10.1016/B978-0-323-39311-9.00013-3
Copyright © 2016 Elsevier Inc. All rights reserved.

INTRODUCTION

A device, fundamentally, is a synonym for an information processor or transducer. Other synonyms are machine or automaton. The most advanced devices are able to adapt to their environment. Adaptation implies sensing (gathering information), processing (information) and actuating (transducing information into physical action) functions. In the analysis of living organisms, or in robotics, or more generally in cybernetics, these different functions are often considered as separate units connected via information channels. In the ultraminiature realm of nanotechnology, separation might not be realizable, nor indeed necessary, except possibly as an aid to conceptual thinking. Analogous situations are encountered in integrated circuits. A traditional a.c. to d.c. converter is a little circuit with a number of clearly separable components. In an integrated device, a piece of pyroelectric material suffices to fulfill the same function. Sensorial nanomaterials [176] already exist (sometimes they are called "smart" materials), combining the sensing, processing and actuating functions without clearly separating them physically. An excellent example of such an adaptive material, actually belonging to the realm of chemistry rather than of nanotechnology, is viscostatic motor oil, containing some randomly coiled long chain hydrocarbons. If the temperature rises the chains unfurl, thereby increasing the viscosity so as to compensate for the generic viscosity-decreasing effects of temperature. Adaptation, of course, confers on the device a certain independence of being; hence, it deserves to be called an entity rather then a mere object.

Apart from this ultimate integration of the macroscopically inseparable functions involved in adaptation, it is especially appropriate to miniaturize the hardware associated with information processing, because there is no definite lower bound to the physical embodiment of one bit of information, considered as its ultimate, irreducible quantum. A single electron or photon can perfectly well carry one bit, even though it lacks extension. This is the basis of the special strength of the connexion between nanotechnology and IT.

No clue as to the extent of a device is given by the fundamental definition. A single neuron or logic gate is as much a device as a brain or a computer. In this chapter, we shall, however, interpret "device" as essentially a rather basic part of an information processor, up to the complexity of a logic gate. A nanodevice is a device with at least one overall dimension in the nanoscale, or comprising one or more nanoscale components essential to its operation. A nanosystem (cf. Chapter 10) is a system of nanodevices, or any system, the nanoscale features of which are essential to its function. Fig. 7.1 summarizes the main functional categories of devices.

In this chapter, logic gates and data storage devices are also roughly classified according to the nature of their internal information carriers (e.g. electrons or photons). Their applications are mostly either "pure" information processing (e.g. a logic gate) or sensory (e.g. detection of a magnetic field, or a chemical). "Smart" materials are covered in Section 4.2 (e.g. for drug delivery) or in Section 6.6 (e.g. self-repairing structures).

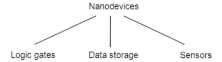

FIGURE 7.1

The main functional categories of devices. Energy transducers might also have been included as a separate category.

A further possible classification is according to the number of terminals: typically either two (e.g. a diode) or three (e.g. a transistor), the latter corresponding to a machine with input in cybernetic terminology. Another one introduces the rôle of internal states and device history: level 1 devices have no internal states and the output only depends on the input (e.g. a resistor); the output of a level 2 device (e.g. a condenser) depends on both input and its internal state; the output of a level 3 device (e.g. a memristor) also depends on the history of its inputs.

One may note that most current activity involving nanodevices is taking place in research laboratories, with the potential for showing dramatic industrial growth in the future.

The first section of this chapter briefly surveys some general consequences of miniaturization. Due to the particular importance of digital information processing as an application of nanotechnology, it is given special consideration. The main alternative is quantum computation, which is summarized in the following section. Specific devices able to execute information-processing operations, based on electron charge or spin, or on photons, are then considered. Finally, nanomechanical (relays and sensors) and fluidic (mixers) devices are described. The concluding sections of the chapter deal with sensors and energy-transducing devices.

7.1 ISSUES OF MINIATURIZATION

This section may be read in conjunction with Section 1.8.2.

Surface relative to the bulk interior. An object is delineated by its boundary. Making an object small has an effect on purely physical processes in which it is involved. For example, suppose a spherical object of radius r is heated by internal processes, and the amount of heat is proportional to the volume $V = 4\pi r^3/3$. The loss of heat to the environment will be proportional to the surface area, $A = 4\pi r^2$. Now let the object be divided into n particles. The total surface area is now $n^{1/3}4\pi r^2$, hence more heat is lost. This is the basic reason why small mammals have a higher metabolic rate than larger ones – they need to produce more heat to compensate for its relatively greater loss through the skin in order to keep their bodies at the same steady temperature. It explains why few small mammals are found in the cold regions of the Earth.

Catalysis and heterogeneous chemistry. Reactions take place at surfaces (i.e. the interface between the catalyst and the reaction medium, the latter considered to be three dimensional). Indeed in some cases the main catalytic effect is due to the reduction in dimensionality imposed on the system [251]. In general, the greater the degree of nanostructuring the greater the preponderance of surface per unit mass.

Ballistic transport (cf. Section 7.4.1). Usually, carriers cannot move through a medium without encountering some resistance, which is caused by the carrier particles colliding with (being scattered by) obstacles (which might be their congeners). The characteristic length associated with this scattering is the mean free path ℓ. If some characteristic size l of the device is less than ℓ, transport will be ballistic and the carrier can move from one end of the device to the other without encountering any resistance. Similar reasoning can be applied to heat conduction, with resistance normally due to phonon scattering. Since mean free paths in condensed matter are usually of the order of nanometers long, l is likely to be a useful candidate for defining the nanoscale (cf. Chapter 2).

How performance scales with size (cf. Section 10.8). Analysis of device performance begins by noting how key parameters scale with device length: area (hence power and thermal losses) as length squared, volume and mass as length cubed, natural frequency as inverse length and so forth. Relationships such as these are used to derive the way a device's performance scales as it is made smaller [123]. This consideration is apart from qualitatively new phenomena that may intervene at a certain degree of smallness (cf. Chapter 2, Fig. 3.1).

Noise in detectors (sensors and dosimeters). Natural phenomena involving discrete noninteracting entities (e.g. the spacial distribution of photons) can be approximated by the Poisson distribution (Eq. (10.15)). A fundamental property of this distribution is that its variance equals its mean. The uncertainty (e.g. of the magnitude of a certain exposure of an object to light) expressed as a standard deviation therefore equals the square root (of exposure).

When objects become very small, the number of information carriers necessarily also becomes small. Small signals are more vulnerable to noise (i.e. the noise is an increasing proportion of the signal). Repetition of a message (e.g. sending many electrons) is the simplest way of overcoming noise. A nanoscale, device using only one entity (e.g. an electron) to convey one bit of information would, in most circumstances, be associated with an unacceptably high equivocation in the transmission of information.

An accelerometer (which transduces force into electricity) depends on the inertia of a lump of matter for its function, and if the lump becomes too small, the output becomes unreliable. Similarly with photodetectors (that transduce photons into electrons): due to the statistical and quantum nature of light, the smallest difference between two levels of irradiance that can be detected increases with diminishing size. On the other hand, there is no intrinsic lower limit to the physical embodiment of one bit of information. One bit could be embodied by the presence of a neutron, for example – indeed distinguishable isotopes have been considered as the basis for ultrahigh density information storage [19]. Information processing and storage is therefore the ideal field of application for nanotechnology. The lower limit of miniaturization is only dependent on practical considerations of "writing" and "reading" the information.

Utility. Considering the motor-car as a transducer of human desire into translational motion, it is obvious that the nanoautomobile would be useless for transporting anything other than nano-objects. The main contribution of nanotechnology to the automotive industry is in providing miniature sensors for process monitoring in various parts of the engine and air quality monitoring in the saloon; nanolubricants for the engine to reduce wear and enhance efficiency; additives in paint giving good abrasion resistance, possibly self-cleaning functionality and perhaps novel esthetic effects; new ultrastrong and ultralightweight composites incorporating CNTs for structural parts; and sensors embedded in the chassis and bodywork to monitor structural health.

7.2 DIGITAL INFORMATION PROCESSING

As already mentioned, a device (or component) is fundamentally an information processor, in other words a transducer that encodes (or decodes) information. As noted by Szilard, the elementary "quantum" of information is the binary digit or bit. Binary or Boolean logic, based on zero or one, true or false, presence or absence and so forth, has very modest physical requirements. There is essentially no intrinsic lower limit to the size of the physical embodiment of "1". Of all technologies, IT is the one most suited to miniaturizing down to the nanoscale. The main residual problem is heat dissipation (see Section 7.5).

The fundamental component of a digital information processor is the switch, or relay (Fig. 7.2). Several relays can be connected together to create logic gates, for example, a not–and (NAND) gate, a fundamental component of a binary logic processor. Its characteristics can be summarized in the following truth table:

input 1	input 2	output
0	0	1
1	0	1
0	1	1
1	1	0

Fig. 7.3 shows a transistor inverter, the simplest logic gate (NOT) and Fig. 7.4 a NOR gate as examples of physical realization.

Someone once remarked, "Give me a relay, and I can build you a computer." The relay has the input–output relationship shown in Fig. 7.5. The earliest digital

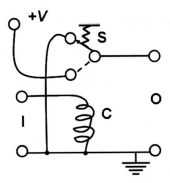

FIGURE 7.2

A relay or switch. When the coil C is energized by applying a voltage across the input terminals I, it pulls the movable contact arm above it to link voltage $+V$ to the output terminals O. If the restoring spring S is present, setting the input I zero will also cause the output to return to zero. Alternatively, a second appropriately placed coil could be used to move the contact arm in the opposite direction. The device is then bistable and would be suitable for use as a memory element.

computers used electromechanical relays. They are large, slow, expensive, energy-hungry (and hence expensive to run) and unreliable. Frequent operational errors during the execution of programs run with such devices provided Hamming with the inspirational motivation for developing error-correcting codes. Thermionic valves (vacuum tubes) are faster and more reliable, but even more expensive and energy-hungry. The first significant step towards miniaturization was taken with the replacement of relays and valves by solid-state transistors (Fig. 7.6). Provided the fabrication process does not pose new difficulties (a clock is usually cheaper to make than a watch), miniaturization uses less material in fabrication and less energy in operation (cf. Section 7.1). At a stroke, the devices became smaller, faster (the electrons carrying the information had less distance to travel), cheaper (not only because the volume of material required was lower, but also because efficient massively parallel fabrication procedures were devised), used less energy and like all solid-state devices were more reliable (the thermionic valve was more reliable than the relay because it had no mechanical moving parts, but the vacuum could leak and incandescent electron-emitting filaments could break). A major step in fabrication technology was the introduction of integration. Miniaturization and, concomitantly,

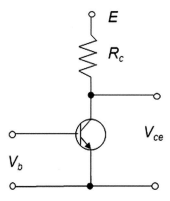

FIGURE 7.3

An inverter. If the voltage V_b on the base is below a certain threshold, corresponding to logic level 0 (typically 0.7 V), the transistor remains in the off state, the current is small and the voltage V_{ce} on the collector is approximately equal to the supply voltage E, typically 5 V, corresponding to logic level 1. If V_b exceeds the threshold then the transistor enters the active region of its characteristic but as V_b increases further (corresponding to logic level 1) V_{ce} saturates at typically around 0.2 V, corresponding to logic level 0.

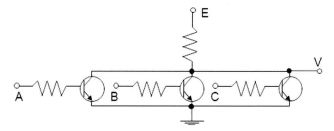

FIGURE 7.4

A NOR logic circuit built according to RTL. Note how it is constructed from the basic inverter (Fig. 7.3).

parallel fabrication now permit millions of integrated transistors to be fabricated on a single "chip" of silicon, with additional gains in operation speed because the electrons have less far to travel, both within and between components.

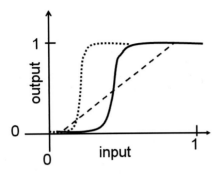

FIGURE 7.5

Input–output relationships approaching the ideal of a step (Heaviside) function. The input might be a voltage (e.g. applied to the coil of a relay, Fig. 7.2, or the gate of a transistor, Fig. 7.6) and the output might be the current flowing through the rest of the circuit. Note that in this example the response characterized by a thick solid line will give an output of one for any input (voltage) exceeding about 0.6. For the response characterized by the dashed line, the input would have to exceed about 0.8, i.e. it is less tolerant to deviations from the ideal (the input of one yielding an output of one). The dotted line marks a possible hysteretic response when decreasing the input from 1 to 0 and beyond to negative values, the existence of which opens up the possibility of using the device as a memory element.

A related device is an information store, or memory. A relay or transistor having the property of bistability could function as an information store (memory), with the disadvantage that it would need to be constantly supplied with electrical power. A more elaborate relay, with two separate coils for switching the current "on" and "off", would be better in this regard, since once its position had been flipped, power could be cut off (see Fig. 7.5). Read-only memories do not even require the flipping to be reversible: an early type of read-only memory was paper tape in which holes were punched. Reading was carried out by passing the tape between pairs of electrically conducting rollers. In the absence of a hole, there would be no electrical contact between the rollers. A later development was the use of ferromagnets, which could be

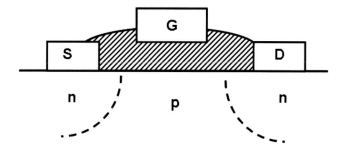

FIGURE 7.6

A FET. Regions marked "n" and "p" are n-type and p-type semiconductors (e.g. appropriately doped silicon). The hatched region is an insulator (e.g. silica). Conductors S, G and D are, respectively, the source, gate and drain. Application of a voltage to G (which plays the rôle of the coil in the relay) increases the concentration of conduction electrons in the p region and allows current to flow from S to D.

poled "up" or "down". Since ferromagnetism cannot exist below a certain volume (cf. Section 2.6), this technology is not suitable for ultimate nanoscale miniaturization, but this limit is still far from being reached – the current limitation is the sensitivity of the magnetic field detector (reading head). Memories based on electrical resistance can be fabricated from materials (e.g. NiO) that can be switched from a conducting to an insulating state by applying a voltage pulse. Other materials can have their phase changed from amorphous to crystalline by exposure to light or by passing an electric current, with corresponding changes in reflectance and resistance, but these materials are not especially "nano".

Apart from the logic gates acting as the components of information processors, the other main types of device to be considered are sensors and actuators. A sensor has a clear transduction function. Examples are a magnetic sensor that registers whether the spin in a memory cell is "up" or "down"; a light sensor such as a photodiode that converts light into electricity; a chemical sensor that converts the presence of a certain chemical compound into electricity or light. The main issue in miniaturization is whether the signal exceeds the noise level. An example of an actuator is the coil in the relay (Fig. 7.2).

7.3 QUANTUM COMPUTING

Extrapolation of Moore's law to about the year 2020 indicates that component size will be sufficiently small for the behavior of electrons within them to be perturbed by quantum effects. This implies a profound perturbation of the proper functioning of the technology, and no solution to this problem within the current framework is in view. Quantum computing can be thought of as "making a virtue out of necessity",

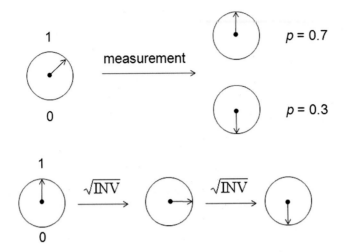

FIGURE 7.7

Qubits. Top: qubits can be represented by spheres (here simplified as disks), a point on the circumference (indicated by an arrow from the center) indicating the net result of the superpositions of all its states. A measurement forces the state into either 1 (with the probability of 0.7 in this example) or 0 (with the complementary probability of 0.3). Bottom: the quantum logic gate. To create an inverter (the NOT operation, cf. Fig. 7.3), the arrow must be rotated by 180 degrees. On the other hand, what might be called $\sqrt{\text{NOT}}$ corresponds to rotation by 90 degrees. This operation, which might be called the square root of inversion, has no classical equivalent. Two successive applications of $\sqrt{\text{NOT}}$ correspond to classical inversion, NOT.

creating computational devices based on the principles of quantum logic. The essence of quantum computing is to use not binary logic bits as the representation of the elementary unit of information, but qubits, which can exist in an infinite number of superpositions of the binary units zero and one (Fig. 7.7).

The key features of quantum objects of interest for computational purposes are superposition – an object can be in several different states simultaneously [60] – and entanglement [6]. Operations can be carried out internally, maintaining superposition, which is only destroyed at the very end of the computation when a single output is required. The system must, however, be kept isolated from its environment during operations. Entanglement with the environment implies decoherence and loss of useful information within the computer. It is easiest to avoid in an isolated small system, hence the interest in realizing quantum computers using nanotechnology.

The architecture of computers needs to be wholly reconceived in order to exploit the peculiarities of quantum mechanics, which means in particular that a particle can exist in two states simultaneously, whereas a cluster of electrons (physically instantiating a bit) in a conventional computer represents either zero or one. The value of a qubit, on the other hand, which might be physically instantiated as a single electron localized on a quantum dot (cf. 7.4.5), depends on its position relative to other electrons. For example, two electrons can exist in four different states – 00, 01, 10, and 11 – depending on their relative positions. If the electrons interact (are entangled) with each other, then any operation carried out on one electron will simultaneously be carried out on the other – implying that one operation is carried out on four different states at the same time. Hence, a computer with just 32 bits could carry out more than a thousand million operations simultaneously.

The physical embodiment of a bit of information – called a qubit in quantum computation – can be any absolutely small object capable of possessing the two logic states 0 and 1 in superposition – e.g. an electron, a photon or an atom. Electron spin is an attractive attribute for quantum computation. Qubits have also been installed in the energy states of an ion, or in the nuclear spins of atoms. A single photon polarized horizontally (H) could encode the state $|0\rangle$ (using the Dirac notation) and polarized vertically (V) could encode the state $|1\rangle$ (Fig. 7.8, upper left). The photon can exist in an arbitrary superposition of these two states, represented as $\alpha |H\rangle + \beta |V\rangle$, with $|\alpha|^2 + |\beta|^2 = 1$. Any polarization can be represented on a Poincaré sphere (Fig. 7.8, lower left). The states can be manipulated using birefringent waveplates (Fig. 7.8, upper right), and polarizing beamsplitters are available for converting polarization to spacial location (Fig. 7.8, lower right). With such common optical components, logic gates can be constructed.

One of the biggest problems was with current supercomputers is energy dissipation. They require tens of kilowatts of energy to run and generate vast amounts of heat. Landauer showed in 1961 that almost all operations required in computation can be performed reversibly, thus dissipating no heat. Reversible computation is possible on a quantum computer.

The capacity of quantum channels has been a matter of conjecture. Hastings has shown that the "additivity conjecture" according to which entangled input states cannot increase capacity is false [118]; hence, the question of the information capacity of quantum channel is still open.

7.4 ELECTRONIC DEVICES

Electronic devices (such as the archetypical FET, Fig. 7.6) depend on the movement of elementary packets of charge (i.e. electrons), carrying their electrostatic charge, that is, current flows. Max Planck already showed in 1914 that an empirical formula for the resistivity ρ of a thin metallic film,

$$\rho = \rho_\infty(1 + A/d) \tag{7.1}$$

accounted satisfactorily for experimental results, where ρ_∞ is the resistivity of the bulk metal, d the film thickness and A an unspecified constant; Weale showed that the constant A was proportional to the electron mean free path [297].

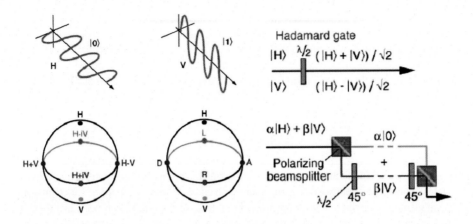

FIGURE 7.8

(Left) Optically encoded qubits. Various combinations of H(orizontal) and V(ertical) polarizations – D(iagonal), A(ntidiagonal), L(eft-circular diagonal), R(ight-circular diagonal) – are represented on the Poincaré sphere. (Right) A birefringent waveplate is used to construct a logic gate, and birefringent waveplates and polarizing beamsplitters can spacially separate photons according to the H ($\equiv$ 0) and V ($\equiv$ 1) polarizations.

-Reproduced with permission from A. Politi, J.L. O'Brien, Quantum computation with photons, Nanotechnol. Percept. 4 (2008) 289–294.

7.4.1 BALLISTIC TRANSPORT

Even a simple electrical wire ranks as a device; a transistor (from several of which a logic gate can be constructed, and hence a digital computer) can be thought of as based on a wire whose resistance depends on an external input. In an ideal binary device, the resistance is either infinite (i.e. the conductance is zero) or has some value R; for a piece of material of length l and cross-sectional area A it is given by

$$R = l\rho/A, \tag{7.2}$$

where ρ is the resistivity, which depends on the material properties,

$$\rho = 12\pi^3 \hbar/(\ell e^2 S_F), \tag{7.3}$$

where S_F is the Fermi surface area. Note in particular the inverse dependence on mean free path, ℓ. As is well known, an electron moving in a perfect lattice experiences no resistance whatsoever; but lattice defects (e.g. impurities) and the thermal fluctuations inevitably present at any temperature above absolute zero result in an effective mean free path of several tens of nanometers. But if a wire is shorter than a critical length

$l_c = \ell$ the resistance becomes ballistic, with a value of $h/(2e^2) = 25.75$ kΩ per sub-band, independent of material parameters. Therefore, any device with a characteristic length smaller than ℓ could justifiably be called a nanodevice.

7.4.2 DEPLETION LAYERS

The operation of an electronic device typically depends on junctions, including internal junctions between different kinds of materials and junctions between the device material and its contact with the external world. Some kinds of transistors depend on the field-dependent barrier arising through the generation of a space charge (depletion layer) depending on the difference V_b between the electron Fermi levels of the two materials juxtaposed at the interface and on the sparse presence of carrier-donating impurities (e.g. electron donor centers). The depletion width may be approximated by

$$W \approx \sqrt{\frac{2\varepsilon_0 \varepsilon V_b}{eN}} \tag{7.4}$$

and N is the impurity concentration, a typical value of which is 10^{16}/cm^3. Therefore, for silicon, the relative dielectric constant ε of which is 11.9, W would be about 35 nm for $V_b = 10$ mV. Rectification cannot occur if the device is significantly smaller than W; hence, the depletion width provides a demarcating length scale below which different behavior occurs.

7.4.3 SINGLE-ELECTRON DEVICES

Even if the device becomes very small, and at any moment only has one electron passing through it, it does not qualify as a single-electron device unless the electron is isolated by substantial tunneling barriers to localize its otherwise extended wavefunction, as in Fig. 7.9.

A key concept is the energy E_a required to add an electron (e.g. to the island shown in Fig. 7.9),

$$E_a = E_c + E_k, \tag{7.5}$$

where E_c is the charging energy required to add an electron to an island of capacitance C $(= 4\pi\varepsilon_0 r$ for a sphere of radius $r)$

$$E_c = e^2/C \tag{7.6}$$

and E_k is the quantum kinetic energy (electron quantization energy); for a degenerate electron gas

$$E_k = 1/(gV), \tag{7.7}$$

where g is the density of states on the Fermi surface and V is the island volume. One would expect thermal fluctuations to swamp single-electron effects (SEEs) unless $E_a \geq 10k_BT$. If the island size is somewhat less than about 10 nm, however, E_a is sufficiently large for SEE to be observable at normal room temperature. Under this condition, however, E_k becomes comparable to or greater than E_c (providing

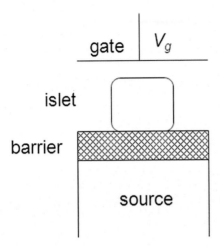

FIGURE 7.9

Single-electron box, consisting of one small
semiconductor island separated from a
larger electrode ("source") by a tunnel barrier (hatched zone). An external electric field
may be applied to the island using another
electrode (the gate) separated from the small
island by a thick insulator preventing tunneling; the field changes the electrochemical
potential of the islet, thereby determining the
tunneling conditions.

the real justification for calling such small islands quantum dots). Not only are such
small objects generally difficult to make, but they have to be made with almost exact
control of size and shape, to which the transport properties are very sensitive. It is,
therefore, preferable to make them large enough for E_a not to be dominated by E_k.

For devices in which information is represented as electrostatic charge, a scalar
quantity, the lower limit of its magnitude is the charge e of a single electron.
Neglecting noise and equivocation issues, single-electron devices can essentially be
achieved by downscaling the components of a conventional transistor. Developments
in fabrication technologies (Chapter 8) have led to nanoscale devices with the
same architecture as their microscopic counterparts. Truly nanoscale devices using
electrons involve single-charge transport in minute tunnel junctions.

The simple but effective "orthodox" theory [182] of electron tunneling assumes
no electron energy quantization inside the conductors. Strictly speaking this is valid
only if the quantum kinetic energy E_k of the added electron is much less than k_BT, but
seems to adequately describe observations provided $E_k \ll E_c$, the charging energy

$[e^2/C$, where C is the capacitance of the object (island) to which the electron is added]. The second major assumption of the orthodox theory is that the tunneling time is assumed to be negligibly small – typically it is of the order of 1 fs – in comparison with other relevant time scales (such as the interval between successive tunneling events). The third assumption is to ignore coherent quantum processes (simultaneous tunneling events), valid if the barrier resistance R is much greater than the quantum unit of resistance

$$R_Q = h/(2e^2) \tag{7.8}$$

which is about 6.5 kΩ. Eq. (7.8) ensures the localization of each electron at any particular instant – hence sufficiently nontransparent tunnel barriers effectively suppress the quantum-mechanical uncertainty of electron location. This suppression makes single-electron manipulation possible and also guarantees that single-electron devices are not quantum devices in the sense of Section 7.3. Tunneling of a single electron is a random event with a rate (probability per unit time) Γ depending solely on the diminution ΔW of the free (electrostatic) energy of the system as a result of the event and expressed as

$$\Gamma = (I/e)[1 - \exp\{-\Delta W/k_B T\}]^{-1}, \tag{7.9}$$

where I is the d.c. current (often satisfactorily given by Ohm's law) in the absence of single-electron charging effects. Useful general expressions for ΔW are [182]

$$\Delta W = e(V_i + V_f)/2, \tag{7.10}$$

where V_i and V_f are the voltage drops across the barrier before and after the tunneling event, or

$$\Delta W = e(V_i - V_t), \tag{7.11}$$

where

$$V_t = e/C_{kl} - e[1/C_{kk} + 1/C_{ll}]/2, \tag{7.12}$$

with C is the capacitance matrix of the system, k and l being the numbers of islands separated by the barrier. Eq. (7.9) implies that at low temperatures (i.e. $k_B T \ll \Delta W$) only tunneling events increasing the electrostatic energy and dissipating the difference are possible; their rate is proportional to ΔW, since an increase in the applied voltage increases the number of electron states in the source electrode, which may provide an electron capable of tunneling into an empty state of the drain electrode (Fig. 7.10, upper diagram). Note that the simplicity of the above equations does not imply that it is necessarily simple to calculate the properties of single-electron systems – several tunneling events may be possible simultaneously, but we can only get the probability of a particular outcome, implying the need for a statistical treatment to yield average values and fluctuations of variables of interest. For large systems with very many possible charge states, a Monte Carlo approach to simulate the random dynamics may be the only practical method.

In a nanosized device, electron confinement engenders significant energy quantization (cf. Section 2.5). Experimental work has revealed features not accounted for by

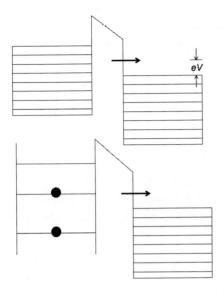

FIGURE 7.10

Top: energy diagram of a tunnel junction with continuous spectra; bottom: energy diagram showing discrete levels of the source electrode.

the orthodox theory, in particular, cotunneling (the simultaneous tunneling of more than one electron through different barriers at the same time as a single coherent quantum-mechanical process) and discrete energy levels (Fig. 7.10, lower diagram), resulting in a different energy dependence of the tunneling rate (cf. (7.9)),

$$\Gamma = \Gamma_0[1 + \exp\{-\Delta W/k_B T\}]^{-1}, \tag{7.13}$$

where Γ_0 is a constant; orthodox rate (7.9) is the sum of rates (7.13) over all the levels of a continuous spectrum of the island. Hence, if $\Delta W \gg k_B$, in the miniature device the tunneling rate is a constant.

The "single-electron box" (Fig. 7.9) is the simplest single-electron device. The Gibbs free energy of the system is

$$W = (ne - Q_e)^2/2C_\Sigma + const, \tag{7.14}$$

where n is the number of uncompensated electrons, C_Σ is the total islet capacitance and the parameter Q_e (the "external charge") is given by

$$Q_e = C_0 V_g, \tag{7.15}$$

where C_0 is the islet–gate capacitance and V_g the gate voltage.

The average charge $Q = ne$ of the single-electron box increases stepwise with the gate voltage (i.e. external charge Q_e); this phenomenon is called the Coulomb staircase, with a fixed distance $\Delta Q_e = e$ between steps (for $E_c \gg k_B T$). The reliable addition of single electrons despite the presence of thousands of background electrons is possible due to the very large strength of the unscreened Coulomb attraction.

Localized states with Coulomb interactions cannot have a finite density of states at the Fermi level, which has significant implications for electron transport within nanoscale material. By definition, at absolute zero, all electronic states of a material below the Fermi level are occupied and all states above it are empty. If an additional electron is introduced, it must settle in the lowest unoccupied state, which is above the Fermi level and has a higher energy than all the other occupied states. If, on the other hand, an electron is moved from below the Fermi level to the lowest unoccupied state above it, it leaves behind a positively charged hole, and there will be an attractive potential between the hole and the electron. This lowers the energy of the electron by the Coulomb term $-e^2/(\varepsilon r)$ where e is the electron charge, ε the dielectric permittivity and r the distance between the two sites. If the density of states at the Fermi level is finite, two states separated by but very close to the Fermi level could be chosen, such that the energy difference was less than $e^2/(\varepsilon r)$, which would mean – nonsensically – that the electron in the upper state (above the Fermi level) has a lower energy than the electron located below the Fermi level. The gap in states that must therefore result is called the Coulomb gap and materials with a Coulomb gap are called Coulomb glasses.

The single-electron box (Fig. 7.9) is unsuitable as a component of an electronic logic device. It has no internal memory, the number of electrons in it is a unique function of the applied voltage. Furthermore, the box cannot carry current; hence, a very sensitive electrometer would be needed to measure its charge state. The latter problem can be overcome by creating a single-electron transistor (Fig. 7.11) [182]. Its electrostatic energy is (cf. Eq. (7.14))

$$W = (ne - Q_e)^2/2C_\Sigma - eV[n_1 C_2 + n_2 C_1]/C_\Sigma + const, \tag{7.16}$$

where n_1 and n_2 are the numbers of electron passed through the two tunnel barriers. This is the most important single-electron device.

If the source–drain voltage is small no current flows because tunneling would increase the total energy ($\Delta W < 0$ in Eq. (7.9)) – this effect is called the Coulomb blockade. Simplistically this is because one cannot add less than one electron; hence, the flow of current requires a Coulomb energy of E_c. The blockade is overcome by a certain threshold voltage V_t. If one gradually increases the gate voltage V_g, the source–drain conductance shows sharp resonances almost periodic in V_g. The quasiperiodic oscillations of the I–V curve are known as Coulomb diamonds and are closely related to the Coulomb staircase.

7.4.4 MOLECULAR ELECTRONIC DEVICES

Another approach to ultraminiaturize electronic components is to use a single molecule as the active medium. Current realizations of such molecular electronic

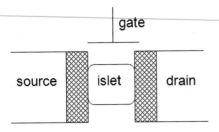

FIGURE 7.11

Single-electron transistor (cf. Fig. 7.9). The
hatched areas represent tunneling barriers.

devices comprise an organic molecule uniting electron donor (D^+, i.e. a cation) and
acceptor (A^-, i.e. an anion) moieties separated by an electron-conducting bridge
(typically π, i.e. a π-conjugated (alkene) chain) between a pair of (usually dissimilar)
metal electrodes $M^{(1)}$ and $M^{(2)}$, mimicking a semiconductor p–n junction. When
a forward bias is applied across the electrodes, chosen for having suitable work
functions, the process

$$M^{(1)}/D^+ - \pi - A^- / M^{(2)} \rightarrow M^{(1)}/D^0 - \pi - A^0 / M^{(2)} \tag{7.17}$$

occurs, followed by intramolecular tunneling to regenerate the starting state. Under
reverse bias, the energetically unfavorable formation of $D^{2+} - \pi - A^{2-}$ that would be
required blocks electron flow; hence, we have rectification (Fig. 7.12).

Organic FETs have been made with such materials [314]. These devices were
made using a benzothiadiazole (BTZ) and cyclopentadithiophene (CDT) copolymer
as the semiconductor. Modifying these molecules enables electron mobilities to be
significantly increased [291].

7.4.5 QUANTUM DOT CELLULAR AUTOMATA (QCA)

The QCA uses arrays of coupled quantum dots to implement Boolean logic functions
[269]. An individual quantum dot can be of the order of 10 nm in diameter; hence,
QCA devices are very small. Fig. 7.13 shows the basic concept. Four quantum dots
coupled by tunnel barriers are placed in a square array. This constitutes one QCA
cell. Electrons can tunnel between the dots but cannot leave the cell. Two electrons
are placed in each cell; Coulomb repulsion ensures that they occupy opposite corners.
Hence, there are two ground state configurations with the same energy; they can be
labeled 0 and 1.

Fig. 7.14 shows several simple QCA devices. On the left we have wires, an
inverter and fanout; on the right a majority gate. In each case, the configuration of
adjacent cells is determined by minimizing Coulomb repulsion. All logic functions
can be implemented using the majority gate and the inverter.

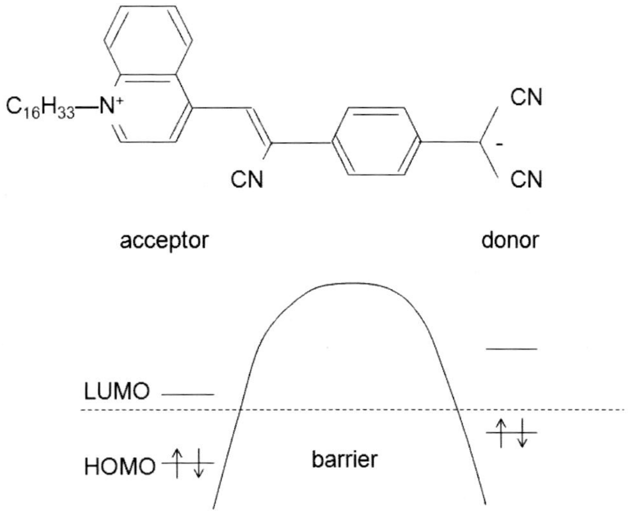

$C_{16}H_{33}$—N$^+$

CN

CN

CN

acceptor   donor

LUMO ——

HOMO ↑↓   barrier   ↑↓

FIGURE 7.12

A typical molecule (Z-β-(1-hexadecyl-4-quinolinium)-α-cyano-4-styryldicyanomethanide) used to create a rectifying junction (see text). The chemical structure is shown at the top. The energetic structure is shown at the bottom.

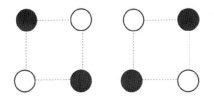

FIGURE 7.13

An illustration of the concept of the quantum cellular automaton. Four quantum dots arranged in a square and occupied by two electrons have two possible ground state polarizations (left and right squares), assigned to represent "0" and "1".

7.4.6 TOWARDS TWO-DIMENSIONAL DEVICES

"Two-dimensional" or layered materials such as molybdenum sulfide have long attracted interest (e.g. [89, 290]). This interest was considerably boosted by the

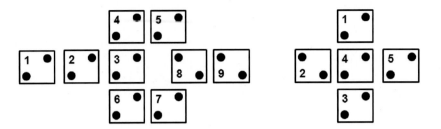

FIGURE 7.14

Some QCA devices. On the left, a wire (cells 1–3) leads input (cell 1) to fanout (cells 3, 4 and 6). When the parallel wires are combined (cells 5 and 7 converging on cell 8), the output (cell 9) is inverted. On the right, we have two inputs (say cells 1 and 2) and a programming input (say cell 3). The central cell 4 cannot minimize all cell–cell Coulomb repulsions but chooses the configuration that minimizes the overall repulsion. Depending on the programming input, the output (cell 5, mimicking the configuration of cell 4) is that of either the AND or the OR logic function.

emergence of graphene as an object of intense scientific interest. Lacking an intrinsic band gap, however, makes graphene unattractive for electronic switching devices, which has refocused interest on the large family of layered materials such as MoS_2 [36]. Among its other remarkable electrical properties, the existence of a metal–insulator transition [44] potentially opens the door for logic and memory devices. The newest such material to be investigated appears to be black phosphorus. Its instability with respect to normal atmospheric conditions meant that progress had to await the development of a way of stabilizing it – such as encapsulating it between layers of BN [45].

Some of these materials, such as Bi_2Se_3 [306], are topological insulators, a fascinating newly discovered quantum state of matter [240]. They have an insulating bulk and gapless surface states.

Despite the attractiveness of many of the intrinsic physical properties of these materials, creating a production industry comparable to what we have for silicon-based electronics is not yet in sight.

7.5 TRENDS IN THE MINIATURIZATION OF ELECTRONICS

Remarkably, Moore's well-known empirical relationship describing the exponential growth of the number of components on a wafer has continued to hold for several decades, although it was based on only a few years of experience when it was formulated. The increase in the number of components has been accompanied by the concomitant decrease in their size and in the switching time of individual transistors. The continuing success of Moore's law is due to the relentless exercise of ingenuity that has enabled new semiconductor technologies to be introduced every two years or so. Some innovations enabling further miniaturization, such as the crucial self-aligned silicon gate process, themselves go back several decades, as does the basic building block of microprocessors and memory chips, namely, the MOSFET (metal oxide–semiconductor FET). The characteristic dimensions of the MOSFET are now in the low tens of nanometers, a scale at which it becomes difficult to maintain their performance. Crucially, with further miniaturization, source–drain current does not fall to zero as the gate voltage is decreased. Nevertheless, within the existing technology framework further innovations to overcome that problem are still possible, such as ultrathin body MOSFETs, which tend to eliminate subsurface leakage paths, and the so-called multigate transistors [78], in which the gate is wrapped around the insulator (this concept comprises several realizations, including the double gate FET, in which a gate above and below the body can be independently controlled, and the finFET, which resembles the tri- or multigate FET). There is also hope that a further diminution of the working wavelength to 13.5 nm (extreme ultraviolet, EUV) will prolong the usability of photolithography. Hence, there is no strong imperative to seriously develop the devices discussed in the preceding sections (based on single-electron operation or molecular electronics) because the impetus of Moore's law can continue with, essentially, the classical technology.

Nevertheless, even the most sanguine semiconductor technologist realizes that there are some ultimate limits that will be reached, hence serious attention is being paid to devices operating on completely new principles, most notably those based on electron spin (Section 7.6).

Although computation-centric applications (especially the "blow-by-blow" methods of modeling systems by high-resolution computational fluid dynamics and finite element analysis) see the bottleneck to expansion as the central processing unit, data-centric applications (such as "mining" vast datasets and analysing global social networks) see the bottleneck as nonvolatile memory. The former is being addressed by continuing innovations in CMOS (complementary metal oxide–semiconductor) technology, including the multigate approach mentioned above and, in the longer term, by migration to spin-based technologies (Section 7.6). The latter is being addressed by a group of innovative technologies known as storage-class memory (SCM), which combines the attributes of existing solid-state random access memory (RAM), such as high performance and robustness, with those of archival media such as conventional hard-disk magnetic storage. The main approaches are resistive RAM (RRAM) and phase-change RAM (Section 7.7).

Interest in how device performance scales with size as very small dimensions are approached is often considered to have begun with paper by Dennard et al. [57]. That paper considered a MOSFET (see Fig. 7.6, where the insulator beneath the gate is typically a metal oxide) channel length of 1 μm as the ultimate lower limit (reflecting then current manufacturing practice). Since then attention has been focused on the fundamental lower limit of miniaturization of information processing, for which the starting point is the Shannon–von Neumann–Landauer (SNL) expression for the smallest energy to process one bit of information [318],

$$E_{\text{bit}} = k_B T \ln 2. \tag{7.18}$$

This equals 17 meV at 300 K. Using Heisenberg's uncertainty principle $\Delta x \Delta p \geq \hbar$ and writing the momentum change Δp as $\sqrt{2m_e E_{\text{bit}}}$ one has

$$x_{\text{min}} = \hbar / \sqrt{2m_e k_B \ln 2} \tag{7.19}$$

amounting to 1.5 nm (at 300 K). Similarly with the other uncertainty relation $\Delta E \Delta t \geq \hbar$ and equating ΔE to E_{bit} one finds that the minimal switching time is 40 fs. The power dissipation per unit area is then

$$P = \frac{E_{\text{bit}}}{x_{\text{min}}^2 t_{\text{min}}}, \tag{7.20}$$

a colossal several megawatts per square centimeter! Although it has been stated that there is no lower limit to the physical embodiment of one bit of information, relation (7.18), and the power dissipation (7.20) derived from it, provides a practical lower limit.

The essential features of classical information-processing technology are the physical carrier of information (e.g. the electron), the creation of distinguishable states within the system of carriers, and the capability for a regulated change of state. Distinguishability and regulated state changes are achievable with energy barriers (Fig. 7.15). This structure constitutes a switch: when "on", electrons can freely move from one well to the other (the transition probability equals unity); when "off" no movement can take place. If the barrier width b is large enough to prevent tunneling, the probability of the thermal transition from one barrier to the other is $\exp(-E_b/k_B T)$, whence $E_b = k_B T \ln 2$ for a probability of one half, equal to the SNL limit (Eq. (7.18)). Roughly speaking x_{min} can be equated to b.

The actual switching operation is represented in Fig. 7.16. It follows from Poisson's equation linking electrical potential ϕ and charge density ρ

$$\nabla^2 \phi = -\rho/\varepsilon_0 \tag{7.21}$$

that a change of barrier height implies a change of charge q; it also implies charging or discharging a capacitance $C = \Delta q/\Delta \phi$. This applies regardless of the detailed nature of the device (FET, QCA, SET etc.) and "changing barrier height" means changing height, width or shape; that is, barrier deformation. Inevitably energy is dissipated in the process (e.g. amounting to CV^2 if it is charged from a constant voltage source V) and this energy will be at least as much as the SNL limit (Eq. (7.18)).

When appraising the minimum heat dissipation of the order of 1 MW/cm^2, one should bear in mind that current heat removal technologies can manage at most about three quarters of a magnitude less than this. *Thus, device miniaturization based on the current paradigm of charge transport-based information processing is limited by heat.* Spin-based logic is one way to eliminate dissipation (Section 7.6.3). Another way might be to introduce new classes of electron charge-based devices and devise

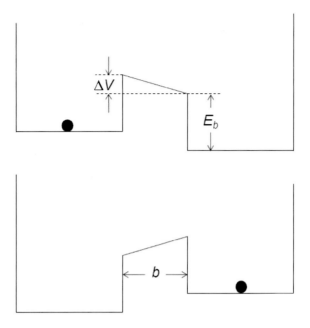

FIGURE 7.15

A very simple model device consisting of two wells separated by a barrier of width b and energy E_b [318]. The electron as the information carrier is represented by the black disk.

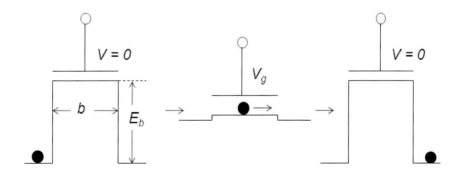

FIGURE 7.16

A model charge transport device with a voltage-controlled barrier of width b. In the absence of an applied voltage the barrier height is E_b, diminishing to practically zero upon application of a voltage V_g [43].

ingenious new ways of carrying out logical operations. The basic variables describing an electronic circuit are current I, voltage V, charge Q and magnetic flux ϕ. Pairs of these basic variables can be permuted in six different ways: $Q = \int I dt$ and $\phi \int V dt$ (definitions of C and ϕ); $R = V/I$ (Ohm's law), $C = Q/V$ ($C = dq/dv$ in differential form) and $L = \phi/I$ (properties of the three basic circuit elements, i.e. the resistor, the condenser and the inductor); symmetry demands the existence of a sixth relation, $M = \phi/V$, where M is memristance (memory resistance)   which implies the existence of a fourth basic circuit element, the memristor, which shows resistance hysteresis as the applied voltage increases and then diminishes, thus endowing it with memory. Memristive components may allow one to create components fulfilling certain logic functions combining several existing classical components. Although the memristor is still considered a curiosity among electronics engineers, the phenomenon of memristance is widely encountered in biological systems [107, 108].

7.6 SPINTRONICS (MAGNETIC DEVICES)

Electrons have spin as well as charge. This is of course the origin of ferromagnetism, and hence magnetic memories, but *their* miniaturization has been limited not by the ultimate size of a ferromagnetic domain but by the sensitivity of magnetic sensors. The influence of spin on electron conductivity was invoked by Nevill Mott in 1936 [216], but remained practically uninvestigated, much less exploited, until the discovery of giant magnetoresistance (GMR) in 1988. Spintronics, sometimes called magnetoelectronics, may be loosely defined as the technology of devices in which electron spin plays a rôle; it has three main directions now:

- The development of ultrasensitive magnetic sensors for reading magnetic memories;
- The development of spin transistors, in which barrier height is determined by controlling the nature of the spins of the electrons moving across it;
- The development of devices in which logical states are represented by spin.

7.6.1 ULTRASENSITIVE MAGNETIC SENSORS

The discovery of the GMR effect in 1988 can be considered as the beginning of the spintronics era. This phenomenon is observed in thin (a few nanometers) alternating layers (superlattices) of ferromagnetic and nonmagnetic metals (e.g. iron and chromium) (Fig. 7.17). Depending on the width of the nonmagnetic spacer layer, there can be a ferromagnetic or antiferromagnetic interaction between the magnetic layers, and the antiferromagnetic state of the magnetic layers can be transformed into the ferromagnetic state by an external magnetic field. The spin-dependent scattering of the conduction electrons in the nonmagnetic layer is minimal, causing a small resistance of the material, when the magnetic moments of the neighboring layers are aligned in parallel, whereas for the antiparallel alignment the resistance is high. The technology is already used for read–write heads in computer hard drives. It is

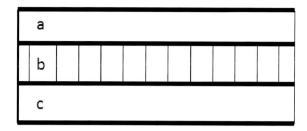

FIGURE 7.17

Diagram for describing spin-controlled electron transport in a thin film multilayer. To exploit the GMR effect, layers a and c are ferromagnetic and layer b is a nonmagnetic metal. In the magnetic tunnel junction layer b is a dielectric.

noteworthy that the discovery of GMR depended on the development of methods for making high-quality ultrathin films (Section 6.3.1). The GMR effect has clearly demonstrated that spin-polarized electrons can carry a magnetic moment through nonmagnetic materials while maintaining spin coherence: this is the meaning of the term "spin transport" nowadays.

A second type of magnetic sensor is based on the magnetic tunnel junction (MTJ). In this device, a very thin *dielectric* layer separates ferromagnetic (electrode) layers, and electrons tunnel through the nonconducting barrier under the influence of an applied voltage (Fig. 7.17). The conductivity depends on the relative orientation of the electrode magnetizations and the tunnel magnetoresistance (TMR): it is low for parallel alignment of electrode magnetization and high in the opposite case. The magnetic field sensitivity is even greater than for GMR. MTJ devices also have high impedance, enabling large signal outputs. In contrast with GMR devices, the electrodes are magnetically independent and can have different critical fields for changing the magnetic moment orientation. The first laboratory samples of MTJ structures ($NiFe–Al_2O_3–Co$) were demonstrated in 1995.

A further direction of development of spintronic devices is based on the development of multilayer structures of ferromagnetic semiconductors, which demonstrate properties not available for their metal analogues (e.g. the possibility to control a magnetic state of material by an electric field) and the giant planar Hall effect, which exceeds by several orders of magnitude the Hall effect in metal ferromagnets. These might be exploited to create memory devices, but they have the disadvantages of poor integratability and the necessity of using additional controlling transistors.

GMR and MTJ are important enablers of size reduction of ferromagnetic memory cells and hence the miniaturization of magnetic random access memory (MRAM), the limiting factor of which is not the ability to make small ferromagnetic islands (which must nevertheless be above the superparamagnetic size boundary, see Section 2.6) but the ability to sense minute magnetic fields.

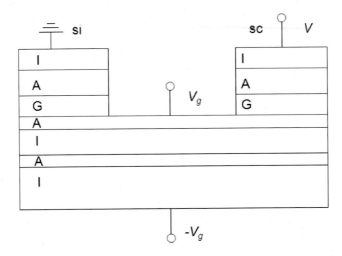

FIGURE 7.18

A spin-dependent transistor. The labels signify: I, InAs; A, AlSb; G, GaSb. The grounded terminal si is the spin injector and the terminal at the potential V is the spin collector. The gate voltage V_g manipulates the spins of the two-dimensional electron gas in the layer I below it. See [114] for more details.

A very important development for increasing storage density is the ability to control the motion of a series of movable domain walls along magnetic nanowires by spin-polarized current pulses [42]. This provides the basis of magnetic domain-wall "racetrack" memory [232].

Permalloy ($Ni_{81}Fe_{19}$) nanoplates with edge length about 1 µm and thickness about 50 nm form magnetic vortices, the core which can project upwards or downwards, potentially enabling one bit of information to be stored. The direction of projection can be switched by an external magnetic field [154].

7.6.2 SPIN-DEPENDENT TRANSISTORS

This kind of spintronics device uses control of electron spin in order to gate electrons passing through a transistor. Fig. 7.18 shows a proposed example. Note that it is made entirely of nonmagnetic materials.

7.6.3 SINGLE SPIN LOGIC (SSL)

Spintronic-based devices have the potential to be very fast and operate with very low power. Spin has polarization and is bistable in a magnetic field (parallel or antiparallel, which can represent the two logic states zero and one). Switching between them requires merely flipping spin without physically displacing the electrons. In contrast,

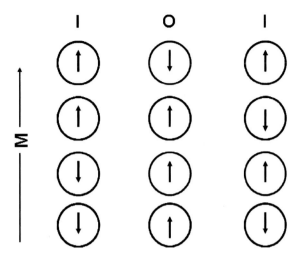

FIGURE 7.19

A spintronic NAND gate. The two inputs are the left and right columns, and the output is the central column. The physical entities are quantum dots. The upspin state parallel to the global magnetic field **M** represents 1 and the downspin state represents 0. Exchange interaction causes nearest neighbors to prefer antiparallel spins, but if there is a conflict, the global magnetic field biases the output to 1.

electronic charge-based devices have the fundamental shortcoming that charge is a scalar quantity possessing only magnitude. Hence, the only way to switch between zero and one (for example, see Fig. 7.3) is to change the magnitude of the charge, which invariably implies current flow I and its associated dissipation I^2R, where R is a resistance. Charged-based devices are therefore intrinsically dissipative.

SSL is the spin analogy of single electron (charge) logic (Section 7.4.3). Fig. 7.19 illustrates a NOT–AND gate based on SSL [16]. A "true" spintronics device encodes binary information as spin, in contrast to the so-called spin transistor (Section 7.6.2), in which spin merely mediates switching.

7.6.4 MAGNON SPINTRONICS

A spin wave is a propagating disturbance in local magnetic ordering; the magnon is the spin-wave quantum, a boson, associated with the flip of a single spin. Spin waves, which propagate in magnetic materials, have wavelengths in the nanometer range and frequencies in the gigahertz–terahertz range. Information transport and processing by spin waves is referred to as magnonics or magnon spintronics [48].

Magnonics overcomes a number of intrinsic disadvantages of electronics, which include the fact that the computational state variable is voltage, a scalar quantity;

the necessary metal interconnects (wires connecting transistors) have no function in modulating the electrical signals they carry; and transistors require a permanent power supply regardless of whether computation is actually being carried out. In contrast, magnetization is a vector state variable (hence both amplitude- and phase-dependent switching are possible). Phase-dependent information enables interconnects to be used as passive logic elements, for they will modulate phase according to their length. Magneto-electric (ME) elements can retain stored information without electrical power.

The basic idea of magnonic logic circuits is to use magnetic thin films to carry the spin waves [157]. A key component is the ME cell, a bistable multiferroic unit with both magnetic and electric polarizations. It could be realized as a layer of a conducting magnetostrictive material (e.g. Ni, CoFe) covered by a layer of piezoelectric material (e.g. $PbZrTiO_3$) in turn covered by a metallic contact (e.g. Al). Information is encoded as the phase of a spin wave and is input as voltage pulses applied to ME cells. They would disturb the magnetic polarization of the magnetostrictive material and launch a spin wave. A possible architecture is to connect the inputs to intersecting spin-wave buses (made from a magnetic material, e.g. permalloy, yttrium iron garnet); where they intersect, the spin waves interfere and the result depends on their relative phases. The result is ultimately passed to output ME cells.

A NOT gate can be achieved simply by choosing an appropriate length for the bus connecting to ME cells. An AND or NAND gate has two input cells connected to one output cell. Larger numbers of input cells can be connected to a single output cell, whose magnetization depends on the majority of the inputs; this is called a MAJ gate. It would appear that magnonic logic circuits can be constructed from fewer components compared with their electronic counterparts [157]. Further elaboration can be undertaken by devising multifrequency magnonic circuits [158]. Another possible approach to data processing is via the magnon equivalent of the Mach–Zehnder interferometer, with electrically controlled phase shifters in each arm [48]. A possible approach to exploiting spin-wave amplitude as the information carrier has been proposed and demonstrated by Jamali et al. [149].

7.7 STORAGE-CLASS MEMORY (SCM)

The essential features of this class of memory combine rapid access with nonvolatility.

7.7.1 FERROELECTRIC MEMORY DEVICES

By analogy with ferromagnetic memory, ferroelectric materials are being investigated for nonvolatile storage (nanocapacitor arrays). Using a nanoporous template, ferroelectric ceramic (e.g. lead zirconate titanate) can be deposited as nanoscale islands on a suitable metal (e.g. platinum). These require ultrasensitive electrometers to be read.

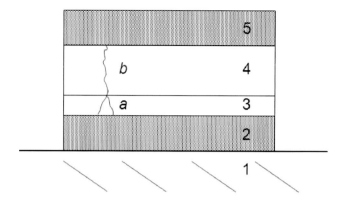

FIGURE 7.20

A RRAM cell. Layer 1 is a silicon substrate, layers 2 and 5 are platinum electrodes, layer 3 is a dielectric oxide (e.g. SiO_2, Gd_2O_3), 2–20 nm thick; layer 4 is a solid electrolyte (ionic conductor), with a resistance about one tenth of that of layer 3, typically a metal-doped nonstoichiometric oxide (with oxygen vacancies) (e.g. MoO_{3-x}:Cu, $0 \leq x \leq 1.5$). See text for explanation of operation.

7.7.2 RESISTIVE RANDOM ACCESS MEMORY (RRAM)

A more promising direction is RRAM. This works as follows (Fig. 7.20) [141]. It is necessary to prepare the device by applying a "foaming" voltage across electrodes 2 and 5 (the polarity is indifferent). The high electric field across layer 3, which has the highest resistance, results in localized electrical breakdown to create a conducting tip with an area of 10 nm² or less. The device is still in a high-resistance state because of layer 4. Now a positive voltage is applied to layer 5. This reduces some of the metal ions in layer 4 and creates a conducting metal filament continuing the path initiated by the conducting tip. The device is now in a low resistance state and remains in it until a positive voltage is applied to layer 5, which oxidizes and hence destroys the metal filament.

Cross-point cell arrays can be made according to this principle to create an addressable memory.

7.7.3 PHASE-CHANGE RAM

Certain materials are rather resistive in their amorphous phase, but relatively conductive in a crystalline phase. If the material is in the molten state, rapidly cooling it results in the amorphous phase, whereas cooling it more slowly allows it to crystallize. A cell made of such a material may be controlled and monitored simply by applying electrical power. A short, high-voltage pulse melts the material and it rapidly cools to the amorphous, high-resistance phase. A longer, lower-voltage pulse allows the cell in the amorphous phase to melt and crystallize. A much lower-voltage short pulse enables the resistance to be determined without significantly changing the temperature [35]. Examples of materials are $Ge_2Sb_2Te_5$ (GST) and silver- and indium-doped Sb_2Te (AIST). The energy barrier to recrystallization to the amorphous phase should be high enough to enable the cell to be stable for at least 10 years at a temperature somewhat above usual room temperature. Operation with pulse durations of the order of ns for recrystallization has been demonstrated, and the cells can be made with sizes of the order of 10 nm.

7.7.4 OTHER POSSIBLE SCHEMES

Although less well-advanced compared to resistive and phase-change RAM, the following approaches to nonvolatile memory may also turn out to be fruitful. Nanotechnology makes memory based on magnetoelectric coupling feasible. The memory cell would be a nanoplate of, say, iron on copper [96]. Application of a strong electric field (using a STM) enables highly local switching of the iron islands, which are only two atoms thick, between antiferromagnetic (face-centered cubic, fcc) and ferromagnetic (body-centered cubic, bcc) states. In a sense this is a (martensitic) phase-change material (the magnetic ground state changes because the fcc phase is more compact than the bcc one).

The ferroelectric photovoltaic effect can be advantageously exploited for non-volatile memory as the reading method. Certain materials, such as $BiFeO_3$, can be polarized by applying an electric field. The polarization state can be read by light [112].

7.8 PHOTONIC DEVICES

The "natural" length scale associated with light is its wavelength, λ: a visible photon of a typical green color has a wavelength of about 500 nm. Macroscopic optics deals with distances $\gg \lambda$ (the far field). There might be some justification for referring to the near field, involving distances $\ll \lambda$ as nanophotonics. The word "photonics" was itself coined in analogy to "electronics" and is particularly used in connexion with integrated optics, in which light is guided in waveguides with at least one characteristic dimension $< \lambda$. Light can be confined in a channel or plate made from a transparent material having a higher refractive index than that of its environment. Effectively, light propagates in such a structure by successive total internal reflexions at the boundaries. The channel (or fiber) can have a diameter, or the plate thickness, less than the wavelength of the light. Below a certain minimum diameter or thickness (the cut-off, typically around one third of the wavelength of the light), propagation is no longer possible, however. The science and technology of light guided in thin structures is called integrated optics (in planar and channel structures) and fiber optics. The main areas of application are in communication (transmission of information) and in sensing (Section 5.7). However, the fact that waveguiding cannot take place below a certain minimum waveguide dimension, typically of the order of 100 nm, means that there can be little justification for using nanophotonics as a synonym for integrated optics. On the other hand, the near field is particularly associated with rapidly decaying evanescent fields, and optical waveguides generate evanescent fields at their surfaces, with a penetration depth (here, into the cover medium) given by (cf. Eq. (5.27))

$$z_C = (\lambda/2\pi)(N^2 - n_C^2)^{-1/2}, \tag{7.22}$$

where N is the effective refractive index of the guided mode and n_C is the refractive index of the cover medium. Hence, the term "nanophotonics" could reasonably be used to describe near-field measurements and processes, especially those associated with evanescent fields generated by optical waveguides. In other words, the nanorégime of photonics is the régime of the near zone and concerns processes taking place within the evanescent field. The term can also be used to describe light sources based on nano-objects (quantum wells in the form of plates, wires or particles) and nanoparticle-based photoelectrochemical solar energy converters (cf. Fig. 7.25).

Semiconductor lasers, in which a voltage is applied across a semiconductor crystal (which in effect constitutes a Fabry–Perot cavity) to create a nonequilibrium population distribution of electrons and holes, whose luminescent recombination generates photons stimulating further emission, were already in existence when Dingle and Henry showed that using quantum wells as the active lasing medium would result in more efficient lasers with lower threshold currents [59], essentially because quantum confinement of the charge carriers and the optical modes enhances carrier–radiation interaction; moreover, the lasing wavelength could be tuned by changing the thickness of the layers. Quantum well lasers are made from alternating ultrathin layers of wider and narrower band gap semiconductors (for example, n-AlGaAs and GaAs, respectively). Improvements in the technology of ultrathin film fabrication, especially with the introduction of techniques such as MBE (Section 6.3.1), enabled quantum well lasers to become a reality.

Reduction of the dimensionality from two to one (quantum wires) and to zero should lead to further improvements (even higher gain and lower threshold current than quantum well lasers). The carriers are confined in a very small volume and population inversion occurs more easily, leading to lower threshold currents for lasing. Furthermore, the emission wavelength can be readily tuned by simply varying the dimensions of the dot (or well). The main difficulty is to ensure that the dots comprising a device are uniformly sized. If not, the density of states is smeared out and the behavior reverts to bulk-like, negating the advantage of the zero-dimensional confinement. Early attempts to produce quantum dots a few tens of nanometers in diameter using electron beam lithography followed by the usual semiconductor processing (etching, see Section 8.1.1) were bedevilled by damage and contamination introduced by the processing. An important advance came through the exploitation of frustrated wetting (Stranski–Krastanov growth): lattice mismatch between the deposited layer and the substratum results in strain, which was found to be relieved by the spontaneous formation of monodisperse islands (Section 8.1.2).

Microcavities have the potential for fabricating ultrahigh-Q and ultralow threshold lasers. Cylindrical or spherical microcavities are particularly interesting in that they have extremely low loss whispering gallery modes (WGMs), which occur when light circumnavigating the cavity is trapped by total internal reflexions and constructively interferes with itself. To construct a laser based on a microcavity it is surrounded by gain medium and pumped by an external light source. The lower limit of miniaturization is exemplified by the laser oscillation demonstrated with a single atom in an optical resonator. Threshold pump energies of the order of 100 μJ have been demonstrated. Quality factors in excess of 10^{10} can be routinely obtained.

An array of plasmonic nanoparticles has been used to create multicolor holograms capable of exceedingly high information storage density [214]. Nano-objects of different shapes and sizes (e.g. particles and rods) can be mixed in the same material. This is an example of light–matter interaction at a scale below the diffraction limit of light, which is where the future of nanophotonics lies [218].

7.9 MECHANICAL DEVICES

The spring constant (stiffness) k of a nanocantilever varies with its characteristic linear dimension l, and its mass m as l^3. Hence, the resonant frequency of its vibration

$$\omega_0 = \sqrt{k/m} \qquad (7.23)$$

varies as $1/l$. This ensures a fast response – in effect, nanomechanical devices are extremely stiff. The figure of merit (quality factor) Q equals ω_0 divided by the drag (friction) coefficient. Q, especially for nanodevices operating in a high vacuum, can be many orders of magnitude greater than the values encountered in conventional devices. On the other hand, under typical terrestrial operating conditions water vapor and other impurities may condense onto moving parts, increasing drag due to capillary effects (cf. Fig. 3.1), and generally degrading performance.

The simplest mechanical device is probably the vibrating cantilever; we may use this to fix ideas (Fig. 7.21; see Fig. 7.22 for a more picturesque representation). From Eq. (7.23), the fundamental (resonant) frequency is the square root of the quotient of the spring constant and the mass, i.e.

$$\omega_0 = \sqrt{Yl/m_{\text{eff}}}, \qquad (7.24)$$

where l is an appropriate length of the device and m_{eff} is an *effective mass* that depends on the geometry and anisotropy of the material of the cantilever as well as on its mass density. The very fact that l/m_{eff} scales as l^{-2} suggests that ω_0 becomes very high as l falls to nanometer size. The quality factor seems to scale with device volume $\sim V^{1/5}$; hence, the power that must be applied to drive the cantilever to an amplitude equal to the thermal fluctuations becomes extraordinarily low, $\sim k_B T/(Q/\omega_0)$. Nevertheless, individual mechanical quanta ($\hbar\omega_0$) have not yet been observed, although it is now feasible for them to be greater than the thermal energy $k_B T$ by appropriately tuning ω_0.

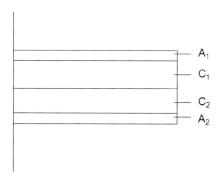

FIGURE 7.21

Schematic diagram of a cantilever. The core structure is represented by layers C and the surface layers are labeled A. The latter may be modified by interaction with the environment. Thus, accrual of mass on C may create A, or very porous pre-existing structure may be stiffened. Any asymmetry between C_1 and C_2 (e.g. only C_1 can accumulate additional material) may lead to static bending of the cantilever.

Envisaged applications of nanoscale cantilevers are as logic devices; in effect ultraminiature versions of the interlocking between points and signals developed for railway systems to prevent a signalman from selecting operationally incompatible combinations; and as sensors.

The advent of ultrastiff carbon-based materials (Chapter 9) has generated renewed interest in mechanical devices that can now be made at the nanoscale (nanoelectromechanical systems, NEMS). Ultrasmall cantilevers (taking the cantilever as the prototypical mechanical device) have extremely high resonant frequencies, effective stiffnesses and figures of merit Q and, evidently, very fast response times $\sim Q/\omega_0$. Therefore, it becomes conceivable that a new generation of relays, constructed at the nanoscale, could again contend with their solid-state (transistor-based) rivals that have completely displaced them at the microscale. Relays have, of course, excellent isolation between input and output, which makes them very attractive as the components of logic gates.

Dissipation, the inverse of Q, is the quotient of the energy ΔW lost per cycle of oscillation and the total mechanical energy W_0 of the resonator. It is approximately equal to the quotient of the damping (friction) γ and the resonant frequency; that is,

$$Q^{-1} = \frac{\Delta W}{2\pi W_0} \approx \frac{\gamma}{\omega_0} \approx \frac{\Delta\omega_0}{\omega_0},$$
(7.25)

FIGURE 7.22

A cantilever.

-From G. Galilei, Discorsi e Dimostrazioni Matematiche, Leyden, 1638.

the last part of this expression representing a practical means of determining the dissipation, $\Delta\omega_0$ being the full width at half maximum of the frequency response.

Nanofriction has, obviously, attracted a good deal of interest. The so-called minimalistic models, including the long-established Prandtl–Tomlinson (PT) model, are useful for gaining some physical understanding of the processes going on [292]. Much work involves dragging an AFM tip over a substrate; the Frenkel–Kontorova (FK) model has been useful for interpreting results [113].

Sensing applications of mechanical nanodevices can operate in either static or dynamic mode. For cantilevers (Fig. 7.21) used as sensors, as the beam becomes very thin, its surface characteristics begin to dominate its properties (cf. Section 2.2). Classically, the adsorption of particles onto such a beam would increase its mass, hence lower its resonant frequency. This effect would be accounted for if the adsorbed adlayer had a higher stiffness than the beam material, thereby *increasing* its resonant frequency. In static sensing mode, the adsorption of particles on one face of a beam breaks its symmetry and causes it to bend.

Manufacturing variability may be problematical for ultrasensitive mass sensors; ideally they should be fabricated with atomic precision using a bottom-to-bottom approach (Section 8.3); in practice nowadays they are fabricated using semiconductor processing technology.

7.10 MOLECULAR MACHINES

The molecular machines considered in this section are typically made by chemists in reactors. They can, therefore, potentially be made in large quantities in the same way as other elaborate molecules, such as some pharmaceutical drugs.

The level of ingenuity required for molecular machines, however, considerably exceeds that of drugs because often the molecular machines consist of independent parts that are threaded into each other. Balzani gives numerous examples [15]. For example, pseudorotaxanes can be threaded and unthreaded by light. The thread is a long molecule containing a –N=N– unit that is stable in the *trans* configuration, in which it can insert itself through the pseudorotaxane ring. Irradiation causes its photoisomerization to the *cis* configuration, which interacts much more weakly with the ring, and hence unthreading takes place.

Another example also involves a rotaxane, which is permanently threaded by a dumbbell, the precise position of which depends on hydrogen bonding between the connecting rod of the dumbbell and the rotaxane ring. These bonds can be destroyed by the addition of a base to the external solution, and the dumbbell shuttles to another stable recognition site. The movement is reversed by the addition of acid.

Whether these phenomena could be harnessed to perform useful work is a moot point. The environment is Brownian, not eutactic, and the movements are statistical: equilibrium is shifted to a new point, but not every molecule in the solution will follow the trend.

7.11 THERMAL DEVICES

Nanoscale junctions can be designed to harness thermopower S, in which a temperature difference ΔT across the two sides of the junction induces a voltage drop ΔV across it. Such junctions may be used to recover waste heat (e.g. from computation) and generate electrical power with no moving parts.

S (the Seebeck coefficient) is defined according to

$$S = -\frac{\Delta V}{\Delta T}\bigg|_{I=0} \qquad (7.26)$$

in the limit of $\Delta T \rightarrow 0$ as well as vanishing current I [65]. The current may be expressed as a linear response

$$I = G\Delta V + L_T \Delta T, \qquad (7.27)$$

where G is the electrical conductance and L_T its analogy, a thermal response. It follows that

$$S = \frac{L_T}{G} \qquad (7.28)$$

which indicates how to, practically, determine S: one needs to measure G and L_T. The latter is approached via the thermal conductance σ_{th}, defined as

$$\sigma_{th} = -\lim_{\Delta T \rightarrow 0} \frac{J_{th}}{\Delta T}, \qquad (7.29)$$

where J_{th} is the total heat current; it is related to the thermal conductivity κ by

$$\sigma_{th} = (A/L)\kappa, \qquad (7.30)$$

where A and L are, respectively, the cross-section and length of the sample; note that Fourier's law links thermal current density to temperature *gradient*,

$$j_{th} = -\kappa\nabla T. \qquad (7.31)$$

A convenient dimensionless figure of merit is [65]

$$ZT = \frac{GS^2}{\sigma_{th}/T} \qquad (7.32)$$

and useful applications should aim to achieve $ZT \gg 1$, although the Wiedemann–Franz law

$$\frac{\kappa}{\sigma} = \frac{\pi^2 k_B^2}{3e^2}T \qquad (7.33)$$

(where σ is the electrical conductivity) militates against it.

Experiments in the nanoscale typically involve placing a gold STM tip on top of a gold substrate covered with selected molecules. The STM tip is lowered until it just touches, and is attached to, a molecule. A thermal gradient is applied and the thermal power is measured by applying a voltage such that no current passes. The field is still in a very early stage of development and we are doubtless a long way from practical devices.

More success is currently being demonstrated with nanogenerators based on piezoelectric materials. A device exhibiting a mechanical-to-electrical conversion efficiency of 11% using polymer nanowires has recently been reported [300].

7.12 FLUIDIC DEVICES

In gases, the dimensionless Knudsen number associated with the motion of the molecules is the ratio of the mean free path ℓ to some characteristic size l,

$$K_n = \ell/l. \tag{7.34}$$

The mean free path depends on the pressure and the simplest possible derivation, assuming the gas molecules to be elastic objects with an effective diameter σ, yields

$$\ell = 1/\pi\sigma^2 n, \tag{7.35}$$

where n is the number per unit volume. Following similar reasoning to Section 7.4.1, one could take the nanoscale as the scale corresponding to $K_n = 1$; it would therefore be pressure-dependent. At atmospheric pressure, the mean free path of a typical gas is several tens of nm. Given that even a moderate vacuum of say 10^{-3} mbar gives a mean free path of about 0.1 m, a low pressure nanodevice would have no unique feature.

In fluids, we typically have molecules or particles of interest dissolved or suspended in the solvent. The particles (of radius r) undergo Brownian motion, i.e. thermal collisions, and the mean free path is effectively the distance between successive collisions with the solvent molecules (formally, it is the average distance from any origin that a particle travels before significantly changing direction) and is related to the diffusion coefficient D and the thermal velocity c of the particle according to

$$\ell = 3D/c; \tag{7.36}$$

the diffusivity depends on the friction coefficient according to the Stokes–Einstein law

$$D = k_B T/(6\pi\eta r), \tag{7.37}$$

where η is the dynamic viscosity of the solvent, and the thermal velocity depends on both the particle size and density ρ,

$$c = \sqrt{6k_B T/(\pi^2 \rho r^3)}. \tag{7.38}$$

Substituting these expressions into Eq. (7.36) gives

$$\ell = \frac{\sqrt{(3/2)k_B T \rho r}}{6\eta}, \tag{7.39}$$

from which it follows that under any practical conditions (e.g. water at 20 °C has a density of 1 g/cm^3 and a viscosity of 10^{-3} Pa s) the particle size exceeds the mean free path; hence, there can be no ballistic transport and hence no unique feature at the nanoscale. In other words, although in analogy to the case of electron conduction (Section 7.4.1), a nanofluidic device could be one for which the mean free path of particles moving within it is greater than the characteristic size of the internal channels, this condition could never be realistically met. If some way could be found of working with nanoparticles of material from a white dwarf star (mass density $\sim 10^9$ g/cm^3), this conclusion would no longer be valid!

For devices that depend on thermal conductivity, it is limited by phonon scattering, but analogously to electron conduction (Section 7.4.1), if the device is small enough ballistic thermal transport can be observed, with the quantum of thermal conductance being $\pi^2 k_B^2 T/(3h)$.

Another approach to defining a nanofluidics scale could be based on the fact that fluidics devices are fundamentally mixers. For reactions in minute subcellular compartments (or taking place between highly diluted reagents) $\langle a \rangle \langle b \rangle \neq \langle ab \rangle$. The nanoscale in fluidics could be reasonably taken to be that size in which $\Delta^2(\gamma_t)$ differs significantly from zero. The scale thus depends on the concentration of the active molecules (A and B).

This calculation ignores any effect of turbulence on the mixing, for which the characteristic scale would be the Kolmogorov length ℓ_K,

$$\ell_K = (\kappa^3/\varepsilon)^{1/4}, \tag{7.40}$$

where κ is the kinematic viscosity and ε is the average rate of energy dissipation per unit mass. But, this length is of micrometer rather than nanometer size. Furthermore, it is essentially impossible to achieve turbulent flow in nanochannels; even microfluidics is characterized by extremely low Reynolds numbers (below unity).

Fluids are often moved in miniature reactors using electroösmosis, in which an electric field is applied parallel to the desired rate of flow. It is assumed that the walls of the device are ionizable. For example, most metal oxides are hydroxylated at the interface with water and can be protonated or deprotonated according to

$$|- OH_2^+ + OH^- \rightleftharpoons |- OH + H_2O \rightleftharpoons |- O^- + H_3O^+. \tag{7.41}$$

Hence, the interface will be charged; the counterions will be dispersed in a diffuse (Gouy–Chapman) double layer, their concentration decaying with a characteristic Debye length $1/\kappa$, given by (for monovalent ions)

$$\kappa^2 = e^2 \sum_i n_i z_i^2/(\epsilon k_B T), \tag{7.42}$$

where n_i is the concentration of the ith ion bearing z_i elementary charges. The decay length is typically of the order of nanometers of tens of nanometers; upon application of the electric field the counterions will move, entraining solvent and engendering its flow. If the radius of the channel is less than $1/\kappa$, electroösmosis will be less efficient. This length could therefore be taken as determining the nanoscale in fluidics devices (provided some electrokinetic phenomenon is present), because below this scale bulk behavior is not followed. Electroösmosis is not of course useful in a macroscopic channel, because only a minute fraction of the total volume of solvent will be shifted.

Note the analogy between the semiconductor space charge (cf. Eq. (7.4)) and the Gouy–Chapman ion distribution.

The capillary length,

$$l_{cap} = (k_B T/\gamma)^{1/2} \tag{7.43}$$

(confusingly, an alternative definition is current in macroscopic fluid systems, in which surface tension is compared with sedimentation) – the surface tension γ is that between the solid wall and the fluid – gives an indication of the predominance of surface effects (cf. Section 2.2). It is typically of the order of nanometers. There is no discontinuity in behavior as the characteristic size of a fluidic device falls to or below l_{cap}. The problem of viscous liquid flow in a nanotube has been formulated and solved by Gogsadze et al. [100].

In summary, as the radius of a fluid channel becomes less than a characteristic length such as $1/\kappa$ or l_{cap}, in effect the properties of the channel wall influence the entire volume of the fluid within the channel. The implications of this are far from having been thoroughly worked out. Especially in the case of aqueous systems, the structure of the water itself (especially the average strength of its hydrogen bonds) is also affected by wall interactions, doubtless with further far-reaching implications. But most discontinuities due to miniaturization of fluidics devices seem to already manifest themselves at the microscale.

7.12.1 MIXERS AND REACTORS

Miniaturizing mixers has been very successful at the microscale, as can be deduced from the huge proliferation of lab-on-a-chip devices for analytical and preparative work in chemistry and biochemistry. Particular advantages are the superior control over flow compared with macroscopic mixers, one very important benefit of which is much more predictable selection of reaction products, wherever several are possible (in consequence, yields can be raised to 100%), and (in principle) great ease of scale up, simply by having many micromixers in parallel (although this does not yet seem to have been convincingly demonstrated for any industrial-scale production facility). It is, however, by no means clear that even greater success will attend further miniaturization down to the nanoscale. On the contrary, performance may be degraded. This needs further investigation. Fig. 7.23 indicates the kind of detail in the nanoscale that is required to understand the course of complex, multistage reactions.

7.12.2 CHEMICAL AND BIOCHEMICAL SENSORS

These devices are included under fluidics since the sample to be analysed is almost invariably a fluid (cf. Section 11.7). The Holy Graal of clinical sensing is continuous, noninvasive monitoring (cf. Section 4.2). Currently, most tests require a sample of the relevant biofluid (e.g. blood) to be drawn from the patient. For most people, this is a somewhat unpleasant procedure, and hence the tests are carried out infrequently. It is, however, recognized that much more insight into a patient's pathological state

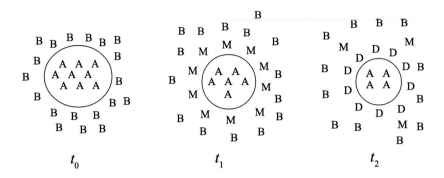

FIGURE 7.23

Schematic diagram of a micromixing scenario, applicable to the pair of reactions A + B → M and A + M → D. At t_0, an eddy of A is present in the solution of B. After an interval $\Delta t = t_1 - t_0$ all the A at the surface of the eddy have reacted with the B in their immediate vicinity to form M; if the reaction of M with a further molecule of B is faster than the diffusion of M, then D will be formed in good yield.

could be obtained by frequent, ideally continuous, monitoring. At present, this is only possible in intensive care stations, where the patient is immobilized, and even then continuous invasive monitoring does not take place (the oxygen content of the blood is monitored noninvasively by analysing the optical reflectance spectrum of the skin covering blood vessels). It seems to be a very difficult technical problem to extend such noninvasive analysis to the plethora of biomarkers currently under intensive study as symptomatic of disease or incipient disease. An alternative approach is to develop sensors so tiny that they can be semipermanently implanted inside the body, where they can continuously monitor their surroundings.

Because of the large and growing number of afflicted people, diabetes has received overwhelmingly the most attention. The sensing requirement is for glucose in the blood. The glucose sensor follows classic biosensing design: a *recognition element* to capture the analyte (glucose) mounted on a *transducer* that converts the presence of captured analyte into an electrical signal (Fig. 7.24). The recognition element is typically a biological molecule, the enzyme glucose oxidase; hence (if small enough), this device can be categorized as both nanobiotechnology and bionanotechnology.

Both components of the biosensor are excellent candidates for the application of nanotechnology. Molecular recognition depends on a certain geometrical and chemical arrangement of atoms in some sense complementary to the analyte molecules, together with coöperative motions to enhance affinity. Atom-by-atom assembly therefore represents the perfect way to artificially fabricate recognition elements. The ultimate goal of the transducer is to detect a single captured analyte molecule, hence the smaller it can be made, the better, provided issues related to noise and detection

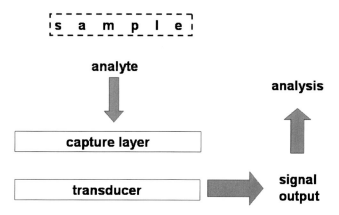

FIGURE 7.24

A prototypical biosensor. The capture layer concentrates the analyte in the vicinity of the transducer, which reports the concentration of analyte in the capture layer, which is directly related to the concentration in the sample.

efficiency (Section 10.8) can be overcome. The goal is to exploit the phenomenon that a minute change in an atomically scaled device (e.g. the displacement of a single atom) can significantly change its properties, including functionality.

7.12.3 ENERGY CONVERSION DEVICES

Originally inspired by the photosynthetic machinery in blue–green algae and plants, in order to overcome the problem that the output of a photovoltaic or thermal solar energy converter (i.e. electricity or heat, respectively) must be subsequently stored, possibly via further conversion and inevitably involving losses, attempts have been made to create devices that convert solar energy directly into chemical form. The basic principle is that the absorption of light by a semiconductor creates highly reducing and highly oxidizing species (respectively, the conduction band electrons and the valence band holes). Instead of separating them in a space charge as in the conventional p–n junction device and collecting the current, if the semiconductor is a nano-object with its characteristic dimension comparable to the Bohr radii of the charge carriers (Section 2.5), they may react with chemical species adsorbed on the nano-object's surface before they have time to recombine (Fig. 7.25).

Challenges that still need to be overcome include: matching the semiconductor band gap to the solar spectrum; matching the positions of the band edges to the redox potentials of the desired chemical reactions; finding catalysts to facilitate the desired chemical reactions; and preventing photocorrosion (i.e. reduction or oxidation of the semiconductor by its own charge carriers).

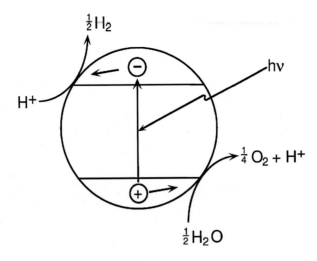

FIGURE 7.25

Radiant energy harvesting using semiconductor nanoparticles. The absorption of a photon raises an electron from the valence band (VB) to the conduction band (CB), leaving a positive hole in the former. The CB electron ⊖ is a strong reducing agent and the VB hole ⊕ a strong oxidizing agent, possibly capable of, respectively, reducing and oxidizing water to hydrogen and oxygen. At present, this concept of nanoparticle photoelectrochemistry is used to sacrificially destroy persistent organic pollutants allowed to adsorb on the particles. This is also the basis of self-cleaning windows and compact water purification devices.

SUMMARY

A device is fundamentally an information processor or transducer. The types of devices considered in this chapter are mainly logic gates and sensors. The relay is considered as the paradigmatical component of the logic gate, which is the foundation of digital processing hardware. Ultraminiature devices based on electron charge as the carrier of information are considered. These include single-electron transistors, which are scaled-down versions of conventional transistors; molecular electronic devices, in which organic molecules replace the inorganic semiconducting media of conventional transistors; and quantum dot cellular automata. As these devices become very small, movement of electronic charge becomes problematic, and using electron spin as the information carrier becomes attractive. This includes both hybrid devices in which spin controls electron current (the main application is for magnetic field sensing), and "all-spin" devices in which spin is the carrier of information. Photons may also be used as information carriers, although mostly not in the nanorealm, except for molecular photonics, which is, however, covered in Chapter 11 because the only actual example uses a biomolecule for the central function. Quantum computation, although not specifically tied to the nanoscale, is briefly described as a completely different approach to increasing computational density, for which photonic devices are well suited. Nanoscale mechanical devices are considered, both as relays and sensors. A brief survey of fluidic devices in the nanoscale is given, including biosensors. Finally, nanoparticulate systems for conversion of solar energy into fuels are considered.

FURTHER READING

1. S. Collin, Nanostructure arrays in free-space: optical properties and applications, Rep. Progr. Phys. 77 (2014) 126402.
2. L.O. Chua, Memristive devices and systems, Proc. IEEE 64 (1976) 209–223.
3. D.K. Ferry and L.A. Akers, Scaling theory of modern VLSI, IEEE Circuits Devices Magazine (September) (1997) 41–44.
4. S. Franssila, Introduction to Microfabrication, second ed., Wiley, 2010.
5. J.A. Hutchby, et al., Emerging nanoscale memory and logic devices: a critical assessment, IEEE Comput. Magazine (May) (2008) 78–82.
6. L. Jacak, Semiconductor quantum dots—towards a new generation of semiconductor devices, Eur. J. Phys. 21 (2000) 487–497.
7. K.K. Likharev, Single-electron devices and their applications, Proc. IEEE 87 (1999) 606–632.
8. P. Lohdahl, S. Mahmoodian, S. Stobbe, Interfacing single photons and single quantum dots with photonic nanostructures, Rev. Modern Phys. 87 (2015) 347–400.
9. S.S.P. Parkin, Giant magnetoresistance in magnetic nanostructures, A. Rev. Mater. Sci. 25 (1995) 357–388.
10. S.S.P. Parkin, et al., Giant tunnelling magnetoresistance at room temperature with MgO (100) tunnel barriers, Nature Mater. 3 (2004) 862–867.
11. A. Politi, J.L. O'Brien, Quantum computation with photons, Nanotechnol. Percept. 4 (2008) 289–294.
12. U. Seifert, Stochastic thermodynamics, fluctuation theorems and molecular machines, Rep. Progr. Phys. 75 (2012) 126001.
13. T. Shibata, Computing based on the physics of nanodevices—a beyond-CMOS approach to human-like intelligent systems, Solid-State Electron. 53 (2009) 1227–1241.

Nanofacture of devices

CHAPTER CONTENTS

Nanotechnology: An Introduction. DOI: 10.1016/B978-0-323-39311-9.00014-5
Copyright © 2016 Elsevier Inc. All rights reserved.

INTRODUCTION

Whereas Chapter 6 considered mainly actual methods of producing nanomaterials with a regular structure (and often with a merely statistical order), in this chapter, we look at more sophisticated technologies, both present and supposed future, capable, in principle, of fabricating functional artifacts of arbitrary complexity. The entire field is encompassed within the three divisions of top–down, bottom-to-bottom, and bottom–up; Fig. 8.1 gives a summary. Top–down requires great ingenuity (and expense) in the fabrication machinery; bottom–up requires great ingenuity in the conception of the building blocks. Bottom-to-bottom, the least developed of the three, requires above all ingenuity in conception at its present stage of development.

8.1 TOP–DOWN METHODS

These share the general feature of requiring large (and also expensive, see Section 1.3.2, requiring considerable concentrations of capital) installations. What might be called the traditional route, that of scaling down processes familiar in macro and micro engineering, appears on the extreme right of the diagram, Fig. 8.1. Concerted incremental improvement in the entire manufacturing process transforms precision engineering into ultraprecision engineering (Fig. 1.2). The stiffness of the parts of mechanical devices used to shape objects is particularly important for achieving precision. These processes are primarily subtractive: material is removed by grinding or etching, although in semiconductor processing this is complemented by deposition.

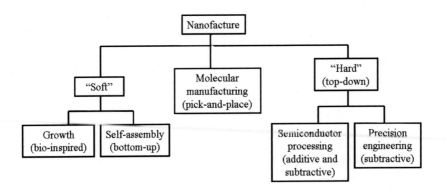

FIGURE 8.1

Different modes of nanomanufacture (nanofacture).

8.1.1 **SEMICONDUCTOR PROCESSING**

This well-established industrial technology refers to the operations of sequentially modifying (e.g. oxidizing), depositing (additional layers on) and removing (parts of) a substratum (e.g. silicon) over areas selected by exposing photoresist coating the working surface through a mask and then dissolving away the unexposed resist (or the converse). Remaining resist either protects from etching or from fresh deposition. This works well at the micrometer scale and is used to fabricate VLSI circuits. Problems of scaling it down to produce features with lateral sizes in the nanorange runs into the diffraction limit of the light used to create the mask (cf. Eq. (5.2)), partly solved by using light of shorter wavelengths or high-energy electrons, and problems associated with etching. Liquid phase processes suffer from the difficulties of precisely controlling mass transfer at the nanoscale; there may be too much or too little etching. Plasma etching (e.g. reactive ion etching, RIE) may be used instead. Insofar as wet chemical processing is used (to dissolve away unwanted resist) hydrodynamics at very small dimensions also becomes important.

In contrast to the difficulties of nanoscaling lateral features, very high quality thin films can be deposited with nanometer control perpendicular to the plane of a substratum. These methods are grouped under the heading of PVD. The material to be deposited is evaporated from a reservoir, or magnetron-sputtered from a target. The most precise control is obtainable with MBE, developed at AT&T Bell Laboratories in the late 1960s: evaporated material is beamed onto the substratum under conditions of ultrahigh vacuum. Deposition is typically very slow (several seconds to achieve 1 nm film thickness) and hence can be epitaxial. Ultrathin layers (of the order of a nanometer) with atomically sharp interfaces can be deposited (Section 6.3.1).

Chemical vapor deposition (CVD) is similar to PVD, except that the precursor of the thin layer is a reactive gas or mixture of gases, and the substratum is typically heated to accelerate chemical reaction to form a solid product deposited as a film. The decomposition can be enhanced with a plasma (this typically allows the substratum to be maintained at a lower temperature than otherwise).

Related technologies are used to modify existing surfaces of materials, such as exposure to a plasma, and ion implantation, in which electrostatically charged high-energy (typically 10–100 keV) ions are directed towards the surface, where they arrive with kinetic energies several orders of magnitude higher than the binding energy of the host material, and become implanted in a surface layer that may be tens of nanometers thick. The change in the structure due to the impact of high-energy ions may be a more important alteration of the material than the incorporation of foreign atoms.

PVD and CVD processes are often able to yield structure at the nanoscale, or the lower end of the microscale, with some structural control achievable via the deposition parameters. The structures obtained emerge from a particular combination of deposition parameters and must be established experimentally. Theoretical understanding of the structure and process is still very rudimentary, although attempts are under way (such as the "structure zone model").

8.1.2 MISMATCHED EPITAXY

In the epitaxial growth of semiconductors (see Section 8.1.1), when material B is evaporated onto a substrate of a different material A, three analogous situations can arise:

- Frank–van der Merwe (B wets A, and a uniform layer is formed);
- Volmer–Weber (no wetting, hence islets distributed in size of B on A are formed);
- Stranski–Krastanov (B can wet A, but there is a slight lattice mismatch between A and B, and the accumulating strain energy is ultimately sufficient to cause spontaneous dewetting, resulting in rather uniform islets of B, which relieves the strain; this explanation is plausible but there is still much discussion regarding the mechanism).

Eq. (3.25) can be applied. If $S > 0$, we have the Frank–van der Merwe régime; the substratum is wet and layer-by-layer growth takes place. If $S < 0$, we have the Volmer–Weber régime; there is no wetting and three-dimensional islands grow. But if $S > 0$ for the first layer at least, yet at the same time there is a geometric mismatch between the lattices of the substratum and the deposited layer [206], strain builds up in the latter, which is subsequently relieved by the spontaneous formation of monodisperse islands (the Stranski–Krastanov régime); it can be thought of as frustrated wetting.

   Its main application is for fabricating quantum dots for lasers [quantum dot lasers are a development of quantum well lasers; the carriers are confined in a small volume and population inversion occurs more easily than in large volumes, leading to lower threshold currents for lasing, and the emission wavelength can be readily tuned by simply varying the dimensions of the dot (or well), see Section 2.5]. The main difficulty is to ensure that the dots comprising a device are uniformly sized. If not, the density of states is smeared out and the behavior reverts to bulk-like. Initially, the quantum dots were prepared by conventional semiconductor processing, but it was very difficult to eliminate defects and impurities, whereas the Stranski–Krastanov self-assembly process does not introduce these problems.

8.1.3 ELECTROSTATIC SPRAY DEPOSITION

Ceramic precursors are dissolved in a suitable solvent and mixed immediately prior to forcing through an orifice maintained at a high potential difference with respect to the substratum. The liquid breaks up into electrostatically charged droplets, which are attracted both by gravity and the Coulombic force to the substratum, which is typically heated to accelerate the reaction that forms the final material. For example, calcium nitrate and phosphoric acid dissolved in butyl carbitol and atomized upon leaving the nozzle at a potential of 6–7 kV with respect to a titanium or silicon substratum maintained at a few hundred °C about 30 mm below the nozzle create coatings of calcium phosphate with intricate and intriguing nanostructures [177].

8.1.4 **FELTING**

The fabrication of texture by felting has been known for centuries (in Europe, and much longer in China) in the guise of papermaking, reputed to have been invented in A.D. 105 by Ts'ai Lun in China. It arrived in Europe, first in Italy, by the end of the thirteenth century, probably via Samarkand, Baghdad, Damascus, Egypt, the Maghreb and Muslim Spain; it appeared in England around 1490. The vegetable fibers (typically based on cellulose) are macerated until each individual filament is a separate unit, mixed with water, and lifted from it in the form of a thin layer by the use of a sieve-like screen, the water draining through its small openings to leave a sheet of felted fiber upon the screen's surface. The first papermaking machine was invented by Robert in France at around the end of the eighteenth century and was later perfected by two Englishmen, the Fourdrinier brothers. The machine poured the fibers out in a stream of water onto a long wire screen looped over rollers; as the screen moved slowly over the rollers, the water drained off and delivered an endless sheet of wet paper.

With the advent of various kinds of nanofibers, it is now possible to make paper-like materials at the nanoscale. This has been attempted most notably using CNTs, when it is sometimes called buckypaper. Mixtures of fibers would be especially interesting for creating nanotexture. The process is different from the shaking and/or stirring used to prepare a mixed powder. Randomly oriented fibers (highly elongated objects) are rapidly placed on top of each other, with their long axes parallel to a substratum (which is, in the case of cellulose-based writing paper, removed later on) to form a random fiber network (RFN) (Fig. 8.2). The network coheres because of numerous fiber–fiber contacts, but the structure is different from that formed by the entanglement of very long randomly coiled polymers, such as the giant glycoproteins such as mucin constituting the mucous films lining the epithelial surfaces of multicellular organisms, which are indeed large and flexible enough to be entangled with one another.

Fiber–fiber contacts can be enhanced by the presence of certain substances (e.g. divalent cations). An example of a naturally felted structure is provided by the

FIGURE 8.2

Sketch of a two-dimensional RFN used to model a felted assembly.

basement membranes assembled from extracellular matrix proteins such as laminin, used in multicellular organisms to support cells and tissues.

Simplified models, such as the RFN, into which the only input is the distribution of fiber lengths and their surface chemical properties, are useful for calculating basic properties of felted materials, for example, mass distribution, number of crossings per fiber, fractional contact area, free-fiber length distribution and void structure.

8.1.5 ULTRAPRECISION ENGINEERING

Current ultrahigh-precision ("ultraprecision") engineering is able to achieve surface finishes with a roughness of a few nanometers. According to McKeown et al. it is achievable using 11 principles and techniques [192], which include

- dynamic stiffness; thermal stability; seismic isolation,
- rigid body kinematic design; three-point support,
- measuring system isolated from force parties and machine distortion,
- high accuracy bearings,
- quasi-static and dynamic error compensation.

Although some of the attributes implied by these principles and techniques would appear to call for large, massive machines, the need for a further, twelfth, principle has become apparent, namely, that of miniaturization – ultraprecision machines should be as small as is feasible taking account of the principles. This is because smaller load paths and metrology and thermal loops directly improve dynamic performance and accuracy. There should be, therefore, ultimately convergence between assemblers and ultraprecision machine tools (UPMT).

On a different scale from UPMT is the exploitation of STM for surface nano machining. In this case the tool is typically a diamond tip (Fig. 8.3). Fig. 8.4 shows an example of a silicon surface machined with such a tip. Ultrahigh precision is achievable, albeit with low throughput. Should there be a demand to increase it, the "millipede" multiple scanning technology developed by IBM for data storage (memory) applications may be applicable. An advantage of this approach is that the instrument can be used to monitor the results of the machining.

8.2 BOTTOM–UP METHODS

The increasing difficulty of continuing the miniaturization of classical photolithography and its derivatives in semiconductor processing (Section 8.1.1), and the extreme laboriousness of current mechanosynthesis (Section 8.3), coupled with the observation of self-assembly in nature, generated interest in alternative fabrication technologies. The primitive idea of self-assembly ("shake and bake") is to gather precursors in random positions and orientations and supply energy ("shaking") to allow them to sample configuration space. The hugeness of this space suggests that a convergent pathway is inherent in the process in order to allow it to be completed in a reasonable time, as in protein "folding" (Section 8.2.11). Once the precursors are

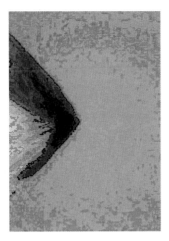

FIGURE 8.3

Single-crystal synthetic diamond tip used in a STM for high-precision machining. The STM is modified to have a loading measurement unit.

-Reproduced with permission from O. Lysenko, et al., Surface nanomachining using scanning tunneling microscopy with a diamond tip, Nanotechnol. Percept. 6 (2010) 41–49.

in position, "baking" is applied to strengthen the bonds connecting them and fix the final object permanently.

8.2.1 SELF-ASSEMBLY

Assembly means gathering things together and arranging them (fitting them together) to produce an organized structure. A child assembling a mechanism from "Meccano" captures the meaning, as does an assembly line whose workers and machines progressively build a complicated product such as a motor-car from simpler components; an editor assembles a newspaper from copy provided by journalists (some of which might be rejected). In all these examples, the component parts are subject to constraints – otherwise they would not fit together – and their entropy S must necessarily decrease. Since the free energy $\Delta G = \Delta H - T\Delta S$ must then necessarily increase, in general, the process will not, by itself, happen spontaneously. On the contrary, a segregated arrangement will tend to become homogeneous. Hence, in order for self-assembly to become a reality, something else must be included. Typically enthalpy H is lost through the formation of connexions (bonds) between

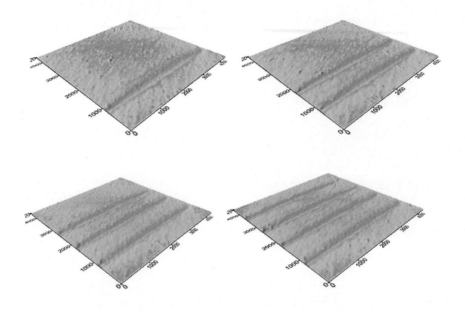

FIGURE 8.4

STM image of grooves on a silicon surface made by the pyramidal diamond tip shown in Fig. 8.3. The depth of the grooves is 14 ± 2 nm.

-Reproduced with permission from O. Lysenko, et al., Surface nanomachining using scanning tunneling microscopy with a diamond tip, Nanotechnol. Percept. 6 (2010) 41–49.

the parts, and provided $|\Delta H|$ exceeds $|T\Delta S|$ we have at least the possibility that the process can occur spontaneously, which is presumably what is meant by self-assembly. The final result is generally considered to be in some sort of equilibrium. Since entropy S is always multiplied by the temperature T in order to compute the free energy, it should be noted that the relevant temperature is not necessarily what one measures with a thermometer; one should only take the (typically small number of) relevant degrees of freedom into account; the temperature is the mean energy per degree of freedom. If slightly sticky granules are thrown into a rotating drum, or stirred in a mixing bowl, typically they will all eventually clump together to form a "random" structure (e.g. Fig. 8.5(a)), but one which is evidently less random than the initial collection of freely flowing granules; hence, the entropy is lower, but every time this experiment is repeated the result will be different in detail, and one feels that the system might be really ergodic if one had enough time to make all the repetitions, and the ergodicity is only broken because one has insufficient time.

The goal of an ideal self-assembly process suitable for manufacturing devices is for the same structure to be formed each time the constituent particles are mixed together (Fig. 8.5(b)) – as was imagined by von Foerster [86]. Of course, energy needs to be put into the system. In the "purest" form of self-assembly, the energy is

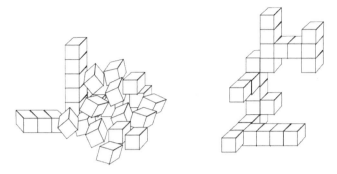

FIGURE 8.5

(a) The result of mixing isotropically sticky cubelets. (b) Putative result of mixing selectively sticky cubelets.

thermal (random), but it could also be provided by an external field (e.g. electric or magnetic). If we are satisfied by the constituent particles being merely joined together in a statistically uniform fashion and, moreover, the process happens spontaneously, then it is more appropriate to speak of self-joining or self-connecting. The mixing of concrete, in which the dry ingredients of concrete (sand and cement) are thrown into a "cement mixer" to which a small amount of water is then added is, at least at first sight, an example of such a self-connecting process; more generally, one can refer to gelation. Approaching the phenomenon from a practical viewpoint, it is clear that gelation is almost always triggered by a change of external conditions imposed upon the system, such as a change of temperature, or hydration, and the spontancity implied by the prefix "self-" is absent. The only action we can impose upon the system without violating the meaning of "self-" is that of bringing the constituent particles together. Note that here we diverge from the everyday meaning of the term "assembly", which includes the gathering together of people, either spontaneously as when a group of people wish to protest against some new measure introduced by an authoritarian government, or by decree.

If the process in which we are interested is deemed to begin at the instant the constituent particles are brought together, then we can indeed put the mixing of concrete in the category of self-joining, because we could (although it is not usually done in that way in practice) bring the wetted particles of sand and cement together, whereupon they would spontaneously join together to form a mass.

The meaning of self-joining (of which self-connecting is a synonym) is then the property possessed by certain particles of spontaneously linking together with their neighbors when they are brought to within a certain separation. One can also imagine there being kinetic barriers to joining, which can be overcome given enough time. Note that the particles each need more than one valency (unit of combining capacity), otherwise dimers would be formed and the process would then stop. A good example is the condensation of steam to form water. We can suppose that the steam is first supercooled, which brings the constituent particles (H_2O molecules) together; the transition to liquid water is actually a first-order phase transition that requires an initial nucleus of water to be formed spontaneously. "Gelation" (cf. Section 3.7) then occurs by the formation of weak hydrogen bonds (a maximum of four per molecule) throughout the system.

Strictly speaking, it is not necessary for all the constituent particles to be brought together instantaneously, as implied in the above. Once the particles are primed to be able to connect themselves to their neighbors, they can be brought together one by one. This is the model of diffusion-limited aggregation (DLA). In nature, this is how biopolymers are formed: monomers (e.g. nucleic acids or amino acids) are joined sequentially by strong covalent bonds to form a gradually elongating linear chain. The actual self-assembly into a compact three-dimensional structure involves additional weak hydrogen bonds between neighbors that may be distant according to their positions along the linear chain (see Section 8.2.11); some of the weak bonds formed early are broken before the final structure is reached.

In the chemical literature, self-assembly is often used as a synonym of self-organization. A recapitulation of the examples we have already discussed shows, however, that the two terms cannot really be considered to be synonymous. The DLA is undoubtedly assembled, but can scarcely be considered to be organized, not least because every repetition of the experiment will lead to a result that is different in detail, and only the same when considered statistically. "Organized" is an antonym of "random"; therefore, the entropy of a random arrangement is high; the entropy of an unorganized arrangement is low. It follows that inverse entropy may be taken as a measure of the degree of organization; this notion will be further refined in the next section. The DLA differs from the heap of sand only insofar as the constituent particles are connected to each other. An example of organization is shown in Fig. 8.5(b). The impossibility of self-organization has been proved by Ashby, as will be described in Section 8.2.3.

Before discussing self-organization, we must first discuss organization, of which self-organization is a part. If the elements in a collection (here we shall not say "system", because that already implies a degree of organization) are organized to some degree, that implies that they are in some way connected to each other, which can be considered as a kind of communication, and are hence subject to certain constraints. In other words, if the state of element B is dependent on the state of A to some extent, then we can say that B's state is conditional on that of A. Likewise, the relation between A and B may be conditional on the state of C. Whenever there is conditionality, there is constraint: B is not as free to adopt states as it would be in a totally unorganized system [11].

8.2.2 THERMODYNAMICS OF SELF-ORGANIZATION

Consider a universe U comprising a system S and its environment E, i.e. $U = S \cup E$ (Fig. 8.6). Self-organization (of S) implies that its entropy spontaneously diminishes,

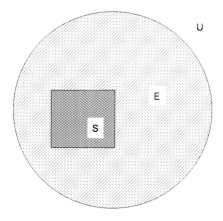

FIGURE 8.6

Universe U comprising system S and its environment E.

that is,

$$\delta S_S/\delta t < 0. \tag{8.1}$$

Accepting the second law of thermodynamics, such a spontaneous change can only occur if, concomitantly,

$$\delta S_E/\delta t > 0, \tag{8.2}$$

with some kind of coupling to ensure that the overall change of entropy is greater than or equal to zero. If all processes were reversible, the two changes could exactly balance each other, but since (inevitably, we may suppose) some of the processes involved will be irreversible, overall

$$\delta S_U/\delta t > 0. \tag{8.3}$$

Therefore, although the system itself has become more organized, overall it has generated more disorganization than the organization created, and it is more accurate to call it a self-disorganizing system [86]. Hence, the "system" must properly be expanded to include its environment – it is evidently intimately connected with it; without it there could be no organization. Despite its true nature as a self-disorganizing system having been revealed, nevertheless we can still speak of a self-organizing *part* S of the overall system that consumes order (and presumably energy) from its environment. It follows that this environment must necessarily have structure itself, otherwise there would be nothing to be usefully assimilated by the self-organizing part.

The link between entropy (i.e. its inverse) and organization can be made explicit with the help of relative entropy R (called redundancy by Shannon), defined by

$$R = 1 - S/S_{max}, \tag{8.4}$$

where S_{max} is the maximum possible entropy. With this new quantity R, self-organization implies that $\delta R/\delta t > 0$. Differentiating Eq. (8.4), we obtain

$$\frac{dR}{dt} = \frac{S(dS_{max}/dt) - S_{max}(dS/dt)}{S_{max}^2}. \tag{8.5}$$

Our criterion for self-organization (namely, that R must spontaneously increase) is, therefore, plainly

$$S\frac{dS_{max}}{dt} > S_{max}\frac{dS}{dt}. \tag{8.6}$$

The implications of this inequality can be seen by considering two special cases [86]:

1. The maximum possible entropy S_{max} is constant; therefore, $dS_{max}/dt = 0$ and $dS/dt < 0$. Now, the entropy S depends on the probability distribution of the constituent parts (at least, those that are to be found in certain distinguishable states); this distribution can be changed by rearranging the parts, which von Foerster supposed could be accomplished by an "internal demon".
2. The entropy S is constant; therefore, $dS/dt = 0$ and the condition that $dS_{max}/dt > 0$ must hold, that is, the maximum possible disorder must increase. This could be accomplished, for example, by increasing the number of elements; however, care must be taken to ensure that S then indeed remains constant, which probably needs an "external" demon.

Looking again at inequality Eq. (8.6), we see how the labor is divided among the demons: dS/dt represents the internal demon's efforts, and S is the result; dS_{max}/dt represents the external demon's efforts, and S_{max} is the result. There is therefore an advantage (in the sense that labor may be spared) in cooperating – e.g. if the internal demon has worked hard in the past, the external demon can get away with putting in a bit less effort in the present.

These considerations imply that water is an especially good medium in which self-assembly can take place because, except very near its boiling point, it has a great deal of structure (Section 3.8) that it can sacrifice to enable ordering in S. Hence, it is not surprising that biological self-assembly of compact protein and nucleic acid structures takes place in an aqueous environment (or, given the emergence of an aqueous environment through geophysical processes, it is not surprising that conditions were then favorable for the building blocks of life to emerge). Presumably thermophilic microbes that live at temperatures close to 100°C have some difficulties on this score because their environment is less structured.

8.2.3 THE "GOODNESS" OF THE ORGANIZATION

Examining again Fig. 8.5, it can be asserted that both putative results of mixing slightly sticky cubelets together are organized, but most people would not hesitate to call the structure in (b) better organized than that in (a) [bear in mind that (b) is unique but (a) is not]. Evidently, there is some meaning in the notion of "good organization", even though it seems difficult to formulate an unambiguous definition. Can a system spontaneously (automatically) change from a bad to a good organization? This would be a reasonable interpretation of "self-organization", but has been proved to be formally impossible [11]: Consider a device that can be in any one of the three states, A, B or C, and the device's operation is represented by some transformation, e.g.

$$\downarrow \quad \begin{matrix} A & B & C \\ B & C & A \end{matrix}$$

Now suppose that we can give an input f into the device, and that the output is determined by the value of f, e.g.

$$\begin{array}{c|ccc} \downarrow & A & B & C \\ \hline f_A & B & C & A \\ f_B & A & A & A \\ f_C & A & B & C \end{array}$$

Spontaneous (automatic) operation means that the device is able to autonomously select its input. The different possible input values are here represented by a subscript indicating the state of the device on which the input now depends. However, this places severe constraints on the actual operation, because $f_A(B)$ (for example) is impossible; only $f_A(A)$, $f_B(B)$ and $f_C(C)$ are possible, hence the operation necessarily reduces to the simple transform, lacking any autonomy,

$$\downarrow \quad \begin{matrix} A & B & C \\ B & A & C \end{matrix}$$

Any change in f must therefore come from an external agent.

8.2.4 PARTICLE MIXTURES

Consider the shaking and stirring of a powder – a collection of small (with respect to the characteristic length scale of the final texture) particles of type A and type B initially randomly mixed up. The interaction energy v is given by the well-known Bragg–Williams expression

$$v = v_{AA} + v_{BB} - 2v_{AB}, \tag{8.7}$$

where the three terms on the right are the interaction energies (enthalpies) for A with itself, B with itself, and A with B. One would guess that the mixture is miscible if $v < 0$, and immiscible otherwise. If the interaction energies are all zero, there will still be an entropic drive towards mixing (the case of the perfect solution). Edwards and Oakeshott [69] introduce the *compactivity X*, analogous to the temperature in thermodynamics, defined as

$$X = \partial V / \partial S, \tag{8.8}$$

where V is the volume (playing the rôle of energy) and S the entropy, defined according to Boltzmann's equation as $S = k_B \ln \Omega$, where Ω is the number of (mechanically stable) configurations. If the number of particles of type A at a certain point r_i is $m_A^{(i)}$ (equal to either zero or one), and since necessarily $m_A^{(i)} + m_B^{(i)} = 1$, by introducing the new variable $m_i = 2m_A^{(i)} - 1 = 1 - 2m_B^{(i)} = \pm 1$. Defining $\phi = \langle m_i \rangle$, which Bragg and Williams give as $\tanh \phi v / k_B X$ [69], three régimes are identified depending on the interaction parameter $v/k_B X$ [69]:

- miscible: $v/k_B X < 1$, $\phi = 0$;
- domains of unequal concentrations: $v/k_B X > 1$, ϕ small ($X = v/k_B$ emerges as a kind of critical point);
- domains of pure A and pure B: $v/k_B X \gg 1$, $\phi = \pm 1$.

8.2.5 MIXED POLYMERS

The entropy of entities being mixed is

$$S_{\text{mix}} = -k_B \sum_i \phi_i \ln \phi_i, \tag{8.9}$$

where ϕ_i is the volume fraction of the entities of the ith kind. The per-site (where each site is a monomer unit) free energy $\mathcal{M}$ of mixing two polymers A and B is

$$\mathcal{M} = \phi_A \ln \phi_A / N_A + \phi_B \ln \phi_B / N_B + \chi \phi_A \phi_B, \tag{8.10}$$

where N_B and N_B are the degrees of polymerization, and χ is the Flory–Huggins interaction parameter, given by (cf. Eq. (8.7))

$$\chi = (v_{AA} + v_{BB} - 2v_{AB})z/2k_B T, \tag{8.11}$$

where z is defined (for a polymer on a lattice) as the number of lattice directions. The first two terms on the right-hand side of Eq. (8.10), corresponding to the entropy of mixing (Eq. (8.9)), are very small due to the large denominators; hence, the free energy is dominated by the third term, giving the interaction energy. If $\chi > 0$, then phase separation is inevitable. For a well-mixed blend, however, the separation may take place exceedingly slowly on laboratory timescales, and therefore for some purposes nanotexture might be achievable by blending two immiscible polymers. However, even if such a blend is kinetically stable in the bulk, when prepared as a thin film on the surface, effects such as spinodal decomposition may be favored due to the symmetry-breaking effect of the surface (for example, by attracting either A or B). Separation can be permanently prevented by linking the two functionalities as in a block copolymer (Section 8.2.6).

8.2.6 **BLOCK COPOLYMERS**

One of the problems with mixing weakly interacting particles of two or more different varieties is that under nearly all conditions complete segregation occurs, at least if the system is allowed to reach equilibrium. This segregation is however frustrated if the different varieties are covalently linked together, as in a block copolymer. A rich variety of nanotexture results from this procedure. If A and B, assumed to be immiscible ($\chi > 0$), are copolymerized to form a molecule of the type AAA$\cdots$AAABBB$\cdots$BBB (a diblock copolymer), then of course the A and B phases cannot separate in the bulk; hence, microseparation results, with the formation of domains with size ℓ of the order of the block sizes (that is, at the nanoscale), minimizing the IF energy between the incompatible A and B regions. One could say that the entropy gain arising from diminishing the A–B contacts exceeds the entropic penalty of demixing (and stretching the otherwise random coils at the phase boundaries). The block copolymer can also be thought of as a type of supersphere [247]. If χN is fairly small (less than 10) we are in the weak segregation régime and the blocks tend to mix, but in the strong segregation régime ($\chi N \gg 10$ the microdomains are almost pure and have narrow interfaces. As the volume fraction of one of the components (say A) of the commonly encountered coil–coil diblock copolymers increases from zero to one, the bulk morphologies pass through a well characterized sequence of bcc spheres of A in B, hexagonally packed cylinders of A in B, a bicontinuous cubic phase of A in a continuous matrix of B, a lamellar phase of alternating lamellae of A and B, a bicontinuous cubic phase of B in a continuous matrix of A, hexagonally packed cylinders of B in A and bcc spheres of B in A.

When block copolymers are prepared as thin films (thickness d less than 100 nm) on a substratum (e.g. by spin-coating or dip-coating), the symmetry of the bulk system is broken, especially if one of the blocks of the copolymer is preferentially attracted to or repelled from the surface of the substratum. If $d < \ell$, the surface may be considered to have a strong effect on the structure of the thin film. For example, poly-2-vinylpyridine does not wet mica, and a polystyrene–polyvinylpyridine block copolymer thin film on mica has a structure different from that of the copolymer in the bulk [272]: structures such as antisymmetric surface-parallel lamellae, antisymmetric hybrid structures (cf. Stranski–Krastanov film growth, Section 8.1.2) and surface-perpendicular lamellae or columns are typically formed. There is at present considerable interest in such processes for fabricating photolithography masks in the nanoscale range more conveniently than by electron beam writing. Reticulated structures seem to have been investigated the most extensively: block copolymer micelles can be formed by dissolving the polymer in a fluid that is a solvent for only one of the components, and using it to coat surfaces, yielding a more or less regular array. This process has attracted interest as a route to making nanoporous membranes. For example, polystyrene–polymethylmethacrylate (PMMA) copolymers prepared with the volume fraction such that there are cylindrical microdomains of the PMMA in the bulk, can be coated on a suitable substratum (silicon or silica) such that the cylinders are oriented normal to the surface. Exposure to ultraviolet light crosslinks the polystyrene but degrades the PMMA, which can then be selectively dissolved out of the film, leaving a nanoporous polystyrene membrane with pore size controllable by varying the molecular weight of the copolymer [307].

One advantage of these polymer-based processes is the tremendous variety of starting materials available (by the same token, the systematic experimental investigation of the effects of compositional variation across the whole range of possibilities represents a huge undertaking). As well as changing the chemical nature of the monomers and the degrees of polymerization, the block copolymers have also been mixed with homopolymers as a way of modifying the characteristic scale of the texture [116].

8.2.7 THE ADDITION OF PARTICLES TO THE SOLID/LIQUID INTERFACE

Many self-assembly processes are based on the addition of nano-objects to a substrate. As noted in Section 8.2.5, a surface can have a symmetry-breaking effect. Consider a chemically and morphologically unstructured surface (medium 1) brought into contact with a fluid (medium 2) in which objects (medium 3) are suspended (i.e. their buoyancy is such that they move purely by diffusion (Brownian motion) and are not influenced by gravity). Suppose that the materials are chosen such that the IF free energy is negative (Section 3.2). On occasion those in the vicinity of the interface will strike it. The rate of arrival of the particles at the substratum is proportional to the product of particle concentration c_b in the suspending medium and the diffusion coefficient D of a particle, the constant of proportionality depending on the hydrodynamic régime (e.g. convective diffusion); this rate will be reduced by a factor $1/\int_{\ell_0}^{\infty}[\exp(\Delta G_{123}(z)/k_B T) - 1]dz$ in the presence of an energy barrier; the reduction factor could easily be several orders of magnitude. Once a particle of radius r adheres to the substratum, evidently the closest the center of a second particle can be placed to the first one is at a distance $2r$ from the center of the first; in effect the first particle creates an *exclusion zone* around it (Fig. 8.7).

A corollary of the existence of exclusion zones is that the interface will be jammed (i.e. unable to accept a further particle) at a surface coverage of substantially less than 100%. The actual value of the jamming limit depends on the shape of the particle; for spheres it is about 54% of complete surface coverage This process is known as random sequential addition (RSA). Although a random dispersion of particles in three dimensions is thereby reduced to a two-dimensional layer, the positions of the particles remain random: the radial distribution function is totally unstructured. Even if the particles can move laterally, allowing the interface to equilibrate in a certain sense, it is still jammed at a surface coverage of well below 100%.

Numerically simulating RSA. The process is exceptionally easy to simulate: for each addition attempt one selects a point at random: if it is further than $2r$ from the center of any existing particle a new particle is added (the available area for less symmetrical shapes may have to be computed explicitly) and if it is closer then the attempt is abandoned. The success of this simple algorithm is due to the fortuitous

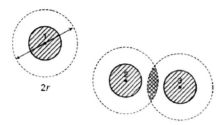

2r

FIGURE 8.7

The concept of exclusion zone. The particles' projected area is hatched. The area enclosed by the dashed lines is the exclusion zone and has twice the radius of the actual particle. The exclusion zone is defined as that area within which no center of any particle can be placed without violating the condition of no overlap of hard bodies. The cross-hatched area marks the overlap of the exclusion zones of particles numbered 2 and 3. If the particles interact with each other with longer range forces than the hard body (Born) repulsion, then the radius is increased to an effective radius equal to the distance at which the particle–particle interaction energy $\Delta G_{323}(z)$ equals the thermal energy $k_B T$.

cancellation of two opposing processes: correlation and randomization. In reality, if a particle cannot be added at a selected position because of the presence of a previously added one, it will make another attempt in the vicinity of the first one, because of the Rabinowitch ("cage") effect [187]; successive attempts are strongly positionally correlated. On the other hand, as a particle approaches the interface through the bulk fluid, it experiences hydrodynamic friction, which exerts a randomizing effect; the two effects happen to cancel out each other [14].

Functions for characterizing nano-object addition. We have the fraction of occupied surface θ, equal to the product of the number of objects ν per unit area and the area a occupied by one object, and the fraction of surface ϕ available for adsorption (sometimes called the available area function). In general, we have for the rate of addition,

$$d\theta/dt = k_a c^* \phi(\theta), \qquad (8.12)$$

where k_a is the addition rate coefficient (dependent upon the IF free energy function $\Delta G_{123}(z)$, see Section 3.2) and c^* is the effective bulk concentration (subsuming hydrodynamic and other factors). Much RSA research concerns the relation of ϕ to θ. An early theory relating them was Langmuir's: if small objects adsorb to discrete substrate sites larger than the particles,

$$\phi = 1 - \theta. \tag{8.13}$$

Substituting this into Eq. (8.12) and integrating, we see that in Langmuir adsorption the surface is completely filled up ($\theta \to 1$) exponentially in time (for a uniform rate of arrival of particles at the surface). In the absence of discrete sites (or in the presence of such sites that are smaller than the particles), the particles adsorb wherever they happen to arrive (assumed to be random locations). If a particle arrives such that its center would fall within the exclusion zone of a previously adsorbed particle (Fig. 8.7) its adsorption attempt is rejected. Since the exclusion zone is four times as large as the particle, we should have

$$\phi = 1 - 4\theta \tag{8.14}$$

but as θ increases, exclusion zones will overlap (Fig. 8.7), and compensating terms have to be added, proportional to θ^2 for two overlapping particles, and so on up to six (the maximum number that can fit; in practice terms up to second or third order will usually be found to be sufficient),

$$\phi = 1 - b_1\theta + b_2\theta^2 + b_3\theta^3 + O(\theta)^4 \tag{8.15}$$

with $b_1 = 4$ and the coefficients b_2 and b_3 determined by purely geometrical considerations; $b_2 = 6\sqrt{3}/\pi$ is identical for both irreversible and equilibrium adsorption, whereas the coefficient b_3 varies from about 1.4 for irreversible (RSA) to about 2.4 for equilibrium (reversible, whether via desorption and readsorption or via lateral movement) adsorption. In this case $\phi \to 0$ for $\theta < 1$; the "jamming limit" at which $\phi = 0$ is $\theta_J \approx 0.55$ for spheres adsorbing irreversibly.

The RSA formalism was developed in the context of particles interacting predominantly via hard body repulsion. The particle radius r is implicitly considered to be the hard body radius r. "Soluble" (stably suspended) particles must have repulsive particle–particle interactions and cannot in fact approach each other to a center-to-center distance of $2r$, but will behave as particles of an effective radius r', where r' is that value of z at which the total IF interaction energy (see Section 3.2) $\Delta G_{(323)}(z) \sim k_B T$, the thermal energy.

Generalized ballistic deposition (GBD). The ballistic deposition (BD) model describes objects falling onto a surface under the influence of gravity. Whereas in RSA if a particle attempts to land with its center within the exclusion zone around a previously adsorbed particle it is rejected, in BD the particle is not eliminated, since it is not buoyant enough to diffuse away, but rolls along on top of previously adsorbed particles until it finds space to adsorb. The coefficients of Eq. (8.15) are then different, namely, $b_1 = b_2 = 0$ and $b_3 \approx -9.95$. BD and RSA are combined linearly in GBD, where

$$\phi(\theta) = \phi^{RSA}(\theta) + j\phi^{BD}(\theta) \tag{8.16}$$

with the parameter j defined as

$$j = p'/p, \tag{8.17}$$

where p' is the probability that a particle arriving via correlated diffusion ("rolling") at a space large enough to accommodate it will remain (i.e. will surmount any energy barrier), and p is the probability that a particle arriving directly at a space large enough to accommodate it will remain. p is clearly related to the lateral interaction ("stickiness") of particles for each other; $j = 0$ corresponds to pure RSA and as $j \to \infty$ the model describes nanoparticle aggregation at a surface. Essentially, the exclusion zones are thereby annihilated, and ϕ can be simplified to Eq. (8.13).

Two-dimensional crystallization. If, however, the particles can adhere to each other on the interface, the possibility for organizing arises. This has been very clearly demonstrated when lateral mobility was expressly conferred on the particles by covering the substrate with a liquid-crystalline lipid bilayer and anchoring the particles (large spherical proteins) in the bilayer through a hydrophobic "tail" [253]. The particles structure themselves to form a two-dimensional ordered array (crystal). When such an affinity exists between the particles trapped at the interface, the exclusion zones are annihilated. From this fact alone (which can be very easily deduced from the kinetics of addition [253]) one cannot distinguish between random aggregation forming a DLA (cf. reaction-limited aggregation, RLA) and two-dimensional crystallization; they can generally be distinguished, however, by the fact that in the latter the crystal unit cell size is significantly bigger than the projected area of the particle (cf. a three-dimensional protein crystal: typically about 70% of the volume of such a crystal is occupied by solvent). The process of two-dimensional crystallization has two characteristic time scales: the interval τ_a between the addition of successive particles to the interface

$$\tau_a = 1/[aF\phi(\theta)], \tag{8.18}$$

where a is the area per particle, F is the flux of particles to an empty surface (proportional to the bulk particle concentration and some power <1 of the coefficient of diffusion in three dimensions), and ϕ is the fraction of the surface available for addition, which is some function of θ, the fractional surface coverage of the particles at the interface; and the characteristic time τ_D for rearranging the surface by lateral diffusion (with a diffusion coefficient D_2)

$$\tau_D = a/(D_2\theta). \tag{8.19}$$

If $\tau_D \gg \tau_a$ then lateral diffusion is encumbered by the rapid addition of fresh particles before self-organization can occur and the resulting structure is indistinguishable from that of RSA. Conversely, if $\tau_a \gg \tau_D$ there is time for two-dimensional crystallization to occur. Note that some affinity-changing conformational change needs to be induced by the interface, otherwise the particles would already aggregate in the bulk suspension. In the example of the protein with the hydrophobic tail, when the protein is dissolved in water the tail is buried in the interior of the protein, but partitions into the lipid bilayer when the protein arrives at its surface.

Another intriguing example of IF organization is the heaping into cones of the antifreeze glycoprotein (AFGP), consisting of repeated alanine–alanine–glycosylated threonine triplets, added to the surface of a solid solution of nanocrystalline $Si_{0.6}Ti_{0.4}O_2$ [175]. Under otherwise identical conditions, on mica the glycoprotein adsorbs randomly sequentially. Despite the simplicity of the structure of AFGP, its organization at the silica–titania surface appears to be a primitive example of examples of programmable self-assembly (PSA, Section 8.2.8).

8.2.8 PROGRAMMABLE SELF-ASSEMBLY (PSA)

The addition of monomers to a growing crystal is self-assembly (Section 6.4), but the result is not useful and a nanofacturing process because there is nothing to limit growth, except when the imposed conditions are carefully chosen (Section 6.1.3). Crystallization is an example of passive or nonprogrammable self-assembly. Its ubiquity might well engender a certain pessimism regarding the ultimate possibility of realizing true self-assembly, the goal of which is to produce vast numbers of identical, prespecified arbitrary structures. Yet in biology, numerous examples are known (see also Section 1.6): the final stages of assembly of bacteriophage viruses, of ribosomes, and of microtubules, which occur not only *in vivo*, but which can also be demonstrated *in vitro* by simply mixing the components together in a test-tube. As apparent examples of what might be called "passive" self-assembly, in which objects possessing certain asymmetric arrangements of surface affinities are randomly mixed and expected to produce ordered structures [86], they seem to contradict the predictions of Sections 8.2.2 and 8.2.3.

It has long been known that biomolecules are constructions: that is, they have a small number of macroscopic (relative to atomic vibrations) degrees of freedom, and can exist in a small number (≥ 2) of stable conformations. Without these properties, the actions of enzymes, active carriers such as hemoglobin, and the motors that power muscle, etc., are not understandable. Switching from one conformation to another is typically triggered by the binding or dissociation of a small molecule, for example, the "substrate" of an enzyme, or adenosine triphosphate (ATP). The initial collision of two particles is followed by a conformational change in one or both of them, e.g.

$$A + B + C \rightarrow A - B^* + C \rightarrow A - B^* - C, \tag{8.20}$$

where the asterisk denotes a changed conformation induced by binding to A; C has no affinity for B, but binds to B^*. This process is illustrated in Fig. 8.8 and is called PSA. Graph grammar, which can be thought of as a set of rules encapsulating the outcomes of interactions between the particles [164, 165] (cf. stigmergic assembly, Section 8.2.12), is useful for representing the process. The concept of graph grammar has brought a significant advance in the formalization of PSA, including the specification of minimal properties that must be possessed by a self-assembling system (e.g. the result implying that no binary grammar can generate a unique stable assembly [166]).

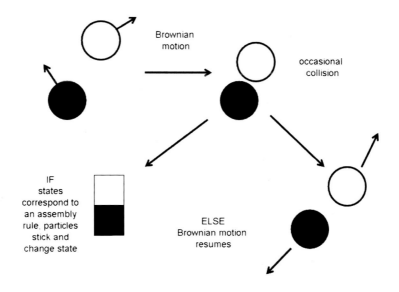

FIGURE 8.8

Illustration of PSA, with a primitive local rule.

While models of PSA robots have been created in the macroscale, artificially synthesizing molecules with the required attribute remains a challenge. Biology, however, is full of examples (e.g. the "induced fit" occurring when an antibody binds an antigen). Microscopically, these are manifestations of coöperativity (Section 3.6). However, the coöperativity is systemic in the sense that entire macromolecules may be acting in concert as a system (cf. Sections 8.2.10 and 8.2.11).

8.2.9 SUPERSPHERES

If the competing interactions have different sign and range, ordered structures of definite size can assemble spontaneously. This provides a simple example of PSA. Consider nanoparticles suspended in water and weakly ionized such that they all carry the same electrostatic charge. When the suspension is stirred, suppose that the repulsive electrostatic force is too weak to overcome the attractive van der Waals force when two particles happen to collide. Therefore, every collision will lead to sticking, and aggregates will slowly form. The van der Waals force is, however, very short range and can only act between nearest neighbors. The electrostatic force, on the other hand, has a much longer range, and can therefore be summed over the entire aggregate. Ultimately, the aggregate will become large enough for the summed electrostatic repulsion to exceed the van der Waals nearest neighbor attraction. The result is monodisperse "superspheres" [i.e. aggregates of small (maybe spherical) particles].

Weakly electrostatically charged quantum dots (nanoparticles) suspended in water aggregate to form uniformly sized superspheres containing several hundred nanoparticles. Nearest neighbors interact with weak, short-range van der Waals interactions, which easily dominate the slight electrostatic repulsion between them. Because, however, the electrostatic interaction is long range (it can be tuned by varying the ionic strength of the solution), the overall electrostatic repulsion within a supersphere gradually accumulates, and when a certain number of nanoparticles have been aggregated, the electrostatic repulsion exceeds the attractive van der Waals force between nearest neighbors [247]. To form superspheres, the attractive interaction should be short range, and the repulsive interaction should be long range.

An interesting kind of nanostructure was shown in Fig. 6.8(d). The small spheres (called micelles) have polar heads that are ionizable in water, resulting in q elementary charges on the surface of the sphere, which exert an expanding pressure

$$\Delta P^{(el)} = (qe)^2/(8\pi\epsilon R^4) ; \qquad (8.21)$$

the size of the micelle adjusts itself to exactly compensate the Laplace contraction (Eq. (2.2)); in consequence such micelles are highly monodisperse because they are at a local energy minimum.

8.2.10 BIOLOGICAL SELF-ASSEMBLY

It has long been known that many biological systems exhibit remarkable capabilities of assembling themselves starting from a randomly arranged mixture of components. These include the bacteriophage virus (the final stages of assembly), and proteins and ribonucleic acids (RNAs), which can be spontaneously transformed from a random coil of the as-synthesized linear polymer to a compact, ordered three-dimensional structure (Section 8.2.11). It is clear that the starting precursors of the final structures have to be very carefully designed – this is a carefully tuned example of PSA in action (Section 8.2.8).

Although appreciation of self-assembly in biology has played a hugely important inspirational rôle, the highly specialized chemistry of living systems, the fragility of many of its products and its inherent variability at many levels have made it unsuitable for mimicking directly and incorporating into our present industrial system (cf. Section 8.2.13). This is particularly so in the case of the food industry. The extreme complexity, both structural and chemical, of its products and the relative ease of letting them grow renders efforts to manufacture food synthetically largely superfluous.

More debatable is whether self-assembly offers a viable route to creating artificial cells for photovoltaic solar energy conversion. The natural system comprises the photosystems embedded within the chloroplast, whose maintenance requires the rest of the machinery of the cell, and whose effective operation requires a macroscopic structure of the plant (stem and branches) to support the leaves in which the chloroplasts are embedded. The classical artificial system is the semiconductor photovoltaic cell. Can its efficiency, as well as ease of manufacture, can be enhanced by using nanostructured photoactive components? Most appraisals of the photovoltaic cell as a "renewable" or "sustainable" energy source pay scant regard to the entire manufacturing cycle, and the key question of working lifetime under realistic conditions is scarcely addressed by laboratory trials. Given the history of considerable efforts to more closely mimic the molecular machinery of the natural photosystems in a nanoconstruction, it has been natural to look at extending the mimicry beyond discrete components to systems. Nevertheless, except for the ultimate, and still hypothetical, stage of molecularly manufactured nanosystems, none of the proposed solutions come anywhere near the performance (considered as an overall system) of natural photosynthesis, which can simply be left to grow over vast areas.

The observation that preassembled bacteriophage components (head, neck and legs) could be mixed in solution exerted a profound inspiration on the world of "shake-and-bake" advocates. These components are essentially made up of proteins – heteropolymers made from irregular sequences chosen from the 20 natural amino acids of general formula $H_2N–C^{(\alpha)}HR–COOH$, where R is naturally one of 20 different side chains (residues), ranging from R=H in glycine, the simplest amino acid, to elaborate heterocycles such as $R=CH_2–[C_8NH_6]$ in tryptophan. The conformation and hence affinity of a protein depend on environmental parameters such as the pH and ionic strength of the solution in which it is dissolved, and this is one mechanism for achieving programmability in assembly, since the local pH and ion concentration around a protein molecule depend on the amino acids present at its surface. These factors also determine the conformation of the highly elongated, so-called fibrous proteins such as fibronectin, now known to consist of a large number of rather similar modules strung together (the "daisy chain" or "pearl necklace" model). Some other examples of biological self-assembly have already been mentioned in Section 8.2.8. A further one is provided by the remarkable S-layers with which certain bacteria are coated. One should also mention the oligopeptides found in fungi (e.g. alamethicine) and the stings of bees (mellitin) and wasps (mastoparan) that self-assemble into pores when introduced into a bilayer lipid membrane. But, biological self-assembly and self-organization is by no means limited to the molecular scale (see Section 8.2.12).

8.2.11 BIOPOLYMER FOLDING

Biopolymer "folding" means the transformation of a linear polymer chain, whose monomers are connected only to their two nearest neighbors, and which adopts a random coil in solution, into a complex three-dimensional structure with additional (hydrogen) bonds between distant monomers.

Predicting the final three-dimensional structure is *prima facie* a difficult problem. Energetics are clearly involved, because bonds between distant monomers form spontaneously (if geometric constraints are satisfied) releasing enthalpy and hence lowering the free energy. On the other hand, this raises the entropy because the chain becomes constrained. Finding the free energy minimum by systematically searching configuration space is a practically impossible task for a large molecule with thousands of atoms – it would take longer than the age of the universe. Since the protein molecule can fold within seconds, it seems clear that the solution to the problem lies in determining the pathways. The *principle of least action* (PLA) is useful for this purpose: the most expedient path is found by minimizing the action.

Action is the integral of the Lagrangian $\mathcal{L}$ ($=L-F$ for conservative systems, where L and F are, respectively, the kinetic and potential energies). Minimization of the action is an inerrant principle for finding the correct solution of a dynamical problem; the difficulty lies in the fact that there is no general recipe for constructing $\mathcal{L}$.

A solution leading to a successful algorithm has been found for the folding of RNA [81]. Natural RNA polymers are made up from four different "bases", A, C, G and U (see Section 4.1.4). As with DNA, multiple hydrogen bonding favors the formation of G–C and A–U pairs, which leads to the appearance of certain characteristic structures. Loop closure is considered to be the most important folding event. F (the potential) is identified with the enthalpy, that is, the number n of base pairings (contacts), and L corresponds to the entropy. At each stage in the folding process, as many as possible new favorable intramolecular interactions are formed, while minimizing the loss of conformational freedom (the principle of sequential minimization of entropy loss, SMEL). The entropy loss associated with loop closure is ΔS_{loop} (and the rate of loop closure $\sim \exp(\Delta S_{loop})$); the function to be minimized is therefore $\exp(-\Delta S_{loop}/R)/n$, where R is the universal gas constant. A quantitative expression for ΔS_{loop} can be found by noting that the N monomers in an unstrained loop ($N \geq 4$) have essentially two possible conformations, pointing either inwards or outwards. For loops smaller than a critical size N_0, the inward ones are in an apolar environment, since the nano-enclosed water no longer has bulk properties, and the outward ones are in polar bulk water. For $N < N_0$, $\Delta S_{loop} = -RN \ln 2$ (for $N > N_0$, the Jacobson–Stockmayer approximation based on excluded volume yields $\Delta S_{loop} \sim R \ln N$).

In summary, SMEL applied to biopolymer folding is a least-action principle that involves sequentially maximizing the number of contacts while minimizing entropy loss.

A similar approach can be applied to proteins [82]. However, in proteins the main intramolecular structural connector (apart from the covalent bonds between successive amino acid monomers) is the backbone hydrogen bond, responsible for the appearance of characteristic structures such as the alpha helix, but which, being single, are necessarily weaker than the double and triple hydrogen bonds in DNA and RNA. They therefore need to be protected from competition for hydrogen bonding by water, and this can be achieved by bringing amino acids with apolar residues to surround the hydrogen bonds [83]. This additional feature, coupled with the existence of multiple conformational states already referred to in Section 8.2.8 means that proteins are particularly good for engaging in PSA, a possibility that is, of course, abundantly made use of in nature.

8.2.12 BIOLOGICAL GROWTH

The development of an embryo consisting of a single cell into a multicellular organism is perhaps the most striking example of self-organization in the living world. The process of cell differentiation into different types can be very satisfactorily simulated on the basis of purely local rules enacted by the initially uniform cells (see [186] for the modeling of neurogenesis). On a yet larger scale, it is likely that the construction of nests by social insects such as ants and wasps relies on simple rules held and enacted by individual agents (the insects) according to local conditions; this process has been called stigmergy and is evidently conceptually related to PSA (Section 8.2.8).

Reproducibility is interpreted somewhat differently by living processes in comparison with the mass standardized manufacturing processes of the Industrial Revolution paradigm. Although the basic building blocks (e.g. proteins) of living organisms are identical, templated from a master specification encoded in DNA (see Section 4.1.4), organisms are not identical in the way that VLSIs are. What is specified (genetically) is at most an algorithm (subject to local environmental influence) for constructing an organism, or maybe just an algorithm for an algorithm. Reproducibility exists at the functional level: every termite's nest is able to protect its denizens from a wide range of hazards under varied local conditions of topography, soil type and vegetation; every dog has a brain able to ensure its survival for a certain period, and so forth. This concept of an algorithm specifying how the construction should take place is what is used for building the nests of social insects, which are constructed stigmergically – each insect is armed with rules specifying what to do in a variety of local circumstances.

There are some hitherto relatively unexploited niches for creating nano-objects via biological growth. For example, the magnetic protein ferritin, which is constituted from an iron oxide core surrounded by protein, can be made on a large scale by low-cost biotechnological manufacturing routes and due to its strict monodispersity can be used, after eliminating the protein matter, in magnetic memory devices [142].

8.2.13 SELF-ASSEMBLY AS A MANUFACTURING PROCESS

Practical, industrial interest in self-assembly is strongly driven by the increasing difficulty of reducing the feature sizes that can be fabricated by semiconductor processing technology. Pending the ultimate introduction of productive nanosystems based on bottom-to-bottom fabrication (Section 8.3), self-assembly is positioned as a rival to the "top–down" processes that currently constitute the majority of nanofacture. For certain engineering problems, such as membranes for separating valuable resources (such as rare earth ions) from a matrix in which they are present in very diluted form, self-assembly may already be useful to manufacture regular structures such as dots or stripes over an indefinitely large area; passive self-assembly might be able to produce such structures with feature sizes at the molecular scale of a few nanometers.

As well as the mimicry of natural surfaces for biological molecular recognition for sensing and other nanomedical applications (Chapter 4), self-assembly fabrication techniques should be generally applicable to create feature sizes in the nanometer range (i.e. 1–100 nm), which is still relatively difficult to achieve, and certainly very costly (limiting its application to ultrahigh production numbers) using conventional top–down semiconductor processing techniques. Furthermore, self-assembly is more flexible regarding the geometry of the substrata to which it can be applied, e.g. there is no restriction to planar surfaces. This is particularly advantageous if it is (for example) desired to engineer the surfaces of intricately curved objects such as vascular stents in order to increase their biocompatibility.

A current challenge of "bottom–up" nanotechnology is the formulation of design rules. Generally we ask "how to design X to carry out a desired function Y?" *Design* means essentially specifying structure, hence the question can be reworded as "what structure will give function Y?" – "function" being interpreted as properties and performance; and structure being constituted from certain *numbers* of different *types* of entities, *connected* together in a certain way. Micro (and macro) engineering benefits from vast experience, i.e. a look-up table with structure in the left-hand column and function in the right. There is less experience in the nanoworld, but if a nanostructure is simply a microstructure in miniature (often it is not), this experience can be transferred. Undoubtedly one needs to ask whether the properties of matter change at the nanometer scale (Chapter 2), i.e. do we need a new set of structure–property relations? These relations may also affect the fabrication process.

Regarding the formulation of assembly rules, again current experience is exiguous and rules come mainly from knowledge of certain biological processes. Typically they are either very general (such as the PLA, see Section 8.2.11), without any specific indication of how to apply them to a particular case, or very specific and possibly quite esoteric (e.g. relying on chemical intuition). Challenges lie in the adaptation of general principles to specific cases, and in the formalization and generalization of known specific heuristic rules (or intuition). Indeed, the main disadvantage of bottom–up is that the process is not well understood theoretically. Hence, although we need to be able to, at present we cannot design the starting objects (precursors) to achieve a specified final device.

8.2.14 SURFACE SELF-ASSEMBLY

Nanoparticles deposited on a surface in the presence of a solvent may form controllable patterns (superlattices) in two dimensions. Several rules useful for predicting the outcome have been established. Monodisperse spheres form regular hexagonal close-packed monolayers. If spheres of two different sizes are mixed, an alloy is formed according to geometric rules comparable to those established for three-dimensional alloys (in which case the spheres are the constituent metal atoms). For example, if gold nanoparticles of types A and B are mixed, with the number ratio $n_A/n_B = 0.5$ and the radius ratio $0.482 < r_A/r_B < 0.624$, then a structure of stoicheiometry AB_2 is formed with a structure of the larger particles arranged hexagonally, each surrounded by six of the smaller particles [160]. This structure is analogous to that of the alloy AlB_2. If, instead, particles with $n_A/n_B = 1.0$ and $0.27 < r_A/r_B < 0.425$, then a structure of stoicheiometry AB is formed, resembling that of NaCl [160]. If the substrate on which these nanoparticle arrays are assembled is itself structured, a further degree of control is attainable. Exploiting solvent dewetting via this route is a powerful adjunct to this control [200]. It is extremely encouraging that actual devices are now being surface-assembled from heterogeneous colloids [47].

8.3 BOTTOM-TO-BOTTOM METHODS

Essentially, the contribution of nanotechnology to the effort of ever-improving machining accuracy is simply to take it to the ultimate level, in the spirit of "shaping the world atom-by-atom" (the subtitle of a report on nanotechnology prepared under the guidance of the US National Science and Technology Council Committee on Technology in 1999). Rather like the chemist trying to synthesize an elaborate multifunctional molecule (e.g. adding conjugated olefins to provide color, and hydroxyl groups to provide Lewis acid–base interactions), the materials nanotechnologist aims to juxtapose different atoms to achieve multifunctionality. This approach is known as mechanosynthetic chemistry or, in its large-scale industrial realization, as molecular manufacturing. The essential difference from chemistry is the eutactic environment, in which every atom is placed in a precise location.

The famous experiment of Schweizer and Eigler, in which they rearranged xenon atoms on a nickel surface to form the logo "IBM" [266], represented a first step in this direction. Since then, there has been intensive activity in the area, both computational and experimental, but it still remains uncertain to what extent arbitrary combinations of atoms can be assembled disregarding chemical concepts such as affinity and potential energy surfaces, and whether the process can ever be scaled up to provide macroscopic quantities of materials.

Also known as molecular manufacturing or mechanosynthesis or "pick and place" chemistry, bottom-to-bottom methods literally construct things atom by atom. In other words, it is chemistry with positional control, i.e. taking place in a eutactic environment. This is what is sometimes called "hard" or "true" nanotechnology (in the sense of being uncompromisingly faithful to Feynman's original vision [85]).

A specific realization based on carbon (mainly diamondoid) structures has been elaborated by Drexler and others. This envisages an appropriately functionalized molecular tool driven by mechanical forces (such as the tip of a SPM), which abstracts hydrogen from passivated surfaces to form radicals ("dangling bonds"), where other atoms can be added. Hence, this approach is also called "tip-based nanofabrication".

Bottom-to-bottom methods are essentially simply scaled-down versions of additive manufacturing (also known as 3D printing) technologies. The first operational 3D printer was created in the 1980s, but it is only since around 2010 that they have become widely available commercially. The conventional raw material is a metal powder; a sophisticated variant is to use a continuous wire [205]. Software control of additive manufacturing is obviously of great importance. It is to be hoped that more

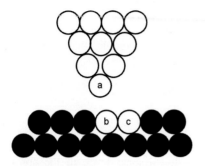

FIGURE 8.9

Illustration of tip-based nanofacture (in cross-section). Tip and substrate can move relative to each other. Each circle or disc corresponds to an atom (or superatom, see Section 8.3.2) The working tip is marked a. It can engage in vertical interchange (of atoms a and b) and lateral interchange (of atoms b and c). Each of these concerted actions involves an intermediate state in which two or more atoms are simultaneously bonded to the tip [275].

convergence between additive manufacturing and bottom-to-bottom nanoassembly will mutually advance the two fields.

8.3.1 TIP-BASED NANOFABRICATION

The use of SPMs to characterize structure at the nanoscale suggests the reciprocal use of SPM-based methods to generate such structure; that is, by picking up atoms from a store and placing them exactly where required, as was first proposed by Drexler [62] (Fig. 8.9).

Regardless of the elements involved, the current molecular manufacturing paradigm involves tips – in fact it is the technology of the SPM (Section 5.1.1). Schweizer and Eigler's manipulation of xenon atoms on an ultracold nickel surface to form the letters "IBM" was an iconic demonstration of the possibilities [266], but also of the extreme laboriousness of the procedure (it took almost 24 hours – and had to be carried out at 4 K). Curiously advance has meanwhile been very slow, possibly because of a lack of long-term investment and clear direction, despite the large overall funds spent on nanotechnology in the USA and elsewhere (too much funding has been allocated to incremental work with short-term goals). A more recent landmark was the demonstration of atomic extraction using purely physical forces

[227, 228] – 13 years after Schweizer and Eigler's demonstration [266]. Clearly this approach is still very much in its infancy. Even the issue of automatic control of the tip (Schweizer and Eigler manipulated their tip manually), which would enable one to simply specify the desired coördinates of each atom, and which would also undertake the necessary checks to ascertain whether each atom had been correctly placed, has received minimal attention. What has been undertaken is extensive calculations, mainly using density functional theory (DFT), of mechanosynthetic reactions; the slowly advancing experimental work is lagging far behind.

Although nanoscale miniature tools work extremely fast (typically in the GHz range), high throughput of nanofacture can only be achieved by massive parallelization, which in turn is only feasible if the required tools can make themselves – this was also part of Feynman's original vision. The IBM "Millipede" project, in which large numbers of SPM tips work in parallel, still falls many orders of magnitude short of achieving the throughput that would be required for materials fabrication yielding macroscopic quantities.

In summary, the main challenges of mechanosynthesis are as follows:

1. Experimental verification of the calculations of rudimentary processes;
2. Extension of work on C, Si to cover all atoms;
3. Software control [189];
4. Strategies to increase throughput.

8.3.2 NANOBLOCKS

Significant acceleration of the process could take place if "nanoblocks" – preassembled (possibly by self-assembly, Section 8.2.1) units that may comprise dozens or hundreds (or more) atoms – are manipulated via bottom-to-bottom methods. This immediately increases throughput (in terms of the volume of artifacts produced) by a factor n_b, the number of atoms per block. The principles of software control should remain unchanged. The basic idea of nanoblocks is to retain the quality, specificity and individuality of quantum objects while having the relative ease of manipulation (compared with individual atoms) of nanoparticles. Regarding the choice of blocks, there are three main possibilities:

1. Multi-atom clusters (also known as superatoms) [50]; examples are fullerenes (see Section 9.1), $Al_{13}K$ and K_3As_7. Some of these have properties analogous to an atom (Section 2.5);
2. Nanoparticles or nanorods prepared conventionally (Sections 6.1 and 6.2);
3. Biopolymers, especially globular proteins or RNA.

An advantage of 1 and 3 is strict identity of the blocks, whereas with 2, one will generally have a distribution of sizes and possibly some other properties too. An advantage of 2 and 3 is that they can readily be prepared in large quantities [this is also true for some kinds of clusters, notably fullerenes (Section 9.1)]. In all cases doping or modification is generally possible. Some of these blocks spontaneously self-assemble into regular structures (superlattices) [50], but their greatest potential use is in bottom-to-bottom fabrication. Chiutu et al. have demonstrated the precise orientation of a fullerene on an SPM tip [46].

In some sense, nanoblocks are the hardware equivalent of finite element analysis.

8.3.3 DIP-PEN NANOLITHOGRAPHY (DPN)

At a length scale slightly larger than that of atoms, DPN has been developed by Chad Mirkin and others as a way of placing molecules in finely resolved zones of a substratum. The technique works by coating the SPM tip with a weakly adherent molecular ink (e.g. proteins, relatively weakly adhering to a hydrophilic tip). When it is desired to write, the tip is lowered to the vicinity of the substratum, to which the affinity of the ink is much stronger, and the ink's molecules are transferred to the substratum through a water meniscus – the molecules ("ink") are transferred to the substratum by capillary action (Section 3.3) [151]. It follows that writing is strongly dependent on the ambient relative humidity. Although not atomically precise manufacturing, it allows features of the order of 100 nm to be written, and may be considered as featuring an approximately eutactic environment.

DPN was presumably inspired by printed electronic devices. Circuits can be fabricated at extremely low cost by printing onto a suitable substrate. Conventional processes such as screen printing and inkjet are suitable, with inks formulated using "pigments" that are conductive or semiconductive nanoparticles. This technology is especially attractive for radio frequency identification tags (RFID), which are expected to become widely used in packaging, and as security devices on products and even on documents if they can be produced at sufficiently low cost.

8.4 TOP–DOWN MANUFACTURABILITY

Many of the nanoscale devices made in the laboratory are unique. Their laborious fabrication is usually a *tour de force* of the artisanal approach. Such a device suffices for the study of their interesting properties, but does not even begin to address the problems of low-cost, high-volume manufacture [155]. This requires:

1. Prespecified performance achieved reproducibly (uniformly) and reliably;
2. High yield to acceptable tolerance;
3. A simulator for right-first-time design and for reverse engineering during development.

These conditions are, of course, fulfilled by semiconductor manufacturing, which has taken many decades to evolve to its present position of being able to reliably create devices in the nanoscale with extremely high throughput.

The fundamental processes of semiconductor manufacturing are the top–down processes of epitaxy, lithography and etching (Section 8.1.1). In these processes, atoms and photons arrive stochastically with a Poisson distribution. Consider an array consisting of 3 nm diameter features on a 6 nm pitch [155]. Each layer of each pillar contains 80 atoms. Since the variance of a Poisson process equals its mean, the coefficient of variation of the pillar area is $\sqrt{80}/80$, that is, about 12%. In contrast, modern VLSIs require a figure of less than 2% in terms of transistor performance.

It is, therefore, sensible to consider other technologies, such as spin-wave devices (Section 7.6.4), which do not require particularly small devices, when trying to increase the density of processing elements.

The protagonists of bottom-to-bottom manufacture are, of course, mindful of this limitation, which is why they are careful to specify that processing requires a eutactic environment (Section 8.3.1). Nevertheless, it should be pointed out that no such environment has hitherto been actually realized in the nanoscale, although at least simulators for the design exist and have been extensively tested.

SUMMARY

Top–down methods (exemplified by ultraprecision engineering and semiconductor processing) constitute the bulk of current industrial nanotechnology. Due to the enormous expense of the capital equipment required, however, it is impractical for use other than for very high-volume products (such as computer or cellular phone chips) or very unique products for which a high price is affordable (giant astronomical telescopes; spacecraft).

The "real" vision of nanotechnology (especially associated with Feynman and Drexler) is based on mechanosynthesis (chemistry with positional control), possibly facilitated by using pre-constructed nanoblocks as the elementary units of fabrication. A productive nanosystem is based on assemblers, devices that are themselves in the nanoscale; hence, the method is also known as bottom-to-bottom. Because of their minute size, the only practical way to fabricate large or large quantities of entities is for the assemblers to first assemble copies of themselves, which then all work in parallel. The practical realization of this vision is focused on tip-based methods inspired by the scanning probe instruments used in nanometrology; at present single objects comprising of the order of ten atoms can be made in this way.

Originally inspired by biology, a third approach is based on creating objects (which could be nanoblocks) capable of spontaneously assembling into useful structures. This method is known as bottom–up or self-assembly. This has been quite successful for creating regular structures (e.g. nanoporous membranes for separating vapors) but the creation of arbitrary geometries requires PSA. In other words bottom–up is presently good at creating materials but not for creating devices, although progress in the latter is occurring [47]. Biological nano-objects have this ability, but it is extraordinarily difficult to reverse-engineer them and use the knowledge to create synthetic analogues.

These three methods can sometimes be advantageously designed, for example, self-assembly could be used to make a mask for photolithography more cheaply and quickly than electron beam writing, which is then used in top–down fabrication (e.g. to create superhydrophobic surfaces).

FURTHER READING

1. A.W. Castleman Jr., S.N. Khanna, Clusters, superatoms and building blocks of new materials, J. Phys. Chem. C 113 (2009) 2664–2675.

2. M. Einax, W. Dieterich, P. Maass. Colloquium: Cluster growth on surfaces: Densities, size distributions, and morphologies, Rev. Modern Phys. 85 (2013) 921–939.

3. S. Franssila, Introduction to Microfabrication, second ed., Wiley, 2010.

4. H. Frauenfelder, From atoms to biomolecules. Helv. Phys. Acta 57 (1984) 165–187.

5. E. Kellenberger, Assembly in biological systems, in: Polymerization in Biological Systems, CIBA Foundation Symposium 7 (new series), Elsevier, Amsterdam, 1972.

6. A.G. Mamalis, A. Markopoulos, D.E. Manolakos, Micro and nanoprocessing techniques and applications, Nanotechnol. Perceptions 1 (2005) 63–73.

7. R.C. Merkle, Molecular building blocks and development strategies for molecular nanotechnology, Nanotechnology 11 (2000) 89–99.

8. P. Schaaf, J.-C. Voegel, B. Senger, Irreversible deposition/adsorption processes on solid surfaces, Ann. Phys. 23 (1998) 3–89.

9. J. Storrs Hall, Architectural considerations for self-replicating manufacturing systems, Nanotechnology 10 (1999) 323–330.

Carbon-based nanomaterials and devices

CHAPTER CONTENTS

INTRODUCTION

Carbon-based materials and devices are dealt with in this separate chapter because of their unique importance and versatility. They are widely considered to represent the epitome of nanotechnology.

Carbon has long been an intriguing material because of its two very well-known allotropes, diamond and graphite. They are so different from one another that if one did not know that they have the same elemental composition one might imagine that they are different elements. Although researchers working on carbon have long been aware of other forms, possibly dubbed "carbon filaments" and "atypical char" (along with "amorphous carbon"), typically these were regarded as a nuisance and discarded if their formation could not be avoided. The importance of the recent "discoveries" of fullerenes ("soluble carbon") and CNTs ("carbon filaments") resides in the fact that their structures were elucidated for the first time. Even nowadays, with ever-improving methods of industrial synthesis, yields of CNT and fullerenes are significantly less than 100%; the desired product is embedded in a gangue of

Nanotechnology: An Introduction. DOI: 10.1016/B978-0-323-39311-9.00015-7
Copyright © 2016 Elsevier Inc. All rights reserved.

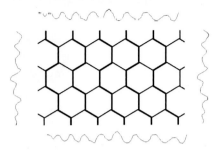

FIGURE 9.1

Part of a graphene sheet, showing the sp^2 chemical bonds. There is a carbon atom at each intersection. The length of each bond is about 0.3 nm.

uncharacterized or defective material that still merits the name "amorphous carbon". The most recent "discovery" is that of graphene, which is simply one of the many parallel sheets constituting graphite; here, the ingenuity resided in the preparation of a single isolated sheet, which opened the possibility of examining experimentally what was already a well-studied material theoretically.

The electron configuration of carbon is $1s^1 2s^2 2p^2$. Its four outer-shell electron orbitals are 2s, $2p_x$, $2p_y$ and $2p_z$; the four valence electrons may hybridize them into sp^1, sp^2 and sp^3, corresponding to a carbon atom bound to 2, 3 and 4 neighboring atoms, respectively. Diamond is composed entirely of sp^3 orbitals: it is ultrahard, a wide band-gap dielectric, has good thermal conductivity and is rather inert chemically. Graphite is composed entirely of sp^2 orbitals: it is soft, an electrical conductor, has highly anisotropic thermal conductivity and is very reactive chemically.

The three new materials, graphene, CNTs and fullerenes, can be called "nanocarbon". Like graphite, they are composed entirely of sp^2 orbitals, but fullerenes contain 12 pentagons and have some sp^3 character. They correspond to nanoplates, nanofibers and nanoparticles, respectively (see Fig. 6.2). Rolling up graphene (Fig. 9.1) into the smallest possible tube makes a single-walled CNT (SWCNT), and curling it up into the smallest possible sphere makes a fullerene – conceptually, that is; the materials are not actually fabricated that way. These nanomaterials have no bulk equivalents, discounting the fact that graphite is made up of an endlessly repeating stack of graphene.

CNTs have hitherto attracted the most commercial interest. Table 9.1 summarizes some of the remarkable properties of these materials. At present, they are most commonly commercially exploited by embedding them in a matrix to form some kind of composite.

Table 9.1 Some Properties of Bulk and Nanoscale Carbon Materials[a]

Property	Unit	Diamond	Graphite	CNT
Young's modulus	$N\,m^{-2}$	10^9	10^{10}	10^{12}
Thermal conductivity	$W\,m^{-1}\,K^{-1}$	2000	20	3000
Electrical resistivity	$\Omega\,m$	10^{12}	10^{-5}	$<10^{-6}$

[a]*The given values are only approximate, given in order to enable a rough comparative idea of the material properties to be formed. Actual measured values still depend on many experimental details.*

9.1 CARBON NANOPARTICLES (FULLERENES)

Fullerenes (also known as buckyballs) exist as C_{60}, C_{70}, C_{80}, etc. (Fig. 9.2). The canonical structure, C_{60}, has icosahedral symmetry and an electronic structure similar to that of graphene. They are soluble in organic liquids like toluene, each kind of fullerene giving a solution of a different color (e.g. C_{60} is violet, C_{70} is reddish-brown). This feature allows them to be purified, most popularly by high-performance liquid chromatography (HPLC).

Fullerenes are insulators like diamond. As for their thermal properties, in air they will start to oxidize when heated to about 300 °C but in an inert atmosphere they will sublime, the temperature depending on the pressure.

Fullerenes can be made in a carbon arc, but burning a hydrocarbon feedstock with strict control of the oxygen supply is a more controllable method. In the arc method, two graphite rods are placed with a small gap between them in an inert gas below atmospheric pressure. When the arc is struck, the carbon of the rods evaporates and migrates away, cooling and accumulating on the inner surface of the reaction chamber, from which the fullerene is collected [10], mixed with sort, from which it must be separated by dissolution. This method is only suitable for a small scale of production.

In the combustion method, a hydrocarbon fuel is burnt imperfectly under oxygen-poor conditions, generating soot. The flame conditions can be adjusted to maximize the yield of fullerene in the soot. This method has a high production capacity [10]. The soot is collected preferably continuously by filtration from the gas.

The possibility of modifying fullerenes by substitutionally doping the carbon lattice (e.g. with boron), adding an endohedral atom, or functionalizing the surface (e.g. with an amine, generally using standard methods of organic synthesis) greatly expands the ways in which fullerenes can be used [10, 50].

9.2 NANODIAMONDS

Nanodiamonds are small diamonds with a typical diameter from 4 to 5 nm. They readily aggregate and agglomerate into much larger powders. Nanodiamonds retain

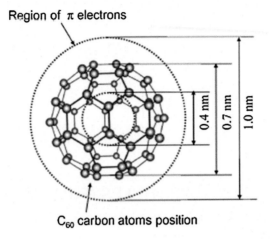

FIGURE 9.2

The structure and size of buckminsterfullerene, C_{60}. Note the presence of both hexagons and pentagons.

-Reproduced with permission from M. Arikawa, Fullerenes–an attractive nano carbon material and its production technology, Nanotechnol. Percept. 2 (2006) 121–114.

most of the properties of bulk material, such as extreme hardness, high thermal conductivity and high electrical resistivity. Typically nanodiamonds contain appreciable quantities (hundreds of ppm) of substitutional nitrogen. When the N is next to a vacancy in the diamond lattice, an NV center is formed, which is fluorescent.

Nanodiamonds were first produced by detonation in the 1960s. The explosives provide both the carbon source and the energy for converting it into diamond [210]. The explosion chamber can be filled either with an inert gas ("dry" synthesis) or with water (" wet" synthesis). The detonation soot may contain up to 75% w/w of diamond. Thanks to the chemical stability of the diamond, the nondiamond carbon can be removed by quite harsh conditions, such as nitric acid, chromic acid, potassium hydroxide solution and the like.

The aggregates of nanodiamond may be disintegrated using methods such as micromilling, although great care must be taken with the choice of media and conditions to prevent contamination. Nanodiamonds can be readily surface-functionalized [210].

9.3 CARBON NANO-ONIONS

Nano-onions (onion-like carbon, OLC) are, essentially, nested fullerenes. They were clearly visible in electron micrographs published by Iijima in 1980 [144] and can be synthesized by annealing nanodiamonds at a high temperature (2000°C), at which an inert atmosphere is required [191]. Another method is arc discharge between two graphite electrodes in water. Hybrid metal–carbon structures can be synthesized using arc discharge; exposure to an electron beam causes the metal core to migrate through the carbon shell, leaving a hollow OLC particle. Yet another method is laser excitation to decompose a hydrocarbon precursor such as ethylene. This is probably the highest-throughput method currently available.

Post-synthesis processing is required for applications. Chemical activation in order to prepare the nano-onions for use in supercapacitors can be accomplished using potassium hydroxide [93].

9.4 CARBON NANOTUBES (CNT)

The synthesis and characterization of CNTs was first reported in the scientific literature by Iijima in 1991, although they are clearly visible in electron micrographs

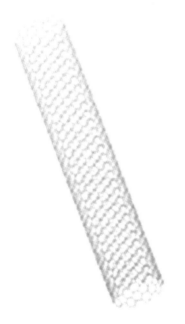

FIGURE 9.3

A SWCNT: a single graphene layer rolled into a seamless tube.

-*Reproduced with permission from B.O. Boscovic, Carbon nanotubes and nanofibres, Nanotechnol. Percept. 3 (2007) 141–158.*

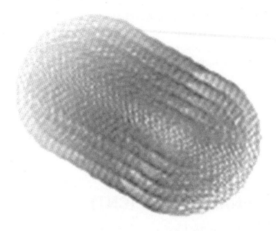

FIGURE 9.4

A MWCNT: concentric single-wall nanotubes of different diameters nested within each other.

-Reproduced with permission from B.O. Boscovic, Carbon nanotubes and nanofibres, Nanotechnol. Percept. 3 (2007) 141–158.

published much earlier [242]. Since the 1990s they have been intensively studied both theoretically and experimentally.

It was long known that carbon filaments are formed by passing hydrocarbons over hot metal surfaces, especially iron and nickel. The actual nature of CNTs was, however, only established relatively recently (by Iijima in 1991): they are seamlessly rolled up tubes of graphene (Fig. 9.3). The ends of the tubes may be open or "capped" with what is essentially a hemisphere of fullerene. Multiwall CNTs (MWCNTs) most typically consist of several concentric tubes of graphene nested inside each other (Fig. 9.4). A form in which graphene is rolled up to give a spiral cross-section is also known.

An important feature of SWNTs is the chiral (or wrapping or rollup) vector $\mathbf{C}_\chi$, defined as

$$\mathbf{C}_\chi = n\mathbf{a}_1 + \mathbf{a}_2 \equiv (n, m), \tag{9.1}$$

where the unit vectors $\mathbf{a}_1$ and $\mathbf{a}_2$ have, respectively, the directions from the ith to the $i + 2$th carbon on any hexagon, and from the ith to the $i + 4$th carbon; $|\mathbf{a}_1| = |\mathbf{a}_2| = \sqrt{3}a_{C-C} = 0.246$ nm, where a_{C-C} is the nearest-neighbor distance between two carbon atoms, and thus the magnitude of $\mathbf{C}_\chi$ is $0.246\sqrt{n^2 + nm + m^2}$; n and m are integers with $m \leq n$. The chiral vector connects crystallographically equivalent sites on the graphene lattice; n (≥ 3) ("twist") and $0 \leq m \leq n$ ("countertwist") are the integers required to ensure translational symmetry of the lattice. $\mathbf{C}_\chi$ defines the disposition of the hexagonal rings of the CNT or in other words describes how the graphene sheet is wrapped. If $n = m$ the SWNT is called "armchair"; if $m = 0$ then the SWNT is called zig-zag; otherwise, it is simply called chiral. The diameter in nanometers is given by

$$d = C_\chi/\pi;$$ (9.2)

the chirality by

$$\chi = \sin \pi(1 - a_2/a_1);$$ (9.3)

and the chiral angle is

$$\theta_\chi = \arcsin \frac{\sqrt{3}m}{2\sqrt{n^2 + nm + m^2}}.$$ (9.4)

The ratio a_2/a_1 provides a measure of helicity; it can range from zero (zig-zag) to 1 (armchair). The chirality ranges from zero (for zigzag and armchair nanotubes) to 1 if $a_2/a_1 = 0.5$. The corresponding chiral angles are 0 for armchair and 30 degrees for zigzag.

The distance between the walls in MWCNTs is about 0.34 nm, similar to the distance between graphene layers in graphite.

SWCNTs change their electrical properties depending on $\mathbf{C}_\chi$. If $n = m$, then the nanotube is metallic and can bear an electrical current density more than 1000 times greater than metals such as silver and copper. Conductance is ballistic, i.e. electrons can pass through the nanotube without heating it. If $n - m = 3$, then the nanotube is semiconducting. Otherwise the nanotube is a moderate semiconductor.

Unlike nanotubes made from many other materials, which tend to have a lower thermal conductivity than the bulk, CNTs have exceptionally high thermal conductivity. This is presumed to be due to the long-range crystallinity, the long phonon mean free path and the high velocity of sound [199].

The three methods for producing CNTs are the laser furnace, the carbon arc (i.e. vaporizing graphitic electrodes) and (plasma-enhanced) CVD. Fig. 9.5 shows a CVD setup in its simplest form, and Figs 9.6 and 9.7 show some results from plasma-enhanced CVD (PECVD). The last-named has turned out to be the best for large-scale

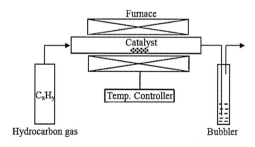

FIGURE 9.5

Schematic diagram of a CVD apparatus, showing only the most basic elements.

-*Reproduced with permission from M. Kumar, Y. Ando, Carbon nanotube synthesis and growth mechanism, Nanotechnol. Percept. 6 (2010) 7–28.*

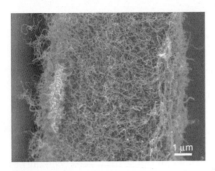 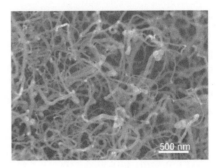

FIGURE 9.6

Scanning electron micrographs of CNTs grown on the surface of a carbon fiber using thermal CVD. The right-hand image is an enlargement of the surface of the fiber, showing the nanotubes in more detail.

-*Reproduced with permission from B.O. Boscovic, Carbon nanotubes and nanofibres, Nanotechnol. Percept. 3 (2007) 141–158.*

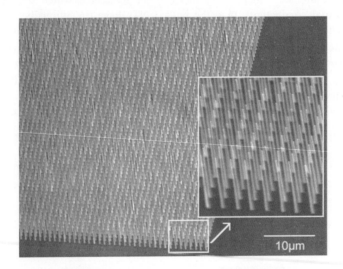

FIGURE 9.7

A forest of CNTs produced by PECVD. The substratum must first be covered with metal (e.g. Fe or Ni) catalyst islands. Hydrocarbon feedstock (acetylene) is then passed over the substratum heated to several hundred °C.

-*Illustration courtesy of Dr Ken Teo, AIXTRON.*

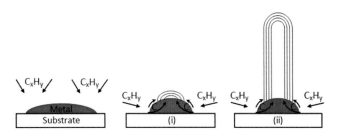

FIGURE 9.8

The base-growth model for catalytically assisted CNT growth.
- *Reproduced with permission from M. Kumar, Y. Ando, Carbon nanotube synthesis and growth mechanism, Nanotechnol. Percept. 6 (2010) 7–28.*

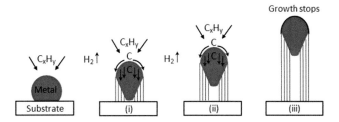

FIGURE 9.9

The tip-growth model for catalytically assisted CNT growth.
- *Reproduced with permission from M. Kumar, Y. Ando, Carbon nanotube synthesis and growth mechanism, Nanotechnol. Percept. 6 (2010) 7–28.*

production of relatively pure material. The process involves passing hydrocarbon vapor (acetylene is a popular choice) through a tubular reactor containing catalyst and maintained at a sufficiently high temperature (600–1200 °C) to decompose the vapor [172]. The method is quite versatile and liquid hydrocarbons can also be used, carried into the chamber by an inert gas. An advantage of using ethanol as feedstock is that the CNTs are almost wholly free from amorphous carbon, since OH radicals are generated and etch impurities away. The molecular structure of the precursor has a surprising influence on the morphology of the CNTs. Whereas linear hydrocarbons such as acetylene generally produce straight CNTs, cyclic hydrocarbons such as benzene or cyclohexane tend to produce curved CNTs with internal bridges.

Two possible mechanisms are favored for the catalytically assisted growth of the CNTs: the hydrocarbon feedstock decomposes at the surface of the catalyst and the CNTs grow up from the catalyst particle (Fig. 9.8), or grow up beneath it, pushing it up (Fig. 9.9) [172]. Metal contamination from the catalyst particles is almost inevitable, and must be removed by post-synthesis purification.

Carefully tuning the reaction conditions can favor one chirality or the other of the nanotubes [263]. Most technical applications require such selectivity. Note that CNTs are often closed at one or both ends by a hemisphere of fullerene. It would be very advantageous if no metallic catalysts were required, since residues of the metal are nearly always present as impurities.

Major problems remain with the large-scale utilization of CNTs. The most severe are as follows:

- Making pure preparations;
- Dispersing them in solvent (since they can scarcely be solvated (cf. Section 3.2), they tend to be strongly aggregated into bundles);
- Reducing their length (a 20 nm diameter tube may be 20 μm long as fabricated, unnecessary for many applications);
- Manipulating them into a desired position.

Post-synthesis processing of nanotube material therefore typically requires the following:

- Purification – methods includes thermal annealing in air or oxygen; acid treatment, microfiltration – which typically reduce the mass by around 50%;
- De-agglomeration to separate the tubes (dispersion) [106]. Methods include ultrasonication (but this can damage the tubes), electrostatic plasma treatment, electric field manipulation and polymer wrapping, ball milling (can damage the tubes); these methods can also reduce their length;
- Chemical functionalization (with electron-donating or electron-accepting groups) to improve interactions with a solid or liquid matrix.

As well as chemical functionalization, the lumen of CNTs may also be filled with different materials. Their use as templates for nanorods or nanotubes has already been mentioned.

9.5 CARBON NANOFIBERS (CNF)

CNFs may be thought of as a precursor to CNTs. It appears that they were already being synthesized in the 1970s, although more careful characterization had to await the 1980s [287]. Similarly to CNTs, they are produced by catalytically assisted pyrolysis of hydrocarbon vapors. The catalyst consists of metallic nanoparticles, which nucleate fiber growth.

Although academic interest has passed from CNF to CNT, the former may be of greater industrial interest, not least because of the persistent difficulties in producing pure CNTs.

CNFs can be produced by electrospinning an organic polymeric precursor such as polyacrylonitrile followed by plasma techniques. In an effort to reduce the existing cumbersome and expensive fabrication routes, Ren et al. have recently proposed a "one-pot" synthesis from carbon dioxide dissolved in molten carbonates [258].

9.6 GRAPHENE

Inspired by learning about naphthalene and anthracene, countless schoolchildren have doubtless doodled endless fused polyaromatic rings. It has long been known that graphite is composed of stacks of such polyaromatic sheets, which are called graphene. Due to convincing theoretical work, it was however long believed that two-dimensional crystals cannot exist. The ostensive demonstration of their existence (graphene sheets) has, *post hoc*, led to the explanation that their stability is due to undulations of the sheet.

The graphene lamellae stacked to make bulk graphite were, from the ease of their detachment (e.g. writing with graphite on paper), known to be only weakly bound to each other. Individual sheets of graphene can actually be peeled off graphite using adhesive tape. Alternatively, a crystal of silicon carbide can be heated under vacuum to 1300 °C; the silicon evaporates and the remaining carbon slowly reorganizes to form some graphene. Graphene appears to have been made in the 1960s by thermal decomposition of acetylene or ethylene on a hot surface [204].

The electronic and thermal properties of graphene are even more remarkable than those of CNT. It is, however, more difficult to manufacture in large quantities. Much effort is currently directed towards high volume (preferably continuous) production of high-quality (with a low concentration of defects) graphene.

Numerous schemes have been investigated. The starting material has often been graphite, from which graphene can be exfoliated. Even ultrasonic agitation in the presence of an organic solvent can yield some usable material. Intercalation of graphite also seems to hold some promise. The into collated material is then decomposed either thermally or with microwaves, followed by chemical and/or mechanical treatment. This route yields "graphene" nanoplatelets. However, they are usually at least one and sometimes several nanometers thick, hence are not true graphene.

CVD using hydrocarbon feedstock (e.g. ethylene) mixed with hydrogen appears to be the most promising route for producing monolayer material. The deposition takes place on a flexible substrate. A concentric tube reactor with limited residence time appears to have been quite successful [239].

9.7 MATERIALS APPLICATIONS

Carbon black. This is a relatively old-established product that even today constitutes the overwhelming bulk of the nanotechnology industry (if it is included as part of it, as is done by many market analysts). It consists of carbon particles ranging in size from a few to several hundred nanometers; the bulk of the mass is therefore present in the form of particles bigger than the consensual upper limit of nanoparticles. It is added to the rubber tires for road vehicles as reinforcing filler; the volume of its manufacture is thus directly related to the popularity of such vehicles. Substitution by fullerenes (its true nano equivalent), cf. Fig. 1.3, would not be an economical proposition (except perhaps in Formula 1), even if the properties of the tires were significantly enhanced, which seems rather unlikely.

Polymers with dispersed CNTs and graphene. Dispersion of CNTs in polymer composites may improve their strength, stiffness, and thermal and electrical conductivities.

Several processing methods are available. Examples such as melt mixing, extrusion, internal mixing, injection molding techniques are all used free of solvents and contamination found in solution processing. There is less fiber breakage than in conventional carbon fibers. High shear mixing is required as the increase in viscosity is greater for nanotubes than for nanoparticles. Controlled alignment is possible using spinning extrusion injection molding. *In situ* polymerization and solution processing have also been used (cf. Section 6.6).

The strength improvement depends on the degree of load transfer and on the degree of dispersion achieved in the matrix. Additions of nanotubes from 1% to 5 vol% can produce increases in stiffness and strength in polypropylene (PP), polystyrene, polycarbonate and epoxies of up to 50% in stiffness and 20% in strength. If the nanotubes are aligned perpendicular to a crack they can slow down propagation by bridging crack faces. This occurs notably in carbon epoxy composites. In some polymers, nanotubes may decrease the coefficient of friction and decrease the wear rate.

Improvements in electrical properties are dramatic even at very low volume fractions since due to their strong affinity for each other the nanotubes tend to form networks rather than being randomly dispersed. Even without that effect, however, their extreme elongation would enable percolation to be achieved at low vol% additions (cf. Section 3.7). Resistivities of 1–100 Ω cm are produced by 1–3 vol% dispersed in PP or nylon. Nevertheless, results are often less spectacular than predicted theoretically because the nanotubes tend to be clustered together in parallel bundles due to inadequate dispersion.

Typical applications for these materials include electrically conducting paint, conducting polymer structures, lighter stiffer structures (the ultimate application of which would be the space elevator), heat sinks for electronics, motor components and smart polymer coatings.

The above trends have been taken up and mimicked by incorporating graphene into polymers. In reality graphene is not actually used, but the relatively low-cost graphite nanoplatelets produced by the intercalation process mentioned in Section 9.6. Since the nanoplatelets are not particularly elongated the percolation threshold is higher than for CNT, but still much lower than for nanoparticles.

9.8 CNT YARN

As fibers, it is potentially possible to spin CNTs into a macroscopic, flexible yarn [315, 316] of great tensile strength and low electrical and thermal resistivity. Within the yarn, each nanotube makes thousands of contacts with other tubes. Electrical resistivity of about 3×10^{-5} Ω m has been achieved, but this is still three orders of magnitude greater than that of copper. The main problem is the mixture of metallic and semiconducting nanotubes. A process of iodine-doping the CNT has been reported [317], which enabled the resistivity to be diminished to about 10^{-7} Ω m; these yarns could carry a current density of about 3×10^4 A/cm^2.

9.9 DEVICE COMPONENTS AND DEVICES

Nano field emitters. Due to their extremely high curvature, CNTs can emit electrons at much lower voltages (a few volts) compared with conventional field emission devices. The principle envisaged application is in flat display screens, competing with liquid crystal technology. They are also attractive as electron guns for SEMs, in high-power microwave amplifiers, and for miniature X-ray sources. CNT-based electron guns for electron microscopes are undoubtedly the best available (but the global market is insignificant in terms of the volume of CNTs required). *In situ* growth on a flat substrate (Fig. 9.7) might diminish the manipulation required to create the final device in this application.

Nanocomposite transparent electrodes. Display technology is inevitably going to have to change in the near future because of the global dearth of indium, presently used for doping tin oxide (at a fairly high level, of the order of 10%) to create electrically conducting, transparent "indium tin oxide" (ITO) thin films on glass, which are used as the counterelectrode in display devices. Annual consumption is around 800 tonnes, yet the total known reserves are less than 3000 tonnes. As much recycling as possible is carried out, but extraction of the indium is becoming increasingly difficult as devices become smaller and more integrated. Due to the very low percolation threshold of highly elongated objects dispersed in a matrix, CNT-doped polymers can be made adequately conductive at levels low enough for the material to remain transparent, which should therefore be able to replace current ITO-based conducting glass technology. CNT-doped polymers are also attractive for organic photovoltaic solar cells and other optoelectronic devices, in which for material compatibility reasons everything should preferably be organic.

Electrical interconnects. The high electrical conductivity of individual CNTs makes them attractive for use in VLSI electronic circuits as connectors between components, especially vertical ones ("vias") to connect stacked layers. Presumably they would be grown in situ.

Thermal interfacing and rectification. Thermal applications of CNT are more attractive than electrical ones, because the extraordinarily high thermal conductivity does not appear to strongly depend on the chirality. Existing technology is based on polymeric greases and metallic solders; a CNT array could have a lower elastic modulus (which is advantageous to accommodate mismatch in the coefficients of thermal expansion of the hot material and the heat sink) than the greases and a lower thermal resistance (hence a better heat sink) than the solders. There is some evidence that the thermal conductivity of CNT depends on the direction of heat transport, potentially enabling rectification, which appears to show a slight dependence on chirality.

Condensers. Use in electrochemical capacitors would be a large-volume application, but supercapacitors are already being made with existing materials, notably the much cheaper carbon black, which already offers performance close to the theoretical limit.

Logic circuits. Logic components, such as single electron tunneling transistors (Section 7.4.3), have been made based entirely on carbon nano-objects. In a veritable tour de force a CNT computer has been demonstrated [267], although the claim that this might be the basis for the next generation of information processors seems overly optimistic, not least since it is difficult to see how the method of making the transistors could be automated.

SUMMARY

Buckminsterfullerene, the SWCNT and graphene epitomize, respectively, nanoparticles, nanofibers and nanoplates. The full potential of their remarkable properties, especially electronic ones, in devices will, however, have to await the development of viable nano-object manipulation technologies. Nevertheless, as fillers in composites these materials are already being used industrially.

FURTHER READING

1. K.H. An, Y.H. Lee, Electronic-structure engineering of carbon nanotubes, NANO: Brief Rep. Rev. 1 (2006) 115–138.
2. D.N. Basov, et al., Colloquium: Graphene spectroscopy, Rev. Moden Phys. 86 (2014) 959–994.
3. Z.F. Wang, et al., Emerging nanodevice paradigm: graphene-based electronics for nanoscale computing, ACM J. Emerging Technol. Comput. Syst. 5 (2009) (1), article 3.

Nanosystems and their design

10

CHAPTER CONTENTS

INTRODUCTION

This chapter comprises a miscellany of topics, because nanotechnology is not yet so far advanced as to comprise systems to any meaningful extent. Key elements in developing a viable industrial system are explored.

10.1 SYSTEMS

The essence of a system is that it cannot be usefully (for the purposes of analysing its function or optimizing its design) decomposed into its constituent parts. Two or more entities (or activities) constitute a system if the following four conditions, enumerated by R.L. Ackoff, are satisfied:

Nanotechnology: An Introduction. DOI: 10.1016/B978-0-323-39311-9.00016-9
Copyright © 2016 Elsevier Inc. All rights reserved.

1. We can talk meaningfully of the behavior of the whole of which they are the only parts;
2. The behavior of each part can affect the behavior of the whole;
3. The way each part behaves and the way its behavior affects the whole depends on the behavior of at least one other part;
4. No matter how we subgroup the parts, the behavior of each subgroup will affect the whole and depends on the behavior of at least one other subgroup.

The word "nanosystem" can be defined as a system whose components are in the nanoscale.

10.2 HAIRY ATTACHMENT DEVICES

An example of a system that might justifiably be called "nano" is the foot of many insects, arachnids, lizards and the like, which can run up vertical walls and upside down across ceilings. The subject has long been studied; it was already considered "hackneyed" in 1862 [299]. Nevertheless, it is only in our present area that a systematic understanding of the mechanism of attachment and detachment has been achieved. Interestingly and intriguingly, the basic principles appear to have evolved separately and independently on several occasions in the history of living creatures (Fig. 10.1).

The feet are hierarchically divided, into ever smaller and larger numbers of fibers. This is a kind of nanification, with the terminal pads being as small as 100 nm in diameter in some cases. Apart from conferring conformability, there is also an elastic modulus benefit due to the splitting [234].

The termination is either a smooth pad or a hairy one (Fig. 10.1) [101]. In some cases (e.g. the gecko and spiders) attachment of the hairs to the surface being walked on is considered to be purely due to Lifshitz–van der Waals forces (Section 3.2), although of course electron donor–acceptor and Coulombic forces may also be present. This is the "dry" adhesive mechanism. On the other hand, many insects secrete an adhesion-promoting fluid through small pores in the terminal plate. This is the "wet" adhesive mechanism [102].

Attempts to mimic the foot with a synthetic nanostructure have only had very limited success, because the real foot is living and constantly adjusted to maintain the close range conformal contact needed for the interaction to be sufficiently strong to bear the weight of the creature, whereas the synthetic foot is static, and typically irreversibly damaged merely upon detachment. Each nanoscale footlet (the smallest subdivision, the part that is actually in contact with the surface) is only part of a system in the living creature, whose brain is presumably involved in maintaining adhesion.

10.3 MATERIALS SELECTION

When confronted with a macroscale design problem, one may use charts of the type shown in Fig. 10.2 to select a suitable material fulfilling the functional require-

FIGURE 10.1

Hairy (fibrillar) adhesion of the fly (upper row) and the gecko (lower row). On the right are scanning electron micrographs of the termination of the feet (adhesive setae). The fly exemplifies a "wet" adhesive system and the gecko a "dry" one (see text).

−Courtesy Prof. Stanislav N. Gorb, University of Kiel.

ments. Complex problems may impose more than two constraints on properties; hence, many such diagrams may be required, since it is visually problematic to construct them in more than two dimensions. However, this procedure imposes some decomposability on the problem, which is contrary to the spirit of it being a system (Section 10.1). Furthermore, the decomposition is arbitrary, there being numerous possible binary combinations of properties, which may influence the design choices in different ways. Finally, we notice that the entire property space is by no means

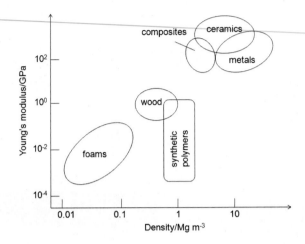

FIGURE 10.2

An example of a materials selection chart used in classical engineering design at the macroscale.

comprehensively covered. Light and strong, or heavy and weak materials remain elusive, for example.

The engineer working in the nanoscale has, in principle, far vaster possibilities available. These will only be fully realized with bottom-to-bottom nanofacture (Section 8.3), but even nanocomposites (Section 6.6) greatly expand the typically small region occupied by classical composites in diagrams like Fig. 10.2.

As components become very small, the statistical nature inherent in present-day "top–down" methods (including the fabrication of a composite by blending nanoparticles into a polymer matrix) becomes more apparent. Ultraprecision engineering mostly achieves its high machining accuracy by rigorous control of the distributions of control errors making additive contributions to the final result, but ultimate nanoprecision is only achievable within a eutactic environment, via bottom-to-bottom nanofacture.

Similar arguments apply to semiconductor processing. Standard semiconductor devices depend on doping (i.e. the introduction of impurities or defects into the pure, perfect semiconductor) for the setting the controlling framework for the mobile electron population. The doping level is typically quite low, hence as working volumes shrink a statistical distribution of dopants is no longer adequate (Section 10.4). Again, the solution is assembly within a eutactic environment, in which each dopant can be placed in a position precisely defined with respect to atomic coördinates.

Charts like Fig. 10.2 really only deal with quantities that are scalars; that is, they can be quantified on a straightforward linear or logarithmic scale, such as density. Anisotropic materials with direction-dependent moduli, electrical conductivities, etc.

require separate charts according to the orientation of the material. Esthetic qualities are not even considered because of their complexity. Even color requires three dimensions; other attributes such as "feel" are more difficult still; indeed there is still no standard way of representing them. Yet the history of technology starting from the earliest known man-made artifacts shows that purely decorative objects designed to be appreciated esthetically have always preceded objects of utility, including weapons and tools, and even entities of a strongly utilitarian nature (such as a motor-car) are often chosen on the basis of esthetic attributes, provided the minimum requirements of the utility function are met. This must be borne in mind by the designer working in the nanoworld.

10.4 DEFECTS IN NANOGRAINS

The simplest kinds of grains (nanoparticles) are made from a single material, e.g. germanium or silicon. Generally, the only issues are whether the surface has a different chemical composition, either due to the adsorption of impurities from the environment or due to the generation of lattice defects at the surface. In the bulk, the two kinds of defects are vacancies and interstitial ions.

In compound materials, formed from a cation C and an anion A, the occurrence of defects is governed by the constraint of the need to maintain electrical neutrality, i.e. vacancies must be compensated by interstitial ions of the same sign (Frenkel type) or by vacancies of the opposite sign (Schottky type). More specifically, there may be cations on interstitial sites and vacancies in the cation half-lattice (Frenkel) or anions on interstitial sites and vacancies in the anion half-lattice (anti-Frenkel); cations and anions on interstitial sites (anti-Schottky) or vacancies in the cation and anion half-lattices (Schottky). Compounds can be created deliberately nonstoicheiometric, e.g. by growth of a granular metal oxide film via PVD in an oxygen-poor atmosphere, thereby assuring an abundance of defects.

Notation (Schottky). The superscript represents the electrical charge of the defect, relative to the undisturbed, overall electrically neutral lattice: $\prime, \cdot, \times$ represent negative, positive and neutral (zero) excess electrical charge. Subscript $\square$ is a vacancy, subscript $\square$ is an interstitial. Hence, $C_{\circ}^{\cdot}$ is an interstitial cation, and $C_{\square}^{\prime}$ is a cation vacancy, "null" signifies the undisturbed lattice, and CA represents the addition (at the surface of the nanoparticle) of a new lattice molecule. A full circle is used to represent the substitution of a native ion with a foreign one, e.g. $X_{\bullet}^{\cdot}(C)$ would be a divalent cation X on a cation lattice site, $Cu_{\bullet}^{\times}(Ag)$ would be (mono)valent copper on a silver lattice site in (say) AgCl, and so forth.

Relations between defects. The following relations are possible [119]:

$$C_{\circ}^{\cdot} + C_{\square}^{\prime} \rightleftharpoons \text{null}, \tag{10.1}$$

$$A_{\circ}^{\prime} + A_{\square}^{\cdot} \rightleftharpoons \text{null}, \tag{10.2}$$

$$C_{\circ}^{\cdot} + A_{\circ}^{\prime} \rightleftharpoons CA. \tag{10.3}$$

The above three equations can be used to derive some more,

$$\text{null} \rightleftharpoons C'_\square + A^{\cdot}_\square + CA, \tag{10.4}$$

$$C^{\cdot}_\circ \rightleftharpoons CA + A^{\cdot}_\square, \tag{10.5}$$

$$A'_\circ \rightleftharpoons CA + C'_\square. \tag{10.6}$$

To each of these there corresponds a mass-action law (MAL), i.e. for the last three, writing x for the mole fraction, identified by its subscript,

$$x_{C'_\square} x_{A^{\cdot}_\square} = K, \tag{10.7a}$$

$$x_{C^{\cdot}_\circ}/x_{A^{\cdot}_\square} = K_1, \tag{10.7b}$$

$$x_{A'_\circ}/x_{C'_\square} = K_2, \tag{10.7c}$$

where the K are the equilibrium constants. The defect concentrations are supposed to be sufficiently small so that the chemical potentials μ depend on the logarithm of x, e.g. for vacancies,

$$\mu_\square - E_\square = -RT \ln x_\square, \tag{10.8}$$

where $E_\square$ is the energy to annihilate one mole of vacancies. Of course the usual electroneutrality condition holds:

$$x_{C^{\cdot}_\circ} - x_{C'_\square} = x_{A'_\circ} - x_{A^{\cdot}_\square}. \tag{10.9}$$

Impurities. This formalism can be powerfully applied to the effect of impurities on defects. As examples let us consider

$$NaCl \rightleftharpoons Na^{\times}_\bullet(Ag) + AgCl \tag{10.10}$$

(silver chloride doped with sodium chloride): here there would be practically no change in defect concentration as a result of the doping, but if it is with divalent cadmium, there would evidently be a significant increase in silver vacancies and a decrease in interstitial silver ions;

$$CdCl_2 \rightleftharpoons Cd^{\cdot}_\bullet(Ag) + Ag'_\square + 2AgCl; \tag{10.11}$$

and

$$Ag^{\cdot}_\circ + CdCl_2 \rightleftharpoons Cd^{\cdot}_\bullet(Ag) + 2AgCl. \tag{10.12}$$

In these mixed phases, the electroneutrality condition becomes

$$x_{C^{\cdot}_\circ} - x_{C'_\square} = x_{A'_\circ} - x_{A^{\cdot}_\square} + y, \tag{10.13}$$

where y is the lattice concentration of the dopant. This relation, together with the three equations (10.7), can be used to derive explicit relations for the concentration of each of the four defect types in Eq. (10.4) on y.

Surfaces. In very small particles, a significant fraction of the atoms may be actually surface atoms. Because of their different bonding, the surface atoms may be sources or sinks of defects, thereby perturbing equilibria (10.1)–(10.3). The adsorption of impurities at the surface of a nanoparticle will generate defects. For example, adsorption of two monovalent Ag^+ onto the surface of a ZnS nanoparticle will require the formation of a zinc vacancy in the cation sublattice. The environment of the nanoparticle (the matrix) thus plays a crucial rôle in determining its properties.

10.5 SPACIAL DISTRIBUTION OF DEFECTS

If p is the probability that an atom is substituted by an impurity, or the probability of a defect, then the probability of exactly k impurities or defects among n atoms is

$$b(k;n,p) = \binom{n}{k} p^k q^{n-k}, \qquad (10.14)$$

where $q = 1 - p$. If the product $np = \lambda$ is of moderate size (~ 1), the distribution can be simplified to

$$b(k;n,p) \approx \frac{\lambda^k}{k!} e^{-\lambda} = p(k;\lambda), \qquad (10.15)$$

the Poisson approximation to the binomial distribution. Fig. 10.3 shows an example. Hence, the smaller the device, the higher the probability that it will be defect-free. The relative advantage of replacing one large device by m devices each $1/m$th of the original size is $m^{1-k} e^{np(1-1/m)}$, assuming that the nanification does not itself introduce new impurities.

10.6 STRATEGIES TO OVERCOME COMPONENT FAILURE

Similar reasoning to that of Section 10.5 can be applied to the failure of individual components (transistors, etc.) on a processor chip, which become "defects". Eq. (10.14) can be used to estimate likely numbers of failures, at least as a first approximation, considering them to all occur independently of each other. As the number of components on a "chip" is increased, instead of applying more rigorous manufacturing standards to reduce the probability of occurrence of a defective component, it may become more cost-effective to build in functional redundancy, such that failures of some of the components will not affect the performance of the whole. One strategy for achieving that is to incorporate the means for their failure to be detected by their congeners, which would switch in substitutes. One of the reasons for the remarkable robustness of living systems is the exploitation of functional redundancy, in neural circuits and elsewhere, although the mechanisms for achieving it are obscure. The oak tree producing millions of acorns each year, of which a mere handful may survive to maturity, provides another example of functional redundancy.

Standard techniques for achieving reliable overall performance when it is not possible to decrease the number of defects include the NAND multiplexing architecture proposed by John von Neumann [221] and simple approaches reminiscent of overcoming errors in message transmission by repeating the message and instructing

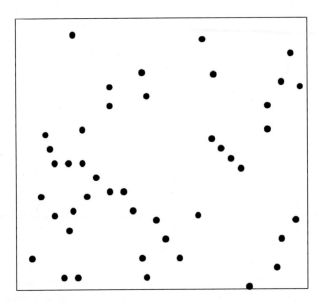

FIGURE 10.3

A two-dimensional piece of matter (white) in which dopants (black) have been distributed at random.

the receiver to apply a majority rule. Thus instead of one logic gate, R gates work in parallel and feed their input into a device that selects a majority output. This is called R-fold modular redundancy (RMR), R being an odd number. For even higher reliability the redundancy can be cascaded; thus, one might have three groups of three modules feeding into a majority selector, feeding into a further majority selector. A more sophisticated approach is to arrange for the circuit to be reconfigurable [221]. In all these approaches, the higher the probability of a component being defective, the greater the redundancy required to achieve reliable computation.

10.7 COMPUTATIONAL MODELING

A general rule of numerical modeling or simulation is that it should only be undertaken if a result can be obtained faster than by an actual physical experiment. The enormous growth in computational power in recent years has greatly reduced what was, in the past, one of the principal barriers to undertaking numerical modeling of materials, devices and processes. At the same time, interest in nanotechnology has engendered a host of possible devices in the nanoscale, which cannot yet be fabricated. Modeling allows their properties to be explored ahead of their physical realization and may help to decide whether the sometimes great investment in hardware required to actually make the device is worth undertaking. This endeavor has been particularly fruitful, with a great proliferation of work and results, because the small size of nanodevices means that in many cases explicit simulation of every constituent atom is possible.

When it comes to advanced manufacturing, it is particularly important to have a simulator of the material or device being manufactured, in order to ensure that the device is, as far as possible, "right first time"; and also to facilitate reverse engineering during development.

10.7.1 METHODS

Materials are typically modeled statically (at 0 K), based on lattice energies computed from interatomic potentials. "Pick-and-place" nanofabrication (Section 8.3.1) in principle enables an almost unlimited variety of materials to be created. At present the technology is still rudimentary and only capable of fabricating minuscule quantities of novel materials; hence, the ability to predict what the properties of the material would be, particularly whether it would be stable, is very attractive.

Of equal importance is the determination of synthetic pathways to novel materials. Especially in the realm of carbon-based molecules, one does not necessarily have to begin with the atoms. The target structure can be broken down into readily available synthons, implying a great saving of synthetic effort [189]. Fig. 10.4 provides an illustration.

In other cases, the target is not a structure but a property or set of properties. Even for molecules of modest size, the chemical compound space (CCS) is too vast to enable a systematic exploration of each point within it. Inspired by the notion of quantitative structure–property relationships (QSPR) the method of using gradients of key attributes with respect to certain dimensions of CCS has been developed to enable the rational selection of a compound exhibiting the target property [183].

When the system under consideration is small, containing only a few dozen or hundreds of atoms, *ab initio* quantum mechanical methods, in which the many-body Schrödinger equation is solved numerically, can be applied. DFT, which defines a molecule's energy in terms of its electron density, is a current favorite for calculating the electronic structure of a system. This may be used for designing a nanosystem – for example, a prototype memory storage system based on magnetoelectric coupling [96]. Fig. 10.5 shows an example of a DFT simulation.

For larger systems, containing hundreds of thousands to a thousand million atoms, the main atomistic simulation technique is molecular dynamics: the physical system of N atoms is represented by their atomic coördinates, whose trajectories $x_i(t)$ are computed by numerically integrating Newton's second law of motion

$$m_i \mathrm{d}^2 x_i / \mathrm{d}t^2 = \partial V / \partial x_i, \tag{10.16}$$

where m_i is the mass of the ith atom and V is the interatomic potential. A general weakness of such atomic simulations is that they use predefined empirical potentials, with parameters adjusted by comparing predictions of the model with available

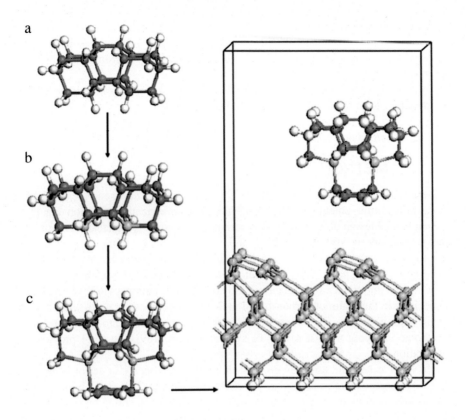

FIGURE 10.4

Diadamantane tool design for the manipulation of benzene, the basic synthon of 4-mercaptophenyl phenylacetylene (MPPA). White spheres represent hydrogen atoms, dark gray spheres carbon, and lightly shaded spheres silicon.

-Reproduced with permission from D.Q. Ly, et al., The Matter Compiler—towards atomically precise engineering and manufacture, Nanotechnol. Percept. 7 (2011) 199–217.

experimental data. For much of the work in nanodevices, it is not possible to give a general guarantee of the validity of this approach because no complete nanodevices of the type being simulated have as yet been constructed; hence, the output of the simulations cannot be verified by comparison with experiment. On the other hand, if a materials formation process, such as a complex crystallization, is being simulated the material is likely to already exist, the purpose of the simulation being to better understand the formation process [76]. Similarly, molecular dynamics can be used to better understand physical properties of nanomaterials [139].

Nevertheless, it is attractive that a nanodevice is small enough for it to be possible to explicitly simulate its operation with atomic resolution (molecular

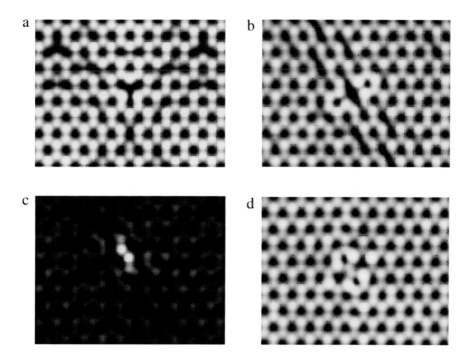

FIGURE 10.5

Simulated STM images (tip height equals 0.80 Å) of a defective graphitic surface: (a) monovacancy; (b) first neighbor divacancy; (c) third divacancy; and (d) Stone–Wales defect.

-Reproduced with permission from D.Q. Ly, et al., The Matter Compiler—towards atomically precise engineering and manufacture, Nanotechnol. Percept. 7 (2011) 199–217.

dynamics), using present-day computing resources. This has been done extensively for the Feynman–Drexler diamondoid devices being considered as a way to realize productive nanosystems. Whereas biological nanodevices (e.g. a ribosome) of a similar size are extremely difficult to simulate operationally because of relaxation modes at many different timescales, extending to tens of seconds or longer, because the structures are "soft", diamondoid devices such as a gear train typically operate in the GHz frequency domain. Furthermore, the biological systems typically have to operate in the presence of bulk liquid water, which is still difficult to deal with because of its highly correlated nature. On the other hand, dynamic mean field theory (DMFT) seems to be capable of successfully dealing with strongly correlated solids, and capable of predicting band gap transitions, for example.

Beyond the capabilities of molecular dynamics, we are really into the meso- or microscale, in which it is not necessary to explicitly simulate each atom. Examples of the technologies in current use include Brownian dynamics, finite element analysis and computational fluid dynamics. For example, if multiatom nanoblocks are used as units of assembly, it may not be necessary to explicitly simulate each atom within the block. In the life sciences, this corresponds to simulating a protein by considering its constituent amino acids as the elementary entities.

Simulation methods for devices are typically *ad hoc*. For example, extensive simulations of magnonic components have been undertaken since the actual physical realization of such components is in the very early stages [157, 158]. Magnetization switching simulation is based on the Landau–Lifshitz equation

$$\frac{d\vec{m}}{dt} = \frac{\gamma}{1 + \alpha^2} \vec{m} \times [\vec{H}_{\text{eff}} + \alpha \vec{m} \times \vec{H}_{\text{eff}}], \tag{10.17}$$

where $\vec{m} = \vec{M}/M_s$ is the unit magnetization vector, M_s is the saturation magnetization, γ is the gyromagnetic ratio and α is the phenomenological Gilbert damping coefficient. The first term in this equation describes the precession of magnetization about the effective field and the second term describes its relaxation towards the direction of the field. Micromagnetic simulations were undertaken to better understand switching in the magnetic vortices being investigated as storage elements [154].

The discrete nanostructures such as CNTs have generated an enormous simulation activity, ranging from *ab initio*, molecular dynamics and Monte Carlo simulation methods to continuum methods for modeling mechanical and thermal properties [243].

10.7.2 EVOLUTIONARY DESIGN

Although the most obvious consequence of nanotechnology is the creation of very small objects, an immediate corollary is that in most cases of useful devices, there must be a great many of these objects (vastification). If r is the relative device size, and R the number of devices, then usefulness may require that $rR \sim 1$, implying the need for 10^9 devices. This corresponds to the number of components (with a minimum feature length of about 100 nm) on a VLSI electronic chip, for example. At present, all these components are explicitly designed and fabricated. But will this still be practicable if the number of components increases by a further two and more orders of magnitude?

Because it is not possible to give a clear affirmative answer to this question, alternative routes to the design and fabrication of such vast numbers are being explored. One of the consequences of vastification is that explicit design of each component, as envisaged in the Feynman–Drexler vision of productive nanosystems, is not practicable. The human brain serves as an inspiration here. Its scale is far vaster than even the largest scale integrated chips: it has $\sim 10^{11}$ neurons, and each neuron has hundreds or thousands of connexions to other neurons. There is insufficient information contained in our genes (considering that each base, chosen from 4 possibilities, contributes $-0.25 \log_2 0.25$ bits) to specify all these neurons (supposing that each neuron is chosen from two possibilities) and their interconnexions (each one

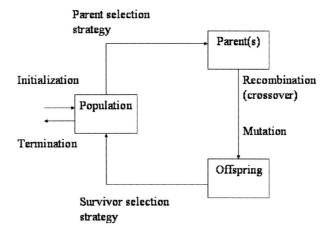

FIGURE 10.6

An evolutionary design algorithm. All relevant design features are encoded in the genome (a very simple genome is for each gene to be a single digit binary value indicating absence (0) or presence (1) of a feature). The genomes are evaluated ("survivor selection strategy") – this stage could include human (interactive) as well as automated evaluation – and only genomes fulfilling the evaluation criteria are retained. The diminished population is then expanded in numbers and in variety – typically the successful genomes are used as the basis for generating new ones via biologically inspired processes such as recombination and mutation.

of which needs at least to have its source and destination neurons specified). Rather, our genes cannot do more than specify an algorithm for generating the neurons and their interconnexions.

In this spirit, evolutionary design principles may become essential for designing nanodevices. An example of an evolutionary design algorithm, or simply evolutionary algorithm (EA) for short, is shown in Fig. 10.6. It might be initialized by a collection of existing designs, or guesses at possible new designs. Since new variety within the design population is generated randomly, the algorithm effectively expands the imagination of the human designer.

Essential requirements of evolutionary design are, then, the following:

1. Encoding the features of the entity to be designed. Each feature is a "gene", and the ensemble of genes is the "genome".
2. Generating a population of variants of the entity; in the simplest genomes, each possible feature may be present or absent, in which case the genome is simply a binary string; more elaborate ones may allow features to take a range of possible values.
3. Selection. If the selection criterion or criteria can be applied directly to the genome, the process is very rapid, especially if the criterion is a simple threshold

(e.g. is the ultimate functional requirement met?). It may be, however, that the genome needs to be translated into the entity it encodes, and the selection criterion or criteria applied to the entity itself; an intermediate situation is if the genome can be translated into a model of the entity, which is then subjected to the selection criterion or criteria. Only those genomes fulfilling the criteria are retained; the rest are discarded.

4. Regeneration of variety. Since selection (3) implies more or less drastic elimination, the survivors (there must be at least one) become the parents of the next generation, which may be asexual ("mutations", i.e. randomly changing the values of genes) or sexual (in which two or more parents are selected, according to some strategy, which may be random, and fragments of the parents' genomes, again selected according to some strategy, are combined into the offspring's genome) or both.

5. Return to 3 and continue iterating. If the selection criterion is graded, one can continue until no further increase takes place. It is almost inevitable that human intervention must then make the final choice (unless there is a single clear "winner"). Indeed, human intervention may be incorporated in every selection stage, although it makes the process very slow. Typically human intervention takes place after, say, one hundred iterations of automatic selection and regeneration; the human selector is presented with a small number ($\sim$10) of the best according to the automatic selection process, and selects a few from those (e.g. [31]). This "interactive evolutionary computation" (IEC) is an excellent way to incorporate expert human knowledge into the design process without having to define and quantify it.

Although the EA design strategy was inspired by biological ("Darwinian") evolution, there is a crucial qualitative difference in that the latter is open-ended, whereas the former is generally applied to achieve a preset goal, even though that goal may be specified at a high level. In fact, the main current use of EAs is multi-objective optimization of designs whose main features are already fixed.

The evolutionary paradigm can operate at various levels. "Genetic algorithms" merely allow the parameters of the system's description to evolve (e.g. the coefficients of equations). "Genetic programming" allows more freedom in that the algorithm itself may evolve. One may also allow the genome itself to evolve, and so forth. EA is generally implemented synchronously (the entire population is simultaneously subjected to the same process), which is of course different from evolution in nature. In a real evolving system, "bad" species may persist, and "good" species may be eliminated for no good reason [5]. These and other biological features [18] will hopefully be implemented in the future as EA themselves continue to evolve. Ultimately one might reach the ability to create self-organizing functionality.

Although this strategy of evolutionary design enables the design size (i.e. the number of individual features that must be explicitly specified) to be expanded practically without limit, one sacrifices knowledge of the exact internal workings of the system, introducing a level of unpredictability into device performance that may require a new engineering paradigm in order to be made acceptable. One should, however, bear in mind that even deterministically designed complex systems (e.g. a motor car, ship or airplane), at current levels of technological sophistication, have a vast behavior space and every possible combination of control parameters cannot be explicitly tested. Given that we already in practice sacrifice complete knowledge of the system (even though in principle it is attainable), and yet still have a high degree of confidence in its ability to function safely, therefore it may be unreasonable to object to using an evolutionarily designed artifact.

10.8 PERFORMANCE CRITERIA

The EA approach (Section 10.7.2) demands "fitness" criteria against which the performance of offspring can be judged. Although a system's performance can only be meaningfully measured by considering its ultimate functional output, it may still be helpful to consider component performance. Thus, a logic gate may be assessed according to its rectification ratio and switching speed. A memory cell can be assessed according to its flip energy (between states) and the stability of states. Sensors may be assessed by (in order of increasing preference) gain, signal-to-noise ratio [123] and detection efficiency [196]. Note that the performance of MEMS such as accelerometers is degraded if they are further miniaturized down to the nanoscale [123].

Criteria applied in IEC are not usually specifiable – if they were, there would be no need to use IEC. It is especially important when considering the esthetic features of a design.

An important criterion applicable to nanobiotechnological devices in medicine (Section 4.2) is resistance to opsonization. Opsonization is that process whereby a foreign object inside living organism becomes coated with proteins that signal to the immune system that the object must be eliminated. Thus, a key parameter for drug delivery nanoparticles is how long they can circulate in the body. Nanoparticles circulating in a complex fluid such as blood typically attract a protein corona around them, a complex multicomponent composite whose structure is usually dynamic, continuously varying. Particularly if protein denaturation takes place (Section 4.1.4 and Fig. 4.5) the particle is likely to be recognized as foreign.

10.9 CREATIVE DESIGN

New ways of connecting known components together could yield new functions. The Nanotechnology Age may usher in a new era of creative leaps, reminiscent of the great Victorian era of engineering. When Richard Trevithick designed and built, in 1803, the first railway locomotive in the world (a replica of which is displayed in the hall of Telford Central Station in the UK) he had to solve design problems that no one had ever encountered before – and many of his solutions have persisted in essence to this day. The same applies to the bridges of totally new structure conceived by

Brunel, the Stephenson brothers, and others. Will we embark on a new panopticon of creative design comparable to the immense creative energy of that earlier era?

One needs to examine how the novel possibilities of nanotechnology can be fully exploited from the design viewpoint. While structure is familiarly a fixed given that determines function, living cells present hints of function retroacting on structure. Can this principle be extended to the inanimate world? Can the "structure" intermediate be short-circuited in nanotechnology? – that is, can we determine the components needed to assemble a device with specified functional properties directly, without considering what structure it should have? This may be especially relevant when considering structures that are not static, but act dynamically (e.g. an enzyme, cf. Section 11.5). The methods described in Section 10.7.2 are especially appropriate for such an approach.

10.10 PRODUCEABILITY

Any viable design must be associated with a practicable route to fabrication. The evolutionary design process described in Section 10.7.2 could readily be extended to encompass manufacturability (cf. Section 8.4). A database of existing manufacturing systems could be a starting point for evaluating the fitness of a design, but the design process could be extended to encompass the production system itself.

Similar considerations apply to marketability. Existing domain knowledge might best be incorporated into the evaluation step interactively, using commercial marketing experts, but the design process could itself devise appropriate and novel approaches to creating a market for the product.

10.11 SCALEOUT

Traditional process design begins with small scale laboratory experiments, continues with a pilot scale fabrication, in some factories fortunate to have the capacity it may then be possible to run an experiment on the existing manufacturing plant, before finally a dedicated production unit is designed and put into operation. At each transition to the next scaleup, it is scarcely to be expected that control parameters can simply be multiplied by the length or volume (etc.) scaling factor. In complicated processes, the behavior may even change qualitatively. Scaleup is therefore very problematical.

On the other hand, a process implemented at the nanoscale requires only to be multiplied (this is called scaleout). Scaling performance up to the level of human utility is simply a matter of massive parallelization. For example, nanoreactors synthesizing a medicinal drug simply need to work in parallel to generate enough of the compound for a therapeutically useful dose. This introduces new problems of connexions and synchronization, but whereas practically each scaleup problem needs to be solved *de novo* in an *ad hoc* fashion, once the general principles of scaleout are established they are universally valid.

With information processors, the problem is the user interface: a visual display screen must be large enough to display a useful amount of legible information, a keyboard for entering instructions and data must be large enough for human fingers, and so forth – these kinds of problems have already been addressed with microsystems technology (e.g. for reading information stored on microfilm).

10.12 STANDARDIZATION

Standardization is the key to any viable system. The richness of bacterial life is due to a high degree of standardization of their genomes, such that genetic material can be readily exchanged between them. Engineers require a standard vocabulary on order to work together and create standard specifications for interchangeable components of industrial systems.

Nanotechnology is no exception. Even though the nanotechnology industry is still in its infancy, the ISO and the International Electrotechnical Commission are jointly preparing a multipart Technical Specification (as a precursor to an International Standard) for the vocabulary of nanotechnology (ISO/TS 80004), some parts of which have already been published. These specify terms and definitions relating to nano-objects and nanostructured materials, carbon nano-objects, the nano/bio interface, nanometrology, nanomanufacturing processes and so forth.

SUMMARY

Systems and nanosystems are defined. Nanotechnology implies many departures from traditional engineering practice. Regarding choice of materials, one should ultimately no longer be constrained to merely select suitable materials; rather materials with exactly the required properties can be specified and produced. Computation is much more intimately connected with nanotechnology than with larger scale technologies, because explicit and reliable simulation of complete systems is often possible. To cope with vastification, evolutionary computation offers a solution, vastly amplifying human capabilities (it can be considered as an "intelligence amplifier", in the same way that muscular power is amplified by a traction engine).

With any nanosystem, it needs to be considered how its function can be made useful for human beings. Scaleout – massive parallelization – eliminates the barriers associated with traditional scaleup.

Any viable, integrated technological system requires standardization, first of vocabulary, than of the components used in constructing the artifacts of the technology.

Nanotechnology offers magnificent opportunities for creative design, possibly allowing a new era of innovation to nucleate and grow. Such growth will be driven not only by possibilities inherent in the technology but also by the fact that a far greater proportion of the population should be able to contribute than hitherto to conventional design and realization.

FURTHER READING

1. P. Hatto, Standardization for nanotechnology, Nanotechnol. Percept. 3 (2007) 123–130.
2. D.C. Sayle, et al., Mapping nanostructure: a systematic enumeration of nanomaterials by assembling nano building blocks at crystallographic positions, ACS Nano 2 (2008) 1237–1251.

Bionanotechnology

CHAPTER CONTENTS

INTRODUCTION

Bionanotechnology is defined as the application of biology to nanotechnology (note that biotechnology is the directed use of organisms to make useful products, typically achieved by genetically modifying organisms); that is, the use of biological molecules in nanomaterials, nanoscale devices or nanoscale production systems. It should be contrasted with nanobiotechnology [Chapter 4; if the bionanotechnology is then applied to human health (nanomedicine or nanobiotechnology), consistency in terminology suggests this should be called bionanobiotechnology].

The discovery of some of the mechanistic details of complicated biological machinery such as the ribosome, which encodes the sequence of nucleic acids as a sequence of amino acids (called "translation" in molecular biology), was happening around the time that Eric Drexler was promoting his assembler-based view of nanotechnology, and these biological machines provided a kind of living proof of principle that elaborate and functionally sophisticated mechanisms could operate at the nanoscale. Some of these biological machines are listed in Table 11.1. There are

Nanotechnology: An Introduction. DOI: 10.1016/B978-0-323-39311-9.00017-0
Copyright © 2016 Elsevier Inc. All rights reserved.

Table 11.1 Examples of Biological Nanosized Machines

Name	Natural Function	State of Knowledge[a]
Muscle (myosin)	Pulling	C, S, T
Kinesin	Linear motion	C, S, T
Nerve	Information transmission	T
ATPase	Synthesis of ATP from proton e.p.g.[b]	C, S, T
Bacteriorhodopsin	Generation of proton e.p.g. from light	C, T
Transmembrane ion pump	Moving selected ions against an adverse e.p.g.	C, T
Hemoglobin	Oxygen uptake and release	C, T

[a]C, crystal structure determined; S, single-molecule observation of operation; T, theoretical mechanism available.
[b]Electrochemical potential gradient.

many others, such as the mechanism that packs viral deoxyribonucleic acid (DNA) ultracompactly in the head of bacteriophage viruses.

The machines listed in Table 11.1 are proteins (polypeptides); some are considered to be enzymes (e.g., ATPase). Enzymes (and possibly other machines as well) can also be constructed from RNA, and some known machines, such as the ribosome, are constructed from both polypeptides and RNA. RNA and polypeptides are synthesized (naturally or artificially) as linear polymers, most of which can adopt their functional three-dimensional structure via a self-assembly process (Section 8.2.11) that occurs spontaneously (and often reversibly).

Some of these machines show consummate scaling out to the macroscopic realm. Muscle is probably the best example: although the actin–myosin pair that is the molecular heart of muscular action develops a force of a few piconewtons, by arranging many "molecular muscles" in parallel, large animals such as elephants can develop kilowatts of power, as humans have known and made use of for millennia.

These natural nanomachines are inspiring in their own right, and their existence and the detailed study of their mode of operation have driven efforts to mimic them using artificially designed and constructed systems – this is called bio-inspired nanotechnology or biomimetic nanotechnology. Many structures, especially devices produced in living systems, are constituted from biopolymers designed to fit to congeners with exquisite specificity and precise stoicheiometry. One of the challenges of biomimetic nanotechnology is to recreate these attributes with simpler artificial systems – without much success until now. Could one, for example, create a synthetic oxygen carrier working like hemoglobin but with a tenth or fewer the number of atoms? Possibly, although one wonders whether such a "lean" carrier would be as resilient to fluctuations in its working environment.

Returning to our definition of bionanotechnology (the incorporation of biological molecules into nanoartifacts), after recalling the basics of biological structure and biomolecular mechanism, we shall survey three example areas in which biological molecules have been used structurally or incorporated in nanoscale devices: DNA as a self-assembling construction material; biosensors; and biophotonic memory and logic gates. Although a rather exotic system of a motile bacterium harnessed to push a tiny rotor has been reported [125], the main current areas of nanotechnological significance are biosensors and biophotonics.

11.1 THE STRUCTURAL NATURE OF BIOMOLECULES

Polypeptides (PP) (proteins) are linear polymers of amino acids (H_2N–CHR–COOH, where R (bonded to the central C) is a variable side chain ("residue") – there are 20 different natural ones. To polymerize them, water is eliminated between –COOH and H_2N– to form the peptide bond, hence there is a common backbone (linked via the "peptide" bonds) with variable side chains – short aliphatic groups, small aromatic groups, carboxylate, amine, hydroxyl, etc. Template-directed synthesis with a very high yield is used in nature, with the templates being closely related to genes via the genetic code (triplets of nucleotide bases encode each amino acid). After synthesis (polymerization), they fold, often spontaneously, to a compact structure according to a least-action principle (see Section 8.2.11). Typical natural proteins have 50–500 amino acids. Depending on their sequence, they adopt a definite remembered conformation (proteins acting as devices, rather than having a passive structural rôle, have two or more stable conformations) and can carry out varied functions, ranging from essentially structural or scavenging to enzymes and motors. Some proteins (called glycoproteins) are branched with oligosaccharides (OS) attached to certain residues.

Nucleic acids (NA)   are polymerized from nucleotides constituted from a sugar, a phosphate group, and a "base" derived from a purine or pyrimidine (aromatic heterocycle). The sugar and phosphate are polymerized by eliminating water to form a linear backbone, with the bases playing the rôle of the residues in PP. There are 4 natural bases, abbreviated A, C, G, T (in DNA) and A, C, G, U (in RNA). The bases pair preferentially: A with T (or U), via 2 hydrogen bonds, and C with G via 3 hydrogen bonds (complementary base-pairing, CBP). Linear polymers are linked via the sugar. Template-directed synthesis with a very high yield is used in nature to create the polymers. The templates are the genes (DNA), and operate according to the principle of CBP. During polymerization RNA spontaneously folds to a definite compact structure according to a least-action principle (see Section 8.2.11), in which base-pairing via hydrogen bonding is equivalent to the potential energy, and loop and hairpin formation is equivalent to the kinetic energy. DNA forms the famous double helix in which genetic information is stably stored in living cells and many viruses.

Polysaccharides (PS) and OS are linear or branched polymers of diverse sugar (cyclic oligoalcohol) monomers, linked via water elimination ("condensation") at any of the several hydroxyl groups. The problem of predicting their structure is not yet solved. Polymerization is not templated (i.e. not under direct genetic control) and there is variability (to a degree that is only poorly characterized) in sequence and length of the PS found fulfilling the same function in comparable organisms.

11.2 SOME GENERAL CHARACTERISTICS OF BIOLOGICAL MOLECULES

The energy contained in a given system can be divided into two categories: (i) the multitude of microscopic or thermal motions sufficiently characterized by the temperature; and (ii) the (usually small number of) macroscopic, highly correlated motions, whose existence turns the construction into a machine (a device). The total energy contained in the microscopic degrees of freedom may be far larger than those in the macroscopic ones, but nevertheless the microscopic energy can usually be successfully neglected in the analysis of a construction (in informational terms, the macrostates are remembered, but the microstates are not).

Biological molecules are constructions. Hence, a statistical approach, in which the motions of an immense number of individual particles are subsumed into a few macroscopic parameters such as temperature and pressure, is inadequate. A construction uses only an insignificant fraction of the Gibbs canonical ensemble and hence is essentially out of equilibrium. This is different from thermodynamic nonequilibrium – it arises because the system is being investigated at time scales much shorter than those required for true statistical equilibrium. Such systems exhibit "broken ergodicity", as epitomized by a cup of coffee in a closed room to which cream is added and then stirred. The cream and coffee equilibrate within a few seconds (during which vast amounts of microinformation are generated within the whorled patterns); the cup attains room temperature within tens of minutes, and days may be required for the water in the cup to saturate the air in the room.

Broken ergodicity may be regarded as a generalization of *broken symmetry*, which leads to a new thermodynamic quantity, the *order parameter* ξ, whose value is zero in the symmetrical phase. ξ may be thought of as conferring a kind of generalized rigidity on a system, allowing an external force applied at one point to be transferred to another. Some protein molecules demonstrate this very clearly: flash photolysis of oxygenated hemoglobin [causing the oxygen molecule to dissociate from the iron core of the porphyrin (the heme) to which it is bound] causes motion of the iron core of the heme, which results in (much larger) movement at the distant intersubunit contacts, leading ultimately to an overall change in the protein conformation involving hundreds of atoms.

The machines that result from these molecules are far larger and more sophisticated than the molecular machines encountered previously (Section 7.10).

11.3 HIERARCHICAL AND COMPOSITE BIOLOGICAL MATERIALS

Many of the most familiar and useful biomaterials are hierarchically structured composites. In the example shown in Fig. 11.1, atoms are first grouped into glucose

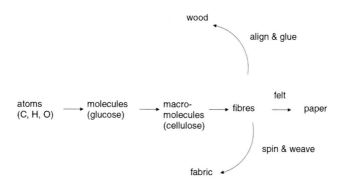

FIGURE 11.1

An example of the hierarchical assembly of a biomaterial, starting with atoms and ending up with wood. Man can intervene before the lignin is added in order to create artificial materials such as paper and textiles.

molecules, which are polymerized into cellulose, which are spun into fibrils, which are bundled into fibers, which are embedded in a lignin matrix to, finally, make wood. At most of the transitions in this hierarchy there is a qualitative change in attributes. From colorless, gaseous elements and a tasteless black solid there emerges a colorless, water-soluble solid with a sweet taste; from this molecule there emerges a strong, insoluble fiber, and finally there emerges a strong engineering material capable of supporting a structure a hundred meters high.

Hierarchical design is also a feature at very small scales for achieving antiwetting (superhydrophobicity), which has already been encountered on the surface of plant leaves (Section 6.3.4). The difficulties insects have in overcoming capillary forces have already been mentioned (Section 3.3); termites, weak flyers, have a multiscale hierarchical surface on their wings to ensure that water drops do not reside on them [296].

Some of the most remarkable materials are found in the most unlikely places. For example, deep on the ocean floor, in the vicinity of hydrothermal vents, one may find the gastropod mollusc *Crysomallon squamiferum*. Its shell is a remarkable trilayered structure [310]. The inner layer is a rigid calcified shell about 250 μm thick with the structure of a type already familiar from common near-surface marine organisms with shells (Section 6.6.6) – calcite nanoplatelets embedded in a protein matrix, the latter only occupying a few percent of the total mass. Next comes a compliant organic, proteinaceous middle layer approximately 150 μm thick, and finally there is a very tough outer layer about 30 μm thick containing greigite (Fe_3S_4) granules of a variety of sizes down to about 20 nm. The overall structure is remarkably hard and tough. One notes that hydrothermal vents are typically rich in sulfides and heavy metals, but one also wonders against what kind of predatory threat this amazing shell is a defcnsc. The structure is, moreover, a veritable system insofar as each part of the shell necessarily contributes to the overall attributes (for example, the organic middle layer is essential for arresting incipient cracks).

11.4 DNA AS CONSTRUCTION MATERIAL

The specific base-pairing of DNA, together with the ease of nucleotide polymerization (it can be accomplished artificially using automated equipment) and the relative robustness of the molecule, has engendered interest in the design and construction of artificial nanoscale artifacts of arbitrary shape made from DNA, as Nadrian Seeman was the first to point out. A drawback is that the design of the required DNA strands is a laborious, empirical process (at least at present) but, in principle, both DNA and RNA could become universal construction materials (provided they are not required to be stable under extreme conditions). The fact that enzymes constructed from RNA are known to exist in nature suggests that ultimately devices could also be made. Once synthesized (according to what are now straightforward, routine procedures – albeit not completely free or errors), however, it suffices to randomly stir a solution of the components together at an appropriate temperature; their assembly then proceeds in a unique (if the sequences have been correctly designed) fashion (Fig. 11.2). This field has recently grown enormously to encompass very elaborate constructions. The process is connected with tile assembly and computation [128].

The specific base-pairing has also been exploited by fastening fragments of DNA to nanoparticles to confer selective affinity upon them, resulting in the formation of specific clusters of particles, although unless the fastening is geometrically precise, rather than merely statistical, the clusters are also statistical in their composition, except for the simplest ones.

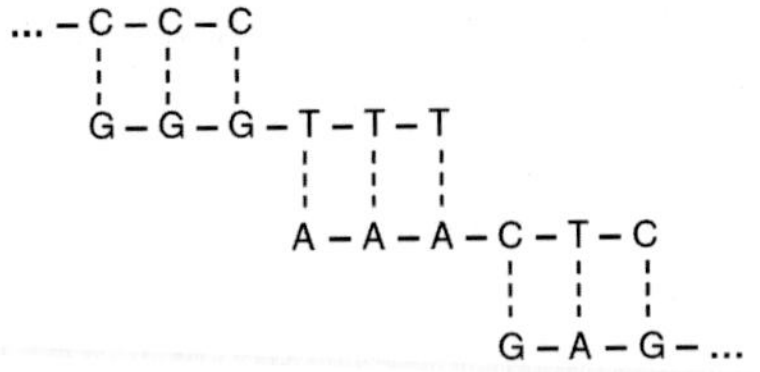

FIGURE 11.2

Four oligonucleotides, that can only assemble in the manner shown. Long dashes represent strong covalent bonds, and short dashed lines represent weak hydrogen bonds.

11.5 THE MECHANISM OF BIOLOGICAL MACHINES

Active proteins (i.e. all proteins apart from those fulfilling a purely passive structural or space-filling rôle) have two or more stable conformations. Unfortunately, there is very little information about these multiple conformations available. Useful as XRD is for determining the structure of proteins, it has the disadvantage of usually fixing the protein in just one of these conformations in the crystal that is used to diffract the X-rays. If a fraction of the protein does happen to be present in one of the other stable conformations, this is usually regarded as "disorder" and any information about it is lost during structural refinement of the diffraction data. However, these multiple conformations are the key to the general mechanism of biological machines, originally formulated by L.A. Blumenfeld to explain enzymatic catalysis [27].

The prototypical example is an enzyme E catalysing (say) the decomposition of a molecule A–B (called the substrate of the enzyme in the biochemical literature) into products A + B. In this case, work has to be done to break the chemical bond A–B; the principle can equally well be applied to any action where work is done, such as pulling on a spring (as in muscle). The binding and release of oxygen to and from hemoglobin also works on this principle.

The enzyme consists of an active site and the rest of the protein, which may be considered to be much bigger than the active site (Fig. 11.3). The "rest" has two stable conformations, E and $\tilde{E}$. The mechanism proceeds in four stages:

1. A complex is formed between the substrate and the enzyme, A–B + E → (A–B)E*. A–B binds to the active site, releasing free energy and resulting in a local conformational change, which creates a strain between the active site and the rest of the protein. Local fast vibrational relaxation takes place in the picosecond time scale, but the active site is no longer in equilibrium with the rest of the molecule and the resulting strain modifies the energy surface on which the enzymatic reaction takes place. The asterisk denotes that the protein is overall in a strained, nonequilibrium state. Strain creation requires energy, but its magnitude must of course be less than the energy of binding.

2. The complex slowly relaxes to a new conformation $\tilde{E}$, releasing the energy to drive the energy-requiring breaking of the A–B bond: (A–B)E* → A$\tilde{E}$B. This is the elementary act of the enzymatic reaction. This conformational relaxation involves making and breaking a multiplicity of weak bonds, but at a slower rate than the reaction being catalysed.

3. The product–enzyme complex is decomposed, i.e. the products are released: A$\tilde{E}$B → A + B + A$\tilde{E}$*. Release of the products from the active site again creates strain between it and the rest of the protein molecule.

4. Finally, the strained enzyme slowly relaxes back to its initial conformation: $\tilde{E}$* → E.

An interesting prediction of this mechanism is that the rate of the overall reaction A–B → A + B should exhibit an inverse Arrhenius temperature dependence, because increasing the temperature accelerates conformational relaxation (step 2), and hence

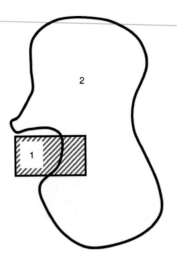

FIGURE 11.3

Model of an enzyme (or motor), consisting of
an active site (1) and the rest (2).

shortens the time during which the strained molecule is able to accelerate the reaction
enzymatically.

11.5.1 BIOLOGICAL MOTORS

Some of the most remarkable devices in nature are not enzymes *per se* – after all,
many efficient synthetic catalysts have been made – but miniature motors, which
perform prodigious feats. Even an entity a simple as a virus possesses a motor – the
device that winds DNA into the head of the bacteriophage. Both linear and rotary
motors are known. Fig. 11.4 shows the working cycle of the myosin linear motor
that powers muscle. Key details have been observed via an experimental approach
based on single-molecule manipulation. For example, myosin is immobilized on the
substratum and allowed to interact with actin tethered to beads held in optical traps
(e.g., [309, 91, 222]). It was long assumed that there was a direct correlation between
the hydrolysis of ATP and the mechanical work performed by the motor; that is,
each hydrolysed molecule resulted in one unit of displacement Δx ("tight coupling
model"). This is, however, rather naïve and essentially treats the biomolecule as a
miniature newtonian construction. Single-molecule experimental observations do not
support this assumption. Simultaneous monitoring of individual hydrolysis events (by
microscopic observation of fluorescently labelled adenosine) and bead displacements
due to the mechanical force exerted on the actin have clearly shown that mechanical
force might be generated several hundred milliseconds after release of ADP [147].

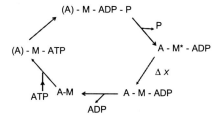

FIGURE 11.4

Simplistic model of the working cycle of muscle. M denotes myosin and A denotes actin; (A) is actin weakly bound to myosin; the asterisk denotes myosin in a strained, nonequilibrium state. Binding of adenosine triphosphate (ATP) to the myosin results in weakened binding to actin, and hydrolysis of the ATP to adenosine diphosphate (ADP) and the subsequent release of phosphate generate strain between the ATP-binding site and the rest of the myosin molecule; the relaxation of this strain drives the movement of the motor (Δx, 8–10 nm) during which mechanical work is done. The hydrolysis of one ATP molecule yields a chemical energy E of 5–20 $k_B T$; the force E/Δ exerted by one myosin motor can therefore be estimated 2–10 pN.

Apparently, myosin can store the chemical energy from several ATP hydrolysis events, and may then subsequently carry out several mechanical work steps ("loose coupling model"). The diagram in Fig. 11.4 is therefore somewhat simplistic. The energy storage demonstrates that the muscle motor follows the Blumenfeld mechanism. Other mechanical motors such as kinesin moving on microtubules also operate on this principle. The general principle of conformational strain between a binding site and the rest of the protein (generated by binding or release of a small molecule) driving (electro)chemical or mechanical work can be expected to be universal in living systems.

11.5.2 MICROTUBULE ASSEMBLY AND DISASSEMBLY

Another example of a process dependent on two conformational substates is the assembly and disassembly of microtubule filaments from the globular protein tubulin.

During the normal state of the eukaryotic cell, these filaments pervade the cytoplasm, acting as tracks for kinesin motors transporting molecules and supramolecular complexes. Prior to eukaryotic cell division (mitosis), the duplicated genome is compactified into chromosomes and the nuclear membrane and the microtubule filaments network are degraded. The duplicated genome must be separated and the two halves relocated in the two halves of the cell that will become separate cells after division. How can this be accomplished? Two centrosomes (protein complexes) form asteriated poles at opposite ends of the cell, and microtubules repeatedly elongate at a speed v_g out from them in random directions, followed by catastrophic disassembly leading to abrupt shrinkage with speed $v_s, v_s \gg v_g$. The process continues until a microtubule filament reaches a chromosome, upon which it attaches itself and drags half of it towards the centrosome. The result is each duplicated genome located in separate halves of the cell, after which the rest of the division process takes place.

The dynamic instability (assembly–disassembly) is characterized by length fluctuations of the order of the mean microtubule length, hinting at a phase transition [130]. Let f_{gs} denote the frequency of switching from growth to shrinkage and f_{sg} the frequency of switching from shrinkage to growth. When $v_g f_{sg} = v_s f_{gs}$ growth switches from unbounded (corresponding to the assembly of the microtubule filament network) to bounded. At this point, the average microtubule length $\bar{\ell} = v_g v_s/(v_s f_{gs} - v_g f_{sg})$ diverges. The molecular origin of growth and shrinkage lies in the fact that tubulin monomers can bind to guanosine triphosphate (GTP), and the complex can spontaneously assemble to form filaments. But the GTP slowly hydrolyses spontaneously to guanosine diphosphate (GDP), thereby somewhat changing the tubulin conformation such that it prefers to be monomeric. However, the monomers can only be released from the end; disassembly can be initiated if the rate of GTP hydrolysis exceeds that of tubulin addition for a while. The overall process is a remarkably effective way of searching a restricted volume for an object when no prior information about the location of the object exists.

11.5.3 THE COST OF CONTROL

The force F which has to be applied to a molecular lever requires accurate knowledge of its position x if reversible work is to be carried out [105]. Specifying the positional accuracy as Δx, the uncertainty principle gives the energy requirement as

$$\Delta E \geq hc/(4\Delta x), \tag{11.1}$$

where h is Planck's constant ($=6.63 \times 10^{-34}$ J s) and c the speed of light in vacuum ($=3.00 \times 10^8$ m/s), ΔE is obviously negligible for macroscopic systems millimeters in size. The uncertainty in the force $F(x)$ generated at x is

$$\Delta F = F(x) \pm \Delta x(dF/dx). \tag{11.2}$$

To compute the work W done by the system, Eq. (11.2) is integrated over the appropriate x interval. The first term on the right hand side yields the reversible work

W_{rev}, and the second term yields $-\Delta x \sum_j |F_j - F_{j+1}|$ for any cycle involving j steps.

The energy conversion factor ϵ is

$$\epsilon = W/(Q + \Delta E), \tag{11.3}$$

where Q is the net energy input during the cycle. With the help of inequality (11.1), the ratio of this to the classical conversion factor $\epsilon_{rev} = W_{rev}/Q$ is

$$\epsilon/\epsilon_{rev} \le (1 - \alpha/z)/(1 + z) \tag{11.4}$$

where

$$\alpha = hc \sum_j |F_j - F_{j+1}|/(4QW_{rev}) \tag{11.5}$$

and the relative energy cost of control is

$$z = \Delta E/Q. \tag{11.6}$$

The maximum possible value of the ratio ϵ/ϵ_{rev} is obtained by substituting z by its optimal value z_{opt}, obtained from the turning point of Eq. (11.4),

$$z_{opt} = \alpha(1 + \sqrt{1 + 1/\alpha}). \tag{11.7}$$

It is

$$\left(\frac{\epsilon}{\epsilon_{rev}}\right)_{max} = \frac{1 - 1/(1 + \sqrt{1 + 1/\alpha})}{1 + \alpha(1 + \sqrt{1 + 1/\alpha})}. \tag{11.8}$$

If more energy than z_{opt} is used, then α decreases because of the energy cost of information; if less, then ϵ decreases because of the irreversibility (dissipation, etc.). For a macroscopic system, these quantities are insignificant. But consider the myosin motor (Fig. 11.4): taking $F_j \approx 2$ pN, the displacement $x \approx 10$ nm, and $Q \approx 0.067$ aJ (the energy released by hydrolysing a single ATP molecule), then the energy cost of optimum control, Qz_{opt}, is equivalent to hydrolysing almost 150 ATP molecules and $(\epsilon/\epsilon_{rev})_{opt} = 0.0033$. Reversible operation is evidently far from optimal; chemical to mechanical conversion occurs at a finite rate which may essentially be uncontrolled; i.e. determined intrinsically. This analysis and conclusion allow the loose coupling model for muscle (Section 11.5.1) to be rationalized.

11.6 BIOLOGICAL SYSTEMS

Even some of the materials mentioned above, such as the iron-plated armor of *C. squamiferum*, are sufficiently elaborate to qualify as systems. Another good example would be the ferritins, a family of large (10–12 nm diameter), self-assembled protein cages that reversibly synthesize Fe_2O_3 nanoparticles containing up to 4500 atoms

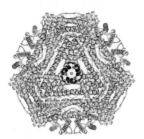 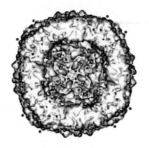 

Cytoplasm
Fe^{2+} In /out

Mineral Growth
Cavity

Ferritin: Protein cage exterior   Protein cage interior Fe^{2+} ion channels

View: Down 3-fold cage axis   Down 4-fold cage axis $\perp$3-fold cage axis

FIGURE 11.5

Ferritin protein cage, showing on the left the exterior surface helices of one of the eight Fe^{2+} channels, in the middle a cutaway view of the interior (from XRD data) showing the mineral growth cavity, and on the right a drawing of one of the Fe^{2+} channels, which are about 1.5 nm long.
- *Reproduced with permission from E.C. Theil, Ferritin protein nanocages—the story, Nanotechnol. Percept. 8 (2012) 7–16.*

[284]. Hence, they are both materials and machines. The protein cage is shown in Fig. 11.5. Because of the uniformity of the hematite particles produced by particular ferritins, they have been perceived as attractive manufacturing routes to magnetic storage media [142].

At a higher scale of complexity are the kidneys, a marvelous example of biological nanoengineering that functions to extract certain substances from highly dilute solutions [286], an operation that may become increasingly attractive to harness nanoindustrially as a way of extracting metals from seawater, as conventionally processable ores become depleted.

Finally, we have the most complex system known to man – the (human) brain [271]. Mimicking the function of the brain is of enormous interest to mankind, especially for creating artificial intelligence. Since it is clear that the VLSI circuit architecture of computing machines does not correspond to that of the brain, work has developed along two lines: attempting to simulate the functioning of the brain using conventional computers (for example, by making each individual microprocessor simulate one neuron); and attempting to understand how living neurons function by scrutinizing their nanoscale components. Both are very difficult undertakings. One problem is the sheer scale of the human brain, the largest on Earth, which contains around 10^{11} neurons, each of which might be connected to as many as 10^4 others. Another problem is that the neurons themselves are far from being simple (like the cells in a model cellular automaton) but are very complex, containing around 10^{11}

atoms that are grouped into an extremely varied hierarchy of molecules, extending from small ions and osmolytes to extremely long filamentous proteins, which are themselves hierarchically assembled from smaller proteins.

11.7 BIOSENSORS

The dictionary definition of biosensor is "a device which uses a living organism or biological molecules, especially enzymes or antibodies, to detect the presence of chemicals" [51] (cf. Section 7.12.2). The classic biosensor is the amperometric glucose sensor, comprising glucose oxidase coating an electrode. The enzyme oxidizes glucose. The main reason for wishing to use an enzyme is to exploit the exquisite specificity of biomolecular binding interactions ("molecular recognition").

The Holy Graal of research in the field is to couple the enzyme directly to the electrode, such that it can be regenerated by passing electrons to it; in current practice, the enzyme concomitantly reduces water to hydrogen peroxide, which is in turn reduced at the electrode, engendering the measured amperometric signal. This is not nanoscale technology, but if the enzyme could indeed be coupled directly to the electrode, this would typically require the active site of the enzyme to be within ∼1 nm of the electrode; hence, it enters the realm of nanoengineering, in which a carbon nanotube might be used as the electrode, and which opens the way to reducing the size of the device, such that ultimately it might incorporate a single enzyme, able to detect single glucose molecules.

Another kind of biosensor exploits the combinatorial uniqueness even of base strings of fairly modest length to fabricate "gene chips" [49] used to identify genes and genomes. In these devices, the sample to be identified (e.g. the nucleic acids extracted from bacteria found in the bloodstream of a patient) is dispersed over the surface of the chip, which comprises an array of contiguous microzones containing known oligomers of nucleic acids complementary to the sought-for sequences (e.g. the fragment GATTACA is complementary to CTAATGA). Binding can be detected by double helix-specific dyes.

11.8 BIOPHOTONIC DEVICES

Apart from the marvelous intricacy of the biological machinery that converts light into chemical energy, which at present only serves to inspire nanotechnological mimics, there are other, simpler, photoactive proteins, robust enough to be incorporated into artificial devices. Molecules based on the chromophore rhodopsin (such as the primary optical receptor in the eye) seem to have a special place here.

One of the most remarkable of these photoactive proteins is bacteriorhodopsin (bR) [230], which constitutes about a third of the outer membranes of the archaeon (extremophilic procaryote) *Halobium salinarum*, living in salt lakes. The optically active site of the protein is the conjugated polyene rhodopsin, and when it absorbs a photon of red light conformational change generating strain between it and the

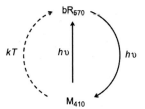

FIGURE 11.6

Simplified view of the bR photocycle. A 570 nm photon absorbed by the ground state bR_{570} (the subscript indicates the wavelength of maximum adsorption of the molecule) rapidly (within a few microseconds) transforms (through a series of intermediate stages) the molecule to the relatively stable intermediate M_{410}. This state slowly relaxes thermally back to the ground state, but it can also be rapidly converted by a 410 nm photon. The thermal stability of the M state can be extended almost indefinitely by genetically modifying the protein.

rest of the protein is engendered, which translocates a proton across the membrane (according to the mechanism outlined in Section 11.5). The process is called the bR photocycle, and a key intermediate state is called M; the altered interaction between the chromophore and its protein environment gives it an absorption maximum of 410 nm (Fig. 11.6).

H. salinarum can be easily grown in a fermenter and the bR harvested in the form of "purple membrane fragments" – pieces of outer membrane consisting of an array of bR with the membrane lipid filling the interstitial space. These fragments can be oriented and dried, in which state they can be kept under ambient conditions for 10 years or more without any loss of activity; they have already generated considerable interest as a possible optical storage medium, using a bR mutant, the M state of which is almost indefinitely thermally stable. In such an optical memory, the ground state would represent "0" and the M state would represent "1".

Nearly all work in biophotonics has been carried out using bacteriorhodopsin. The two main applications are holographic optical memories with ultrahigh data storage density and optical switches. In the former, the biological part is a block of bR and the nonliving part interacting with it is light [115]. Native bR can be used to construct an optically switched optical switch (Fig. 11.7). Not only can the switch operate extremely rapidly (at megahertz frequencies and above), but only

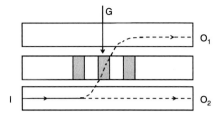

FIGURE 11.7

An optically switched optical switch. The basic construction is a slab of bR-containing purple membrane fragments sandwiched between two optical waveguides. Initially, let us suppose that an optical wave introduced at the input I is guided through the structure to emerge at output O_2. If a grating (indicated by the stripes) is holographically generated in the bR slab by illuminating with light of 570 nm (from G, the "gate"), light will be coupled out from the lower waveguide and coupled into the upper waveguide, emerging at output O_1. Destroying the grating by illuminating with light of 410 nm will cause the output to revert to O_2.

weak light is needed. The remarkable optical nonlinearity of the protein is manifested by exposing it to a single photon! These switches can be used to construct all-optical logic gates [304], which can be combined to construct optical computers.

SUMMARY

Bionanotechnology is defined as the incorporation of biological molecules into nanoartifacts. The highly refined molecular binding specificity is particularly valued and used to facilitate the assembly of unique structures from a solution of precursors and for capturing chemicals from the environment prior to registering their presence via a transducer (biosensors). Further applications involve using the widely encountered ability of biomolecules to easily accomplish actions associated with difficult and extreme conditions in the artificial realm, such as the catalysis of many chemical reactions, and exploiting optical nonlinearity with single photons, a feature that can be exploited to construct all-optical computers.

FURTHER READING

1. F.A. Armstrong, H.A.O. Hill, N.J. Walton, Reactions of electron-transfer proteins at electrodes, Q. Rev. Biophys. 18 (1986) 261–322.
2. L.A. Blumenfeld, Problems of Biological Physics, Springer, Berlin, 1981.
3. J. Chen, N. Jonoska, G. Rozenberg (Eds.), Nanotechnology: Science and Computation, Springer, Berlin, 2006.
4. G.S. Chirikjian, K. Kazerounian, C. Mavroidis, Analysis and design of protein-based nanodevices: challenges and opportunities in mechanical design, J. Mech. Design 127 (2005) 695–698.
5. A. Dér, L. Keszthelyi (Eds.), Bioelectronic Applications of Photochromic Pigments, IOS Press, Amsterdam, 2001.
6. J.J. Ramsden, Bioinformatics: An Introduction, third ed., Springer, London, 2015. See chapters 10 and 11 for a succinct introduction to biology.
7. F. Scheller, F. Schubert, Biosensoren, Akademie-Verlag, Berlin, 1989.
8. J.-M. Valleton, Information processing in biomolecule-based biomimetic systems. From macroscopic to nanoscopic scale, Reactive Polym. 12 (1990) 109–131.
9. M. Yoshida, E. Muneyuki, T. Hisabori, ATP synthase—a marvellous rotary engine of the cell, Nature Rev. Mol. Cell Biol. 2 (2001) 669–677.
10. J. Youell, K. Firman, Biological motors for nanodevices, Nanotechnol. Percept. 3 (2007) 75–96.

The impact of nanotechnology

12

CHAPTER CONTENTS

INTRODUCTION

One might well wonder, after having been confronted by all this marvelous technology and potential technology, whether it has the potential for positively benefiting civilization by contributing to the elevation of society. This final chapter examines the actual and potential impacts of nanotechnology. These introductory paragraphs provide an overview; scientific and technical revolutions are considered, in order to decide whether nanotechnology merits being considered as one; more

Nanotechnology: An Introduction. DOI: 10.1016/B978-0-323-39311-9.00018-2
Copyright © 2016 Elsevier Inc. All rights reserved.

detailed coverage of individual areas is then given: scientific, technical, economic, environmental, social and finally ethical aspects. There is of course overlap between these areas and the division between this to be somewhat arbitrary. Technical impacts, in other words applications, are covered in the greatest detail, focusing attention on the "big three" areas – computing, energy and health. Here too there is overlap between them.

It is necessary to consider what might be called both "soft" and "hard" aspects of nanotechnology, in the sense of unexceptionable and controversial, respectively. The former corresponds to the near-term – in some cases what is already realized; the latter corresponds to the long-term, that is, PN, as embodied by the personal nanofactory. As with any long-term extrapolation of technology, there is a considerable element of speculation regarding the latter, especially regarding timescales.

Applications can be considered as both direct and indirect. An example of the former is a nanoparticle that functions as a medicinal drug and can be injected as such directly into the bloodstream of a patient. An example of the latter is an information processor (computer) based on VLSI chips with individual circuit components in the nanoscale (but the overall size of the device and that of many essential peripheral components is bigger than the nanoscale); the direct application of the nanotechnology is to the realization of the integrated circuit; the many applications of the circuitry count as indirect applications of nanotechnology.

Can nanotechnology help to solve the great and pressing problems of contemporary humanity? Although, if ranked, there might be some debate about the order, most people would include rapid climate change, environmental degradation, energy and other resource depletion, unfavorable demographic trends, insufficiency of food and nuclear proliferation among the biggest challenges. Seen from this perspective, nanotechnology is the continuation of technological progress, which might ultimately be revolutionary if the quantitative change becomes big enough to rank as qualitative. For example, atom-by-atom assembly of artifacts implies, conversely, that discarded ones can be disassembled according to a similar principle, hence the problem of waste (and concomitant environmental pollution) vanishes. More advanced understanding at the nanoscale should finally allow us to create artificial energy-harvesting systems, mimicking photosynthesis, hence the potential penury of energy disappears. If the manufacture of almost everything becomes localized, the transport of goods (a major contributor to energy consumption and environmental degradation) should dwindle to practically nothing. Localized energy production would have a similar effect, eliminating the need for a vast distribution infrastructure. However, the achievement of this ultimate state of affairs depends on the advent of the personal nanofactory, or something resembling it, which is by no means inevitable. Perhaps the miniature medical robot (often called the nanobot) is somewhat closer to realization – in essence it is simply a more advanced version of the responsive drug delivery nanoparticles that are already being deployed. Would indefatigably circulating nanobots inside our bodies enable our lives to be extended almost indefinitely? And, if so, what would be the consequences?

Reports published during the last few years are typically euphoric about nanotechnology and all the benefits it will bring. Many of the examples are, however, of a relatively trivial nature and do not seem to represent sufficient breakthrough novelty to constitute a revolution. Thus, we already have nanostructured textiles that resist staining, self-cleaning glass incorporating nanoparticulate photocatalysts capable of decomposing dirt (Fig. 7.25); nanoparticle-based sun creams that effectively filter out ultraviolet light without scattering it and are therefore transparent; even lighter and stronger tennis rackets made with carbon fiber or even carbon nanotube composites; and so forth. None of these developments can be said to be truly revolutionary in terms of impact on civilization. Indeed, they are rather low-profile; one might not even notice them if they were not pointed out. In contrast, the Industrial Revolution was very visible because of the colossal size of its products: gigantic bridges (e.g. the Forth bridge), gigantic steamships (e.g. the *Great Eastern*) and, most gigantic of all if the entire network is considered as a single machine, the railway. And the steel for these constructions was produced in gigantic works (a modern chemical plant or motor-car factory may cover the area of a medium-sized town). In sharp contrast, the products of nanotechnology are, by definition, very small. Individual assemblers would be invisible to the naked eye. But of course the *products* of the assemblers would be highly visible and pervasive – such as the ultralight strong materials from which our built environment might be constructed.

12.1 TECHNICAL REVOLUTIONS

The anatomically modern human being, *Homo sapiens*, emerged some 200,000 years ago. Over 100,000 years apparently elapsed before his superior mental capacities began to be exploited, as evidenced by sophisticated hunting tools, figurative arts, bodily embellishment (jewelry, etc.) and musical instruments, which have been found in Africa dating from 90,000 years before the Christian era (BCE), although the initial bursts seem to have later died out; about 45,000 years ago one finds similar evidence in Europe and western Asia (the so-called upper Palaeolithic explosion). The first writing (cuneiform script) dates from as recently as around 3000 BCE, and comprehensive written records of the past (i.e. the beginning of history) only emerged around 600 (China)–500 BCE (Greece).

The development of man is marked by technological breakthroughs, especially those concerning materials processing. So important are they that the technologies give their names to the successive epochs of prehistory [the Stone Age (predating the emergence of *Homo sapiens* by over three million years), so-called because of simple stone implements, for example, axes, made by knapping flint (pottery first seems to have been made around 20,000 BCE, but it took another 15,000 years before the potter's wheel was invented); the Bronze Age (starting about 3000 BCE); the Iron Age (starting about 1000 BCE, coincidentally with the emergence of glassmaking)] rather than modes of life such as hunting, agriculture (perhaps starting about 17,000 BCE – somewhat earlier than the Neolithic Age, which is considered to have begun around

8000 BCE) and pastoralism (around 5000 BCE). The beginning of urbanization is often considered to be marked by the Sumerian city of Uruk (around 3500 BCE) with an estimated population of about 50,000, believed to have been the largest settlement in the world at the time. The capital city of Ur covered 15 hectares 500 years later. The roots of urbanization, however, may be traced back to as early as 17,000 BCE, when the site of Kharaneh IV (in Transjordan) was first occupied on a seasonal basis; it was to remain in use for over a millennium and covered an area of about 2 ha.

The most significant change in our way of life during the last two or three millennia was probably that brought about by the Industrial Revolution, which began in Britain around the middle of the eighteenth century and marked the beginning of the Industrial Age; by the middle of the nineteenth century it was in full swing in Britain and, at first patchily but later rapidly, elsewhere in Europe and North America. Note that this revolution, unlike its predecessors, was very much production-oriented – in other words, manufacturability was as much a consideration for what was produced as usefulness. A corollary was that an advertising industry was then required to persuade people to buy the many artifacts that could be cheaply produced in large quantities, but whose usefulness was not immediately apparent. This in turn was replaced in the latter half of the twentieth century by the still ongoing Information Revolution, which ushered in the Information Age, marked by the development of unprecedented capabilities in the gathering, storage, retrieval and analysis of information, and heavily dependent upon the high-speed digital electronic computer. The next revolution already appears to be on the horizon, and it is thought that it will be the Nano Revolution. Insofar as the vast increase of availability of cheap computing power is nano-enabled, one can also say that the current trend in the advertising industry, to direct advertisements to individuals on the basis of their Internet browsing history and so forth, is a nanophenomenon, because it depends on the availability of huge computing and memory resources.

Within a revolution capabilities grow exponentially – one could even say that the duration of the epoch of such exponential growth temporally defines the revolution. This is sometimes quite difficult to perceive, because an exponential function is linear if examined over a sufficiently small interval, and if the technology (or technological revolution) unfolds over several generations, individual perceptions tend to be strongly biased towards linearity. Nevertheless, empirical examination of available data shows that exponential development is the rule (Ray Kurzweil has collected many examples, and in our present epoch the best demonstration is probably Moore's law), although it does not continue indefinitely, but eventually levels off.

Very often a preceding technological breakthrough provides the key to a successive one. For example, increasing skill and knowledge in working iron was crucial to the success of the steam power and steel that were the hallmarks of the Industrial Revolution, which ultimately developed the capability for mass production of the VLSI electronic circuits needed for realizing the Information Revolution.

Why do people think that the next technological revolution will be that of nanotechnology? Because once the technology has been mastered, the advantages of making things "at the bottom", as Feynman proposed [85], will be so overwhelming

Table 12.1 The Infiltration of Science into Industry

Stage	Description	Characteristic Feature(s)
1	Increasing the scale of traditional industries	Measurement and standardization
2	Some scientific understanding of the processes (mainly acquired through systematic experimentation in accord with the scientific method)	Enables improvements to be made
3	Formulation of an adequate theory (implying full understanding of the processes)	Possibility of completely controlling the processes
4	Complete integration of science and industry, extensive knowledge of the fundamental nature of the processes	Entirely new processes can be devised to achieve desired ends

Based on J.D. Bernal, The Social Function of Science, Routledge & Kegan Paul, London, 1939.

it will rapidly dominate all existing ways of doing things. Once iron-making and -working had been mastered, no one would have considered making large, strong objects out of bronze; mechanical excavators now reign supreme on building sites; no one uses a slide rule now that electronic calculators are available, and even domestic appliances such as washing machines are controlled by a microprocessor.

Is it to be expected that IT will be crucial for the realization of nanotechnology? Very probably yes. As explained in Chapter 10, the design of nano materials and systems will be heavily dependent upon computation. Furthermore, nanofacture is scarcely conceivable without computer-enabled automation of (bottom-to-bottom) assembly.

Another consideration is that the Nano Revolution will consummate the trend of science infiltrating industry that began with the Industrial Revolution. As J.D. Bernal has pointed out, this infiltration can be roughly described in four stages of increasing complexity (Table 12.1). Clearly nanotechnology belongs to stage 4, at least in its aspirations. Traditional or conventional technologies (as we can label everything that is not nanotechnology) also have stage 4 as their goal but in most cases are still quite far from realizing it.

Note that stage 4 also encompasses the cases of purely scientific discoveries (e.g. electricity) being turned to industrial use. Nanotechnology is the consummation of stage 4, a corollary of which is that nanotechnology should enable science to be applied at the level of stage 4 to even those very complicated industries that are associated with the most basic needs of mankind, namely, food and health.

Consideration of the anticipated impacts of nanotechnology on society needs to be set in the general context of technology impacts. Bernal has pointed out the difficulties that arise from the discrepancy between the primitive needs of man, which are actually extraordinarily complex from the scientific viewpoint (e.g. the biochemistry of food preparation, and the animal psychology involved in hunting and domestication); the problems in these fields were solved empirically long before understanding was achieved, which needs to proceed from the simple to the complex: what can be understood rationally must necessarily be simple, at least to begin with. Unfortunately, the simplest sciences, astronomy and mechanics, appeared (around 3000 and 400 BCE, respectively) only after the main techniques of human life had already been fixed. As a result, these techniques – encompassing such things as agriculture, cookery, husbandry, metalwork, pottery and textiles – remained almost unchanged for many centuries at least until the early eighteenth century, largely untouched by the scientific method. Subsequent attempts to improve them "scientifically" have actually led to a mixture of benefits and disbenefits, and rational expectations of the impacts of nanotechnology must be tempered by this past history. The Industrial Revolution led to a focus on machine production rather than direct human needs such as food and health. A fundamental difference between the Nanotechnology Revolution and the Industrial Revolution is that the former is supposed to be consumer-oriented, unlike the production orientation of the latter, hence should redress the balance back in favor of fulfilling human needs. The ultimate stage of nanotechnology, PN, in essence abolishes the difference between consumer- and production-orientation. Even now, some large commercial manufacturing corporations, rooted in the Industrial Revolution, are talking about "mass customization" – in which mass production encompasses cost-effective individualization – and "masstige" – prestige products nevertheless made on a large-scale (and possibly benefiting from individualization).

Nanotechnology has the potential of refocusing the way society satisfies its needs on the more human aspects, which is itself a revolutionary enough departure from what has been going on during the last 300 years to warrant the label Nanotechnology Revolution. Furthermore, the Nanotechnology Revolution is supposed to usher in what I.J. Good referred to as the "intelligence explosion", when human intelligence is first surpassed by machine intelligence, which then rapidly spreads throughout the entire universe. This is what Kurzweil calls the singularity – the ultimate revolution.

12.2 SCIENTIFIC IMPACTS

Nanotechnology implies scrutinizing the world from the viewpoint of the atom or molecule, while remaining cognizant of structure and process at higher levels. Practically speaking, this should have a huge impact on applied science, in which domain the most pressing problems facing humanity, such as food, energy and other resource security, fall (and which are covered in Section 12.3). Nanotechnology will hopefully give humanity new impetus to formulating and solving the problems in a more rational way, according to which it is first ascertained whether the problem requires new fundamental knowledge for its solution, or whether it "merely" requires the application of existing fundamental knowledge.

Nanotechnology is often considered not in isolation, but as one of a quartet: nanotechnology, biotechnology, IT and cognitive science (nano–bio–info–cogno, NBIC, Section 1.7). Nanotechnology is itself inter-, multi- and trans-disciplinary and associating it with this quartet of emerging technologies further emphasizes that catholicism.

There is already impact on science from nanometrology, the instrumentation of which is useful are the fields other than that of nanotechnology itself.

12.3 TECHNICAL IMPACTS

This section covers the main anticipated fields of applications. Apart from the "big three" (IT, health and energy), there are other areas that are also due to benefit from nanotechnology, notably chemical processing (e.g. through better, rationally designed and fabricated, catalysts). This will partly be given consideration under energy. General purpose material manipulation capability will doubtless enable the anticipated future shortages of rare elements required for current technologies to be overcome.

12.3.1 INFORMATION TECHNOLOGY (IT)

IT applications are often called indirect, since the main impact does not arise directly from the nanoscale features of a VLSI circuit, but from the way that circuit is used. Nanotechnology is well expressed by the continuing validity of Moore's law, which asserts that the number of components on a computer chip doubles every 18 months. Feature sizes of individual circuit components are already below 100 nm; even if the basic physics of operation of such a nanotransistor is the same as that of its macroscale counterpart, the ability, through miniaturization, of packing a very large number of components on a single chip enables functional novelty.

There is little difference between hard and soft in this case, because there is a continuous drive for miniaturization and whether a nanoscale transistor is made by the current top–down methodology of the semiconductor industry or by atom-by-atom assembly should not affect the function. Other developments, notably quantum computing, that are not specifically associated with nanotechnology but have the same effect of increasing the processing power of a given volume of hardware, will compete.

This indirect nanotechnology is responsible for the ubiquity of Internet servers (and, hence, the World Wide Web) and cellular telephones. The impact of these information processors is above all due to their very high-speed operation, rather than any particular sophistication of the algorithms governing them. Most tasks, ranging from the diagnosis of disease to ubiquitous surveillance, involve pattern recognition, something that our brains can accomplish swiftly and seemingly effortlessly for a while, until fatigue sets in, but which requires huge numbers of logical steps when reduced to a form suitable for a digital processor. Sanguine observers predict that despite the clumsiness of this "automated reasoning", ultimately artificial thinking will surpass that of humans – this is Kurzweil's "singularity". Others predict that it will never happen. To be sure, the singularity is truly revolutionary, but is as much a product of the Information Revolution as of the Nano Revolution, even though the latter provides the essential enabling technology.

Information processing and storage constitutes the most "classical" part of nanotechnology applications, in the sense that it was the most readily imaginable at the time of the Feynman lecture [85]. The physical embodiment of one bit of information could be, in principle, the presence or absence of a single atom (the main challenge is reading the information thus embodied). The genes of the living world, based on four varieties of DNA, come quite close to this ultimate limit and the reading machinery is also nanosized.

Ever since the invention of writing, man has been storing information but the traditional technologies, whether clay tablets or books, are voluminous, whereas the miniaturization of storage that has already been envisaged creates what is essentially unlimited capacity (including, for example, the "life log" – a quasicontinuous record of events, actions and physiological variables of every human being).

12.3.2 ENERGY

Nanotechnology has the opportunity to contribute in several ways to the problem of energy, which can be succinctly expressed as the current undersupply of usable energy and the trend for the gap to get worse. The principle near-term technical impacts of nanotechnology will be as follows:

Renewable energies: There is expected to be direct impact on photovoltaic cells. The main primary obstacle to their widespread deployment is the high cost of conventional photovoltaic cells. Devices incorporating particles (e.g. Grätzel cells) offer potentially much lower fabrication costs. The potential of incorporating further complexity through mimicry of natural photosynthesis, the original inspiration for the Grätzel cell, is not yet exhausted. The main secondary obstacle is that except for a few specialized applications (such as powering air-conditioners in Arabia) the electricity thus generated needs to be stored, hence the interest in simultaneous conversion and storage (e.g. Fig. 7.25). Undoubtedly natural photosynthesis is only possible through an extremely exact arrangement of atoms within the photosystems working within plant cells, and the more precisely artificial light harvesters can be assembled, the more successful they are likely to be. Such precision biomimicry is still some way off, however. Meanwhile, more and more carefully nano-engineered artificial structures are being created for solar cells [171].

Fuel cells: Although the scientific basis of this technology, whereby fuel is converted to electricity directly, was established over 150 years by Christian Schönbein it has been very slow to become established. As with photovoltaic cells, the main primary obstacle is the high cost of fabrication. Nanotechnology is expected to contribute through miniaturization of all components (especially diminishing the thickness of the various laminar elements), simultaneously reducing inefficiencies and costs, and through realizing better catalysts for oxygen reduction and fuel oxidation. A particular priority is developing fuel cells able to use feedstocks other than hydrogen.

Energy storage: The primary means of storing energy is as fuel, but unless photo-electrochemical cells generating fuel from sunlight receive renewed impetus, renewable sources will mainly produce electricity and except for some special cases at least some of the electricity will have to be stored to enable supply to match demand. Supercapacitors based on CNTs have attracted interest, but the impact of nanotechnology is likely to be small, since using ordinary carbon black has already enabled over 90% of the theoretical maximum charge storage capacity to be achieved, at much lower cost. In any case, the classic photoelectrochemical process generates hydrogen (from water), the storage of which is problematical. Through the design and fabrication of rational storage matrix materials, nanotechnology should be able to contribute to effective storage, although whether this will tip the balance in favor of the hydrogen economy is still questionable. An attractive novel approach is nano-engineered materials with ultrahigh thermal capacities.

Energy transmission: The so-called renewable energies, such as wind turbines, are generally associated with high transmission costs because the point of generation is far from the point of use. In some cases, the extra costs incurred by transmission, which implies not only cables but also conversion from one type of electricity to another, makes the concept nonsensical from an engineering viewpoint [236]. This is especially true for offshore wind turbines. Although certain nanomaterials, such as CNTs of metallic chirality, have superior electrical conductivity and could, therefore, potentially reduce transmission losses, the present state of such technology stands some way before industrial realization. It may well be that the best application of photovoltaic cells is to niches such as isolated electricity-consuming devices for which the cost of connecting to the electricity grid would be high, and certain portable or mobile appliances.

Energy efficiency: This heading comprises a very heterogeneous collection of technical impacts. Nanostructured coatings with very low coefficients of friction and extremely good wear resistance will find application in all moving machinery, hence improving its efficiency and reducing the energy required to achieve a given result. Examples include electricity-generating wind turbines and electric motors. For all applications where collateral heat production is not required, nanotechnology-enabled light-emitting diodes can replace incandescent filaments. They can achieve a similar luminous output for much less power than incandescent lamps. The heat produced by the latter may be of value in a domestic context (e.g. contributing to space heating in winter); it is simply wasted in the case of outdoor streetlighting (although the esthetic effect is very agreeable). Note that, however, the actual operational efficiency of a device in a given functional context typically represents only a fraction of the overall effectiveness in achieving the intended function. For example, if the intended function of streetlighting is to reduce road accidents, there is probably an ergonomic limit to the number of lamps per unit length of street above which the reduction becomes insignificant. Although data are hard to come by, it seems that few amenities have been properly analysed in this manner. It may well be a 50% reduction in the number of lamps could be effected without diminution of the functional effect. Such a reduction would be equivalent to a really significant technological advance in the device technology. Miniaturizing computer chips diminishes the heat dissipated per floating point operation (but not by as much as the increased number of floating point operations per unit area enabled by the miniaturization, cf. Section 7.5). Devices in which bits are represented as electron spins rather than as electron charges will dissipate practically no heat.

Resource extraction: Current technologies used to extract metal from ores use vastly more energy than is theoretically required. Nanotechnology can be brought to bear on this problem in many different ways. Biomimicry seems a very attractive route to explore, especially since living organisms are extremely good at extracting very dilute raw materials from their environment, operating at room temperature and, seemingly, close to the thermodynamic limit.

Localized manufacture: Possibly the greatest ultimate contribution of nanotechnology, once the stage of the personal nanofactory has been reached, to energy conservation will be through the great diminution of the need to transport raw materials and finished products around the world. The amount of energy currently consumed by transport in one form or another is something between 30 and 70% of total energy consumption. A reduction by an order of magnitude is perhaps achievable.

The above are realizable to a degree using already available nanotechnology – the main issue is whether costs are low enough to make them economically viable.

12.3.3 HEALTH

Impacts can be summarized under the following headings:

Diagnosis: Many diagnostic technologies should benefit from the contributions of nanotechnology. Superior contrast agents based on nanoparticles have already demonstrated enhancement of tissue imaging. Nanoscale biochemical sensors offer potentially superior performance and less invasiveness, to the extent that implanted devices may be able to continuously sense physiological parameters. More powerful computers able to apply pattern recognition techniques to identify pathological conditions from a multiplicity of indicators provide an indirect diagnostic benefit.

Therapy: The most prominent development has been the creation of functionally rich nanostructured drug packaging, enabling the more effective delivery of awkward molecules to their targets. Implants with controlled drug-eluting capability enabled by nanostructured coatings have also been demonstrated. Indirect therapeutic impacts include more powerful computers accelerating the numerical simulation stage of drug discovery, and nano-enabled microreactors facilitating the affordable production of customized medicines, which depends at least in part on the availability of individual patient genomes.

Surgery: Miniaturized devices are making surgery less and less invasive. The basic surgical tools are unlikely to be miniaturized below the microscale, however, but their deployment may be enhanced by nanoscale features such as ultralow friction coatings and built-in sensors for *in situ* monitoring of physiological parameters. The autonomous robot ("nanobot") is still some way in the future. Superior nanostructured tissue scaffolds will enhance the ability to regenerate tissue for repair purposes. Advanced materials, especially sensorial materials [176], may even allow malfunctioning organs to be replaced (e.g. in the pancreas could be replaced by an insulin-releasing polymer, which can regulate the release according to environmental conditions).

In medicine, scrutinizing the world from the viewpoint of the atom or molecule amounts to finding the molecular basis of disease, which has been underway ever since biochemistry became established, and which now encompasses all aspects of disease connected with the DNA molecule and its relatives. There can be little doubt about the tremendous advance of knowledge that it represents. It, however, is part of the more general scientific revolution that began in the European universities founded from the eleventh century onwards – and which is so gradual and ongoing that it never really constitutes a perceptible revolution. Furthermore, it is always necessary to counterbalance the reductionism implicit in the essentially analytical atomic (or nano) viewpoint by insisting on a synthetic systems approach at the same time. Nanotechnology carried through to PN could achieve this, because the tiny artifacts produced by an individual assembler have somehow to be transformed into something macroscopic enough to be serviceable for mankind.

The application of nanotechnology to human health is usually called nanomedicine and is thus a subset of nanobiotechnology. To recall the dictionary definition, medicine is "the science and art concerned with the cure, alleviation and prevention of disease, and with the restoration and preservation of health, by means of remedial substances and the regulation of diet, habits, etc." Therefore, in order to arrive at a more detailed definition of nanomedicine, one needs to simply ask which parts of nanotechnology (e.g. as represented by Fig. 1.1) have a possible medical application.

Much of nanomedicine is concerned with materials. Quantum dots and other nano-objects, surface-functionalized in order to confer the capability of specific binding to selected biological targets, are used for diagnostic purposes in a variety of ways, including imaging and detection of pathogens. Certain nano-objects such as nanosized hollow spheres are used as drug delivery vehicles. Based on the premise that many biological surfaces are structured in the nanoscale, and hence it might be effective to intervene at that scale, nanomaterials are being used in regenerative medicine both as scaffolds for constructing replacement tissues and directly as artificial prostheses. Nanomaterials are also under consideration as high-capacity absorbents for kidney replacement dialysis therapy. Other areas in which materials are important and which can be considered to be sufficiently health-related to be part of nanomedicine include the design and fabrication of bactericidal surfaces (e.g. photocatalytic coatings for hospital interiors, which under illumination that does not even have to be continuous keep surfaces permanently sterile by destroying any organic matter that arrives – in this case nanotechnology enters into the process for creating the coatings).

The advantages of miniaturized analytical devices for medicine ("labs-on-chips"), especially for low-cost point-of-care devices, are already apparent at the microscale. Further miniaturization would continue some of these advantageous trends, resulting *inter alia* in the need for even smaller sample volumes and even lower power consumption, although some of the problems associated with miniaturization such as unwanted nonspecific binding to the internal surfaces of such devices would be exacerbated. However, whereas microscale point-of-care devices rely on the patient (or his physician) to take a sample and introduce it into the device, miniaturization down to the nanoscale would enable such devices to be implanted in the body and, hence, able to monitor a biomarker continuously. The apotheosis of this trend is represented by the nanobot, a more or less autonomous robotic device operating in large swarms of perhaps as many as 10^9 in the bloodstream.

As well as analysis, synthesis (especially of high-value pharmaceuticals) is also an area where microtechnology is contributing through microscale mixers. This technology is very attractive to the pharmaceutical manufacturing industry. Enough evidence has accumulated for it to be generally recognized that many drugs are typically efficacious against only part of the population. This may partly be due to genetic diversity, and partly to other factors, which have not yet been characterized in molecular detail. In the case of genetic diversity, it is possible that almost all the pharmaceutically relevant DNA sequence variants occur in haplotype blocks, regions of 10,000 to 100,000 nucleotides in which a few sequence variants account for nearly all the variation in the world human population (typically, five or six sequence variants account for nearly all the variation). If a drug that has been found to be efficacious against the majority haplotype variant can be made to be efficacious against the others by small chemical modifications, then the task of the drug developer might not be insuperable.

At present, clinical trials do not generally take account of haplotype variation. Advances in sequencing, in which nanotechnology is helping both through the development of nanobiotechnological analytical devices (although it seems that microtechnological "labs-on-chips" may be adequate to fulfill needs) and through more powerful information processing, should make it possible in the fairly near future for haplotype determination to become routine. It is, however, not known (and perhaps rather improbable) whether small modifications to a drug would adapt its efficacity to other haplotype variants. At any rate, different drugs will certainly be needed to treat different groups of patients suffering from what is clinically considered to be the same disease. Micromixers represent a key step in making custom synthesis of drugs for groups of patients, or even for individual patients, economically viable.

Although there is undoubtedly considerable interindividual variation in drug response, the results from large-scale attempts (such as genome-wide association studies) to link such variation to genetic variation have been disappointing. The large and expensive international HapMap project was designed to quantify linkage disequilibrium relationships among human DNA polymorphisms in an assortment of populations in order to identify markers for disease risk. However, it was found that "genetic variation did not improve on the discrimination or classification of predicted risk" [233] and "treatment based on genetic testing offers no benefit compared to without testing" [68]. Therefore, one should be cautious about asserting the potential benefits from the so-called personalized medicine.

The precise control of the hydrodynamic régime that is attainable (see Section 2.4) typically enables reactions that normally run to what are considered to be quite good yields of 90% say to proceed at the microscale without by-products (that is, a yield of 100%). The advantages for the complicated multistep syntheses typically required for pharmaceuticals are inestimable. Added to that advantage is the ease of scaling up the volume of synthesis from a laboratory experiment to that of full-scale industrial production simply by scaleout; that is, multiplication of output by the addition of identical parallel reactors.

It is not, however, obvious that further miniaturization down to the nanoscale further enhances performance. Similar problems to those afflicting nanoscale analytical labs-on-chips would become significant, especially the unwanted adsorption of reagents to the walls of the mixers and their connecting channels.

Two further aspects of nanomedicine deserve a mention. One is automated diagnosis. It is an inevitable corollary of the proliferation of diagnostic devices enabled by their miniaturization that the volume of data that needs to be integrated in order to produce a diagnosis becomes overwhelming for a human being. Interestingly, according to a University of Illinois study by Miller and McGuire (quoted by Fabb [72]), about 85% of medical examination questions require only recall of isolated bits of factual information. This suggests that automated diagnosis to at least the level currently attainable by a human physician would be rather readily realizable. The program would presumably be able to determine when the situation was too complicated for its capabilities and would still be able to refer the case to one or more human agents in that case. Indeed, medicine has already become accustomed to depending on heavy computations in the various tomographies that are now routine in large hospitals. Diagnosis is essentially a problem of pattern recognition: an object (in this case, the disease) must be inferred from a collection of features. Although there have already been attempts to ease the work of the physician by encapsulating his or her knowledge in an expert system that makes use of the physician's regular observations, significant progress is anticipated when measurements from numerous implanted biosensors are input to the inference engine. This is an example of indirect nanotechnology: the practical feasibility depends on the availability of extremely powerful processors, based on chips having the very high degree of integration enabled by nanoscale components on the chips.

The second aspect is drug design. In many cases, useful targets (enzymes, ion channels, promoter sites on DNA) for therapeutic action are already known through molecular biology work. The key tasks for the designer in current practice are (i) to generate a collection of molecules that bind to the target and (ii) to select those that bind negligibly to structurally related but functionally unrelated or antagonistic targets. Task (i) may nowadays be largely accomplished automatically through molecular modeling (trial-and-error docking), and thus requires high-performance (possibly nano-enabled) computing. Task (ii) is a germane problem, the solution of which is mainly hampered through ignorance of the rival targets. However, the growing importance of scrutinizing biological processes not only from the viewpoint of what is now traditional molecular biology but also from that of the nano-engineer is yielding new insight (a particularly valuable example of such new insight is the characterization of the propensity of biomolecules to bind according to their dehydron density [83]).

Since diet is an important contribution to health, nanotechnology applied to foodstuffs could also be considered to be part of nanomedicine. This is of course a vast area, ranging from nanoscale sensors to discreetly probe food quality to nanoscale food additives (to create "neutriceuticals") to nanoscale field additives such as fertilizers, pesticides and enhancers of agriculturally advantageous natural symbioses. Since water is ingested by humans in greater quantity than any other substance, it might also be considered as food, making nanotechnologies applied to water purification also part of nanomedicine. Furthermore, creating micronutrients in nanoparticulate form permits substances that are normally considered too insoluble to be of nutritional value to be administered (cf. Eq. (2.7)).

Finally, human enhancement (HE) is often considered to be inseparably associated with nanotechnology, or the nano–bio–info–cogno quartet. In contrast to traditional medicine, which is mainly concerned with repairing damage, HE seeks to improve on existing design. Some aspects of such improvement are so commonplace – such as drinking coffee or smoking a cigarette to improve concentration on a particular task – that they would barely be considered as HE, the minimal level of which appears to involve organ replacement. Cosmetic surgery – intervention to change body shape – could be considered as a rudimentary form of HE, even though there is generally no functional change. Nanotechnology is involved because most living organs are so intricate that artificial surrogates require close to atomic-scale precision in order to mimic them. In a few cases function (e.g. of the kidney) can be mimicked by an artificial device (a renal dialysis machine), but it is so large the kidney patient has to go to the machine rather than the machine being incorporated inside the patient. Extreme supporters of HE advocate the systematic replacement of all body parts by artificial surrogates in order to prolong the existence of the individual indefinitely. Any kind of HE raises ethical issues; ethics are part of the human "system", as defined in Section 10.1.

12.4 COMMERCIAL AND ECONOMIC IMPACTS

Although, if the many nanotechnology market reports are to be believed, the technology sector is growing robustly, this growth will be unsustainable unless issues of standardization and risk are addressed. At present, the nanoscale products (e.g. nanoparticles and nanofibers) supplied by many companies are essentially research grades sold on the basis of *caveat emptor*. Standardization implies robust nanometrology (cf. Chapter 5) in order to be able to verify compliance with specifications; the existence of the specifications in turn implies the possibility of freely tradable commodities via an exchange. Once this begins to happen on a significant scale, the nanotechnology industry will have reached a certain stage of maturity and will be able to continue to develop more and more sophisticated commodities, such as sensorial materials.

The economic impacts of the technological innovations considered in the previous section are likely to be relatively minor as far as the "soft" implementations of nanotechnology are concerned. There will be many innovations but these will in turn breed new demands and one cannot therefore expect a dramatic change in work–life balance. At least, this has been the experience of the developed world for the last century or so and there is no reason to expect radical change from what are essentially incremental improvements. For example, many of the new nanostructured drug delivery vehicles will lead to therapeutic enhancement, but once the old ailments are cured, new ones are likely to be found requiring further new drugs. In contrast, the "hard" implementation – the personal nanofactory – will lead to very dramatic changes in terms of personal empowerment. The entire system of the joint stock company concentrating resources into large, central, capital-intensive factories will become obsolete.

It is often stated that the driver behind Moore's law is purely economic. This is not, however, self-evident. It presumably means that growth is dominated by "market pull" rather than "technology push", but this raises a number of questions. It is doubtful whether there are intrinsic markets, and even if there are why should the technology advance exponentially? How could the actual doubling time be predicted? Does it depend on the number of people working on the technology? If they are mostly employed by large corporations (such as semiconductor chip manufacturers) the ability of those corporations to pay the salaries of the technologists indeed depends on economic factors, but this would not hold in the case of open source development, whether of software or hardware. Similar questions apply to all technical innovations, which generally grow exponentially. Technology push seems to require fewer assumptions as an explanation than market pull. However, it may be erroneous to say that the dynamics must reflect either technology push or market pull. Both may be in operation; in analogy to the venerable principle of supply and demand, push and pull may also "equilibrate", as illustrated in Fig. 12.1.

Further aspects connected with "push" and "pull" are the degree to which technology changes society, and the degree to which society is ready for change. The Internet – first e-mail and now social networking sites – has enormously changed

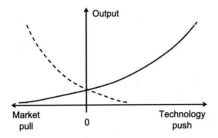

FIGURE 12.1

Proposed quasi-equilibrium between technology push (solid line) and market pull (dashed line). The ideal level of output occurs where they exactly match each other. This diagram neglects consideration of possible temporal mismatch between push and pull.

the way people communicate with one another as individuals. It cannot be said that there was a market for these innovations, or for mobile telephony; the market was somehow created. The latter, in particular, epitomizes many aspects of a sustainable commercial activity; most users are eager to keep up with the latest technological innovations that imply a high degree of personal enthusiasm among the technology developers. Although these technologies are operating and are already nano-enabled (through their ultraminiature information processors), the economic model behind them is basically that of the Industrial Revolution. The new communications media are heavily infiltrated by commercial advertising. Internet search engines and social networking sites are not actually floating somewhere in cyberspace but are firmly anchored to strong financial interests and user preferences can easily be linked to advertising via the cellular phone. All this activity has taken on a strong cultural dimension, since for it to be widely acceptable culture has somehow had to become dominated by its commercial and popular side, virtually squeezing out anything else. Although this general trend appears to be fully in the spirit of Adam Smith's "invisible hand" – every member of society acting in his or her best interest and thereby contributing to the general good – there seems to be a growing asymmetry between "producers"   (production itself is anyway becoming more and more automated) and "consumers". Whereas previous visions extrapolating the trend of ever more efficient production, which implies ever increasing leisure time, have thought of this leisure time being used for personal cultural enrichment and "working out one's own salvation", presumably because of the preponderance of IT today personal leisure is also nowadays characterized by "producers" and "consumers". Even though the technology actually enables anyone to be an author, film-maker, journalist or singer, we have not arrived at increasing micro-production of culture, with works produced for small audiences, maybe of just a few dozen, with everyone more or less on an equal footing. These potentially very enriching aspects of technological development have been, at least until now, largely thrown away.

IT has already enabled the "personal nanofactory" of culture to be realized (but little use has been made of this possibility by the vast majority). If and when the real personal nanofactory capable of fabricating virtually any artifact becomes established, will things be different? Open source software demonstrably works and its products seem on the whole to be superior to those produced in closed, "professional" companies. Attempts are already being made to mimic this model with open source hardware. Nanotechnology implies that everyone can be a designer, or at least contribute to design in the case of complex technology. Among other things, this implies the complete obsolescence of patents and intellectual property laws. At present, nanotechnology is at best seen as an enabler of "mass customization", taking a little further existing technologies that allow the purchaser of house paint, and possibly of a T-shirt or a motor-car, to specify a unique, individual color. Ultimately, though, everyone *could* be as much a producer as a consumer, of hardware as well as software. This would render the present commercial model obsolete.

Although some nanoproduction facilities – most notably for semiconductor processing – are so expensive that even many developed countries cannot afford them, many other aspects of nanotechnology require rather minimal investment in infrastructure. This has rendered nanotechnology in principle attractive for the developing world, offering developing countries a rapid route to technological and economic development. Some indication of what might be achievable is provided by the rapid growth of cellular telephony in Africa, which now has a greater density of cellphone users than developed countries, although its landline infrastructure lags far behind. This technology was not, however, developed indigenously but imported. In order to have a real impact on development, the technology would have to be developed locally. Having some flagship nanotechnology projects in the van would create disequilibrium [126], resulting in development of the entire economy, since the technological demands of having any success at all in an advanced system of production force the rest of the economy to develop in its train. Disappointingly, and for reasons that cannot be gone into in this book, no developing countries have hitherto chosen this route, but the potential is surely there.

12.5 ENVIRONMENTAL IMPACTS

One often hears it stated that nanotechnology will enable the environment to be returned to a pristine state, without any explanation of the process by which this rather vague assertion might be realized being offered. It seems that there are going to be two principal impacts of nanotechnology on the environment. The first is immediate and direct, the second long-term and indirect. The first is concerned with the use of nanoparticles for environmental remediation. In particular, iron-containing nanoparticles are being promulgated as superior alternatives to existing remediation procedures for soil contaminated with chlorinated hydrocarbons using a more or less comminuted scrap iron. This proposed technology raises the number of questions – to start with there does not seem to be any conclusive evidence that it is actually efficacious, and furthermore there is the still open question of the effect of dispersing a significant concentration of nanoparticles (and it has to be significant, otherwise there would be no significant remediation) on the ecosystem, especially microbial life.

Fortunately, there appears to be a general realization of this potential danger, and many countries have now imposed a moratorium on the large-scale deliberate release of nano-objects into the environment, thereby avoiding the same kind of large-scale social experiment that is, regrettably, under way with genetically modified organisms (GMOs). However great the purported benefits, they cannot justify society being used as an experimental laboratory, not least because there is then no control group; hence, the experiment lacks scientific validity. Furthermore, a comprehensive assessment of the benefits is hitherto lacking, despite the availability of the objective and universal J-value technique for making such an assessment (Section 12.6.5) [285]. Attitudes with respect to GMOs seem to have become too entrenched to expect a very progressive development in this sphere; fortunately nanotechnology, being in an earlier stage of development, is associated with more positive and open attitudes.

The long-term and indirect effects follow from the obvious corollary of atomically precise technologies – they essentially eliminate waste. This applies not only to the actual fabrication of artifacts for human use but also to the extraction of elements from the geosphere (should those elements still be necessary, cf. 12.3.2). Furthermore, the localized fabrication implied by the widespread deployment of PN and personal nanofactories should eliminate almost all of the currently vast land, sea and air traffic involved in wholesale and retail distribution; this elimination (and a commensurate downscaling of transport infrastructure) will bring about by far the greatest benefit to the environment.

Nano-engineered "artificial kidneys" can be used not only to extract desirable elements from very dilute sources, such as seawater, but also to extract elements or extractable compounds from natural water polluted with them.

12.6 SOCIAL IMPLICATIONS

The focus in this section is on some of the collective activities of humankind that have not been covered in the preceding sections. Clearly the technical advances sketched out in Section 12.3 also have social implications – or example, localized medicine – diagnosis and treatment – will make large central hospitals obsolete, which will be quite a big social change.

If the present commercial model becomes obsolete, are there political implications? The political arrangements in most countries have co-evolved with commercial developments. If centralized production becomes obsolete, does that imply the same for centralized government?

The implications do not stop there. Cities arose because of the need to exchange goods and, even more importantly, information. If everyone is connected to everyone else via the Internet, and production is decentralized, it may be that the trend of the last few millennia towards ever more urbanization will be reversed.

12.6.1 REGULATION

There are already widespread calls for stricter regulation of the deployment of nanotechnology, particularly nano-objects in consumer products. These calls are driven by a growing awareness of the potential dangers of nano-objects penetrating into the human body, and by the realization that understanding of this process (cf. Section 4.4) is still rather imperfect, and prediction of the likely effects of a new kind of nano-object is still rather unreliable. Furthermore, there have been a sufficient number of cases, albeit individually on a relatively small-scale, of apparently unscrupulous entrepreneurs promoting nano-object-containing products that have turned out to be quite harmful.

These calls, while seemingly reasonable, raise a number of difficulties. One is purely practical: once nano-objects are incorporated into a product they are extraordinarily difficult to trace. Traceability is only generally feasible up to the point of manufacture, and even then only if the manufacturer has sourced materials through a regulated or self-regulated commercial channel such as a commodity exchange. Establishing the provenance of nanoparticles that might turn up in a waste dump, for example, poses a very difficult forensic challenge.

Furthermore, regulation will become essentially meaningless if PN become established: every individual would be producing his or her own artifacts according to his or her own designs and it is hard to see how this could be regulated.

12.6.2 MILITARY IMPLICATIONS

It is a striking fact that the mechanization of war has resulted in a reversion of society's involvement in fighting to an early epoch when every member of a tribe was essentially a warrior. During the last 3000 years, the trend in civilized countries was for fighting to become a highly specialized activity practised by a small minority. This changed dramatically in the First World War. Not only was the actual number of fighters a significant proportion of the active working population, but virtually the entire rest of the nation was actively working for the war effort, supporting the fighters. The nature of the warfare also changed. Military activity was beginning to become highly mechanized in the First World War, but it still saw battlefields of the kind epitomized by the Battle of Waterloo; that is, places where pitched battles between the professionals took place, in a manner analogous to a game played on a sports field. In the Second World War the tendency was rather to speak of battle theaters, and that war was marked by wholesale bombing of great civilian centers, made relatively easy by advances in the design and production of aircraft. More recently, this trend to involve the entire civilian population has continued through urban resistance movements turning civilian cities into battlegrounds.

Against this background, it is highly pertinent to inquire how nanotechnology will affect the social organization implicated in warfare. Clearly many of the technical advances brought about by nanotechnology are useful for military purposes. At present, even though the entire population might be involved in a war, it is still only the professionals who are armed; the civilian population might at best have handguns; they are essentially defenseless. Advances in nanotechnology that are put to belligerent use would appear to mainly serve the professional military services; the performance of munitions can be enhanced or made cheaper. But in the era of the personal nanofactory (PN), any individual could design and fabricate offensive weapons, such as small pilotless flying objects (drones). Drones have already become so cheap and ubiquitous that they are even being used for criminal activities, such as the delivery of supplies of prohibited articles to the inmates of jails. Thus, the ultimate impact of military nanotechnology is much greater for the civilian population, who will in principle be able to defend themselves rather effectively against aggression.

Nanotechnology represents the apotheosis of mechanization. On the other hand, the universal plenty that the Nano Revolution is supposed to usher in (Section 12.4) should largely remove the raison d'être for warfare in the first place.

12.6.3 TECHNICAL LITERACY

The real benefits of nanotechnology, which by changing the way we understand technology and science will also impact on the way we understand our relationship with the rest of the universe, can only be realized if there is a significant – at least tenfold – increase in the technical literacy of the general population. Technical literacy means the ability to understand an intelligible account of the technology (that is, an account in plain language not requiring any prior specialized knowledge). It would be tragic if the technology remains – like so many technologies today – a mysterious, impenetrable black box (cf. Section 12.2).

12.6.4 EDUCATION

There are doubtless implications at all levels, starting with the newborn infant and continuing throughout life. The commercial and political reorganization implied by the growth of nanotechnology provides an opportunity to thoroughly rethink the whole structure of education, especially the formal years of schooling from the ages of six to eighteen. The content of what is currently taught seems very far from what would be required for future designers of nanoartifacts.

During the last century or so, we have become accustomed to highly formalized education in which a growing state bureaucracy has played a large part, not only in building and running schools but also in devising and managing curricula. Yet, if the technical literacy associated with nanotechnology (Section 12.6.3) develops to a significant degree, the need for such tutelage will largely wither away, and parents will be capable, and hopefully willing, to once again take on the principal responsibility for their children's education. For that they will, of course, also have to have the time to be able to do so, but this should follow automatically from increased productivity – fewer hours should have to be spent in working. This is just another example of how the Nanotechnology Revolution will lead to far-reaching social changes that will go very far beyond merely devising new educational materials and objects with the help of nanotechnology, while leaving the fundamental aims and processes unchanged. The paradigm exemplified by every citizen designing their own material artifacts for fabrication on their personal nanofactory applies equally well to the possibility of their designing every aspect of their lives.

One nano-enabled trend (since it requires large IT resources) that is already well established is the existence of massive online open courses (MOOCs). These are lectures recorded by distinguished professors at the world's leading universities and posted on the Internet. Anyone can access them, and they are backed up with tutorial exercises and tests designed to be automatically marked. The learning experience might well be superior to that of being physically present as an undergraduate at some mediocre mass university.

12.6.5 RISK ASSESSMENT

In our present age of rationality, it is fortunate that a universal methodology has been developed [286], initially in order to assess safety systems, with which the benefits of any new technology can be compared with their costs. The benefits can be rationalized in terms of increased longevity, the value of which is quantifiable using actuarial tables, and quality of life, which is more difficult to quantify but work–leisure balance might be an acceptable measure. The costs are generally straightforward enough to identify and quantify, at least in principle.

The life quality index Q combines longevity (as discounted life expectancy, X_d), earnings G from work, and the average fraction of time spent working w. This assumes that people value leisure more highly than work. The index is defined as

$$Q = G^q X_d, \tag{12.1}$$

where q is defined as

$$q = w/(1 - w); \tag{12.2}$$

a value $q = 1/7$ seems to be typical for industrialized countries [286]. Raising G to the power q gives it the form of a utility function – initial earnings, spent on essentials, are valued more highly than later increments spent on luxuries. Averages over the relevant group (population N) under consideration (which might be an entire nation) for all the quantities contributing to Q are taken.

An individual may choose to divert a portion of his income ΔG into a safety measure that will prolong his life by an amount ΔX. Assuming ΔG and ΔX are small, expanding Eq. (12.1) and neglecting higher powers and cross-product terms yields

$$\Delta Q/Q = q\Delta G/G + \Delta X_d/X_d. \tag{12.3}$$

Since it makes no sense to spend more on safety than the equivalent benefit in terms of life prolongation, the right-hand side of Eq. (12.3) should be equal to or greater than zero. The limiting case, equality, may be solved for ΔG and multiplied by N to yield the maximum sensible annual safety spend

$$S_{\text{max}} = -N\Delta G = (1/q)NG\Delta X_{\text{d}}/X_{\text{d}}, \tag{12.4}$$

where the minus sign explicitly expresses the reduction in income. The J-value is defined as

$$J = S/S_{\text{max}}; \tag{12.5}$$

if $J > 1$, the expenditure S cannot be rationally justified.

12.6.6 DEMOGRAPHY

One can scarcely omit this topic, since population growth and other unfavorable demographic trends represent the biggest challenges to future human development, measured in any way other than sheer numbers. The direct link with nanotechnology seems, however, tenuous. In fact, nanotechnology might even promote population growth because by increasing technological efficiency it will take the pressure off resources. At the same time, enhancing longevity removes the *raison d'être* for reproduction. This problem, or suite of problems, has not yet been addressed and even a comprehensive analysis of the economic impact of nanofactories does not touch upon the problem of population [88], although it is considered in an analysis of some of the lesser implications [289].

12.6.7 THE STRUCTURE OF SOCIETY

Of comparable importance to demography is the way society is structured: it is not simply an amorphous mass of people. The universal availability of personal nanofactories [88] is often taken to imply universal social equality. Nevertheless, it is likely that access to nanotechnology, while universal in principle, will be differentiated in practice. There will be premium on technical literacy, which will doubtless not be possessed by everyone. Hence, nanotechnology might end up creating more social division than we have at present.

12.7 IMPACTS ON INDIVIDUAL PSYCHOLOGY

Much of what has already been written in this chapter impacts psychology, both positively and negatively. At present, the negative aspects seem to be slightly dominant. People worry about information overload, which might be quite as harmful as overloading the environment with nano-objects. "Theranostics", therapy combined with diagnostics, as exemplified by ingested nanoparticles that can both diagnose an internal ailment and release a drug to treat it, is seen as disempowerment of the patient (whereas if he or she is prescribed a drug to take as a tablet, it is after all possible to refuse to take it). Nanotechnology may offer a clear solution to solving, say, a nutritional problem, but implementation stalls. Yet the potential of nanotechnology is surely positive, because it offers the opportunity for all to fully participate in society. The answer to the question how one can move more resolutely in that direction would surely be that under the impetus of gradually increasing technical literacy in an era of leisure in which people are as much producers as consumers there will be a gradually increasing level of civilization, including a more profound understanding of nature. The latter, in particular, must inevitably lead to revulsion against actions that destroy nature, and that surely is how the environment will come to be preserved. Elevation of society implies other concomitant advances, such as in the early (preschool) education of infants, which has nothing to do *per se* with nanotechnology, but which will doubtless be of crucial importance in determining whether humanity survives.

12.8 SOME ETHICAL ISSUES

There are quite a few books on the subject of "nanoethics". Is there actually such a subject? The contents of these books mostly deal with matters such as whether it is right to release nanoparticles into the environment when the effects of inhaling those nanoparticles are imperfectly understood. Clearly that is wrong if (for example) one's choice of ethics is based on the dictum "Primum, nil nocere" (often ascribed to Hippocrates but not actually found in his oath). Most of these books seem, however, to overlook the fact that the choice of ethics is ultimately arbitrary. As Herbert Dingle has sagely pointed out, there exists no final sanction for any particular answer to the question, "How shall I choose what to do?"

It follows that books dealing with ethics can be of two types: one type can present a system (possibly with the intention of convincing others that it is worthwhile adopting it). The second type starts with a particular choice of ethical system and examines the consequences for action in a particular area (e.g. nanotechnology). It seems that most of the books dealing with nanoethics are of the second type, but the defect of omitting to clearly state what are the ethics on which they are based seems to be widespread.

Perhaps there is an underlying current of thought that the new era of rationality that might be ushered in by nanotechnology and its way of looking at the universe would lead to an indefeasible and unique set of "natural" ethical principles. Certainly the technology encourages a more consequential attitude: if one knows a structure with atomic precision then its properties should be more predictable than those of conventional or statistically assembled materials; likewise the "atoms" of ethical convictions have certain consequences for human survival, and if they lead to the destruction of humanity those convictions should be discarded. At the same time, the use of evolutionary design principles (Section 10.7.2) means that we shall become surrounded by devices, the mechanisms of whose inner workings may not be known. This will perhaps focus attention on the real features of reality, notably its microdiversity, open-endedness and continuous evolution, which the preponderance of engineering based on fixed, uniform systems and science based on linearity and isolation has tended to eclipse.

It should not be forgotten that morals are essentially stationary, and dependent on the state of intellectual knowledge for their interpretation. Therefore, if there is some agreement on a common set of ethical principles, nanotechnology as part of the corpus of intellectual knowledge could at least contribute to, and hopefully enhance, their interpretation and effectiveness.

12.9 NANOTECHNOLOGY AND HUMANITY'S GRAND CHALLENGES

In this concluding section, we summarize how nanotechnology may enable humanity to meet the grand challenges that mark our present era, such as the US National Academy of Engineering Grand Challenges, or those implicit in the United Nations Millennium Development Goals. The three top themes appear to be health, education and environmental sustainability (including tackling fresh water shortage, climate change and pollution) – in order of descending importance, for ignorant people cannot tackle climate change, and it is hard to educate ill people. How can nanotechnology contribute to tackling these challenges?

It is not necessary to recapitulate the preceding sections in order to see that nanotechnology is able to make significant contributions to all these areas. Only the problem of population growth seems to be left largely untouched. Interestingly, this is not always pronounced to be one of the grand challenges, even though it obviously underpins them; indeed, it is often taken as a given and meeting the grand challenges is required to enable its untrammeled progression.

In the final analysis, preserving humanity really means preserving civilization; hence, one needs to ask how does nanotechnology help to do that. We conclude this book with that thought.

SUMMARY

It is fairly easy to list likely technical impacts of nanotechnology. The focus on information processing, energy and health implies that these are the areas in which the greatest impact is foreseen. The most important impacts will, however, be on a broader plane, because nanotechnology epitomizes a deeper, more rational view of the universe, including all the practical aspects of our environment. Nevertheless, it is far from certain that humanity will take up the challenge of the new opportunities, especially for everyone to participate in shaping his or her own environment, as much a producer as a consumer. For this reason, those that hold this vision, and see such a development as the only one offering a way forward for humanity (by which I mean its very survival) have a special responsibility to promote the vision and its realization.

FURTHER READING

1. J. Altmann, Military Nanotechnology, Routledge, London, 2006.
2. J. Baumberg, et al., Where is nano taking us? Nanotechnol. Percept. 3 (2007) 3–14.
3. J.D. Bernal, The Social Function of Science, Routledge & Kegan Paul, London, 1939.
4. R.A. Freitas Jr., Economic impact of the personal nanofactory, in: N. Bostrom, et al., Nanotechnology Implications: More Essays, Collegium Basilea, Basel, 2006, pp. 111–126.
5. S.L. Gillett, Nanotechnology, resources, and pollution control, Nanotechnology 7 (1996) 177–182.
6. OECD, Nanotechnology: An Overview Based on Indicators and Statistics, STI Working Paper 2009/7.
7. OECD, The Impacts of Nanotechnology on Companies, 2010.
8. J.J. Ramsden, Applied Nanotechnology, second ed., Elsevier, Amsterdam, 2014.
9. J.J. Ramsden, Nanotechnology and Gaia, Nanotechnol. Percept. (2014).
10. T.T. Toth-Fejel, Irresistible forces vs immovable objects: when China develops Productive Nanosystems, Nanotechnol. Percept. 4 (2008) 113–132.

Nano neologisms

Any new technology almost inevitably introduces new words. A great many of the terms that have been invented for nanotechnology are simply existing words with "nano" prefixing them. Their meaning is therefore self-evident; a formal definition is only needed in some cases to remove ambiguity. This alphabetical list describes these and some other new words.

Errification. A change in the parameters of a system that promotes a higher incidence of defective components or errors in operation.

Eutactic. (adj.) Describes the ideal nano-environment, in which every atom is positioned and oriented as desired. The eutactic environment is an essential precursor for carrying out mechanosynthetic chemistry (cf. Section 8.3).

Micro-enabled nanotechnology. A general challenge of nanotechnology is how to scale it up or out in order to provide artifacts usable by human beings. Microtechnology has the potential to act as a key intermediary in this process. For example, nanocatalysts may line in the inner tubes of a microchemical reactor [145].

Nanification. To approach the meaning of the word "nanification", think of miniaturization but in a more all-encompassing fashion. To nanify electronics, for example, is not only to make individual components smaller (right down to the nanoscale) but also to adapt all parts of the industry to that situation, including design aspects. In short, nanification means introducing nanotechnology in an integrated rather than a piecemeal fashion. Hence, to nanify manufacture is, ultimately, to introduce molecular manufacturing, which involves not only the actual assembly devices themselves but also logistics, indeed the entire supply chain, and the (re)organization of the economic system. Nanification of physical systems (design for human use) implies vastification (q.v.) – many systems are required to work in parallel in order to achieve something of practical use. But in each individual system, fluctuations will play an enhanced role (see, e.g. Section 2.4).

Nanobiology. The obvious meaning is the investigation of biological objects at the nanoscale. It is difficult to perceive how this differs from molecular biology. The application of nanotechnology to biological research is called nanobiotechnology (see Chapter 4). We deprecate use of the word nanobiology as a synonym for molecular biology. On the other hand, the mindset of molecular biology, which is rooted in biology and chemistry, is very different from that of nanotechnology, which is rooted in engineering and physics. Nanobiology has a valuable meaning describing the scrutiny of biological processes at the molecular level using the formal scientific method associated with physics, namely, the mapping of the observed phenomena onto numbers and the manipulation of those numbers using the rules of mathematics. The relationship of nanobiology to molecular biology is the same as that of Sommerhoff's analytical biology [270] to biology.

Nanoblock. Nanostructured object, typically made by a bottom-to-bottom assembly process and designed to be capable of self-assembling into a finished object, or

Nanotechnology: An Introduction. DOI: 10.1016/B978-0-323-39311-9.00019-4
Copyright © 2016 Elsevier Inc. All rights reserved.

of being conveniently processed using conventional microscopic or macroscopic techniques.

Nanochemistry. If one allows nanobiology, it seems highly appropriate to also allow this term, defined as chemistry with explicit consideration of processes involving only a few molecules. It could therefore be said to have been founded by Alfred Rényi with his landmark paper on reactions involving very small numbers [259]. Micromixing as studied by chemical engineers, but extended down to the nanoscale (i.e. nanomixing), would also belong to nanochemistry.

Nanocriterion. A criterion that can be applied to determine whether something is nanotechnology or not.

Nanofacture. Contraction of nanomanufacture. Linguistic purists prefer this word because the etymology of manufacture implies the involvement of human hands in the making process (however, manufacture is currently also used to describe the operations taking place in a wholly automated factory).

Nanology. Surprisingly, given its obvious parentage, this word has been rather rarely used so far. It is useful as an umbrella term covering both nanotechnology and nanoscience and could be used wherever it is awkward to distinguish between them.

Nanonutrition. The exact quantification of all elements and compounds (in principle by enumerating them) present in food. The word is contrasted with micronutrition, which is defined as the micronutrients (vitamins, trace elements, etc.), which are required for assimilation and transforming the macronutrients (the energy-rich proteins, carbohydrates and lipids), from which the micronutrients cannot be synthesized by our bodies. This term well illustrates the essential difference between "micro" and "nano" approaches to nutrition. Macronutrition concerns the provision of macronutrients.

Nanophenomenon. An observable object or event in the nanoscale or which depends on a process in the nanoscale for its appearance or occurrence.

Nanoscience. Is there a need for this term? Sometimes it is defined as "the science underlying nanotechnology". If so, is it not those parts of biology, chemistry, and physics that can be grouped under the term of "molecular sciences"? It is the *technology* of designing and making functional objects at the nanoscale that is new; *science* has long been working at this scale and below; for some time in universities and research institutes, a movement has existed to unite departments of chemistry and molecular biology and some parts of physics into a "department of molecular sciences". No one is arguing that *fundamentally* new physics emerges at the nanoscale; rather, it is the new combinations of phenomena manifesting themselves at that scale that constitute the new technology. The term "nanoscience" therefore appears to be superfluous if used in this sense. A possible meaning could, however, be to denote those parts of molecular sciences that are useful (at any particular epoch) for nanotechnology. Another possible meaning is *the science of mesoscale approximation*, which amounts to roughly the same as virtual nanotechnology (see the legend to Fig. 1.1). The description of a protein as a string of amino acids provides a good example: at the mesoscale, one does not need to inquire into details of the internal structure (at the atomic and subatomic levels) of the amino acids. This is perhaps the most useful meaning that can be assigned to the term.

Nanoscope. Device for observing in the nanoscale. The term "microscope" is clearly misleading; an optical microscope cannot resolve features in the nanoscale. Unfortunately, the word was (rather unthinkingly but perhaps understandably) applied to new instruments such as the electron microscope and the STM well able to resolve objects in the nanoscale. It may be too late to change this usage, in which case the term "ultramicroscope" might be used instead, a word which is already established in German.

Scaleout. Massive parallelization of a nanoscale production process.

Sensorization. The embedding or incorporation of large numbers of sensors into a structure (which might be something fixed like a bridge or a building, or a structure such as an aircraft, or even a living organism). It is only feasible if the sensors are in, or close to, the nanoscale, from the viewpoints of both cost and space requirements; that is, nanotechnology is an enabling concept. Sensorization is likely to lead to a qualitatively different way of handling situations in at least four areas:

- Structural (civil) engineering: bridges, walls, buildings, etc. Sensors – typically optical fiber Bragg gratings, the technology of which already exists – will be incorporated throughout the structure (e.g. embedded in the concrete or in the wings of an aircraft). The output of these sensors is indicative of strain, the penetration of moisture and so forth.
- Process (including chemical) engineering: sensors embedded throughout machinery and reaction vessels will monitor physical (e.g. temperature) and chemical (e.g. the concentration of a selected substance) variables.
- Sensors will be incorporated into the human body, continuously monitoring physiological variables (cf. Section 4.2).
- Sensors will be dispersed throughout the environment (e.g. along rivers and in lakes), reporting the purity of water. The concept is somewhat analogous to what is already taking place in agriculture ("microfarming", that is, intervention guided by high-resolution satellite images of fields, indicating local moisture, etc.).

In most, or perhaps all, cases where sensorization is envisaged, at present we simply do not have data with the spacial and temporal densities that will be obtainable. Its availability will almost certainly qualitatively change our views. It will perhaps come to seem primitive to base an assessment of health on a single analysis of key physiological biomarkers. Vehicle health – as appraised by analysing sensor readouts – will become the criterion for the airworthiness of aircraft (etc.).

Apart from making the miniature sensors themselves, the two main challenges of sensorization are (i) how to deal with the vast proliferation of data (an example of vastification) and (ii) what about the reliability of the sensors? The first challenge can presumably be dealt with by automated processing of the data, and human intervention will only be indicated in the event of some unusual pattern occurring. This implies vast data processing capacity, which is, however, anyway an envisaged development of nanotechnology. The second challenge may be dealt with in a similar fashion: the system will come to be able to determine from the pattern of its readouts when sensors are malfunctioning (cf. Chapter 10).

Ultramicroscope. Synonym for nanoscope, that is, an instrument able to reveal features smaller than in the microscale (that is, in the nanoscale).

Vastification. An enormous increase in the number of elementary units or components in a system. Typically, vast numbers of objects are a corollary of their being very small (cf. Moore's law, which is as much about the vast increase in the number of components fabricated in parallel on a chip as the diminishing size of the components). The main qualitatively distinctive consequence of vastification is that explicit specification and control become impracticable. To tackle this problem, an evolutionary approach will be required, notably in design (see Fig. 10.6), but very possibly also in operation (e.g. a stigmergic approach). Biomimicry of the human brain may become the favored approach. Vastification usually implies complexification (although typically not found in dictionaries, this term is already used by mathematicians and does not therefore rank as a neologism) and, sometimes, errification (q.v.).

Abbreviations

AES Auger emission spectroscopy

AFM atomic force microscope or microscopy

AIC algorithmic information content

APED alternating polyelectrolyte deposition

APFIM atom probe field ion microscopy

APM atomically precise manufacturing

APT atomically precise technology

BCE before the Christian era

BD ballistic deposition

CMOS complementary metal-oxide–semiconductor

CNT carbon nanotube

CVD chemical vapor deposition

DFT density functional theory

DLA diffusion-limited aggregation

DNA deoxyribonucleic acid

DPI dual polarization interferometer

DPN dip-pen nanolithography

EA evolutionary algorithm

ECM extracellular matrix

EDS, EDX energy dispersive (X-ray) spectroscopy

ESEM environmental scanning electron microscope or microscopy

FET field-effect transistor

FIB fast atom bombardment

GBD generalized ballistic deposition

GCI grating-coupled interferometry

GMR giant magnetoresistance

HOMO highest occupied molecular orbital

IEC interactive evolutionary computation

IF interface, interfacial

IT information technology

KMAL kinetic mass action law

LB Langmuir–Blodgett

LS Langmuir–Schaefer

LUMO lowest unoccupied molecular orbital

MAL mass action law

MBE molecular beam epitaxy

MOF metal–organic framework

MOO multiobjective optimization

MEMS micro electromechanical systems

MMC metal matrix composite

MOSFET metal–oxide–semiconductor field effect transistor

MRAM magnetic random access memory

MST microsystems technology

MTJ magnetic tunnel junction

MWCNT multiwall carbon nanotube

NEMS nano electromechanical systems

NSOM near-field scanning optical microscope or microscopy

OWLI optical waveguide lightmode interferometry

OWLS optical waveguide lightmode spectroscopy

PDB protein data bank

PECVD plasmon-enhanced chemical vapor deposition

PN productive nanosystems, personal nanofactory

PSA programmable self-assembly

PVD physical vapor deposition

QCA quantum cellular automaton

QD quantum dot

RAM random access memory

RFN random fiber network

RLA reaction-limited aggregation

RNA ribonucleic acid

RRAM resistive random access memory

RSA random sequential addition or adsorption

RWG resonant waveguide grating

SAM self-assembled monolayer

SAR scanning angle reflectometry

SCM storage-class memory

SEM scanning electron microscope or microscopy

SET single electron transistor

SFG sum frequency generation

SICM scanning ion conductance microscope or microscopy

SIMS secondary ion mass spectroscopy

SMEL sequential minimization of entropy loss

SNOM scanning near-field optical microscope or microscopy

SPM scanning probe microscope or microscopy

SPR surface plasmon resonance

STEM scanning transmission electron microscope or microscopy

STM scanning tunneling microscope or microscopy

SWNT single wall carbon nanotube

TEM transmission electron microscope or microscopy

UPMT ultraprecision machine tool

VLSI very large scale integration or integrated

XRD X-ray diffraction

Bibliography

1. E. Abad, et al., NanoDictionary, Collegium Basilea, Basel, 2005.
2. N. Aggarwal, et al., Protein adsorption on heterogeneous surfaces, Appl. Phys. Lett. 94 (2009) 083110.
3. S. Alexandre, C. Lafontaine, J.-M. Valleton, Local surface pressure gradients observed during the transfer of mixed behenic acid/pentadecanoic acid Langmuir films, J. Biol. Phys. Chem. 1 (2001) 21–23.
4. C. Allain, M. Cloitre, Characterizing the lacunarity of random and deterministic fractal sets, Phys. Rev. A 44 (1991) 3552–3558.
5. P.M. Allen, W. Ebeling, Evolution and the stochastic description of simple ecosystems, Biosystems 16 (1983) 113–126.
6. L. Amico, et al., Entanglement in many-body systems, Rev. Modern Phys. 80 (2008) 517–576.
7. P.W. Anderson, More is different, Science 177 (1972) 393–396.
8. J.M. Andreas, E.A. Hauser, W.B. Tucker, Boundary tension by pendant drops, J. Phys. Chem. 42 (1938) 1001–1019.
9. A. Aref, et al., Optical monitoring of stem cell-substratum interactions, J. Biomed. Opt. 14 (2009) 010501.
10. M. Arikawa, Fullerenes—an attractive nano carbon material and its production technology, Nanotechnol. Percept. 2 (2006) 121–114.
11. R.W. Ashby, Principles of the self-organizing system, in: H. von Foerster, G.W. Zopf (Eds.), Principles of Self-Organization, Pergamon Press, Oxford, 1962, pp. 255–278.
12. Report of the Working Party on the Experimental Manipulation of the Genetic Composition of Micro-Organisms (Cmnd 5880), HMSO, London, 1975.
13. S. Auyang, Scientific convergence in the birth of molecular biology, Nanotechnol. Percept. 11 (2015) 31–54.
14. J. Bafaluy, et al., Effect of hydrodynamic interactions on the distribution of adhering Brownian particles, Phys. Rev. Lett. 70 (1993) 623–626.
15. V. Balzani, Nanoscience and nanotechnology: the bottom-up construction of molecular devices and machines, Pure Appl. Chem. 80 (2008) 1631–1650.
16. S. Bandyopadhyay, Single spin devices—perpetuating Moore's law, Nanotechnol. Percept. 3 (2007) 159–163.
17. L. Banyai, S.W. Koch, Absorption blue shift in laser-excited semiconductor microspheres, Phys. Rev. Lett. 57 (1986) 2724.
18. W. Banzhaf, et al., From artificial evolution to computational evolution, Nat. Rev. Genet. 7 (2006) 729–735.
19. A.A. Berezin, Stable isotopes in nanotechnology, Nanotechnol. Percept. 5 (2009) 27–36.
20. C.R. Berry, Effects of crystal surface on the optical absorption edge of AgBr, Phys. Rev. 153 (1967) 989–992.
21. C.R. Berry, Structure and optical absorption of AgI microcrystals, Phys. Rev. 161 (1967) 848–851.
22. W.C. Bigelow, Oleophobic monolayers, J. Colloid Sci. 1 (1946) 513–538.
23. J.J. Bikerman, The criterion of fracture, SPE Trans. 4 (1964) 290–294.
24. J.J. Bikerman, Surface energy of solids, Phys. Status Solidi 10 (1965) 3–26.
25. G. Binnig, H. Rohrer, Scanning tunneling microscopy, Helv. Phys. Acta 55 (1982) 726–735.
26. C. Binns, Prodding the cosmic fabric with nanotechnology, Nanotechnol. Percept. 3 (2007) 97–105.
27. L.A. Blumenfeld, D.S. Burbajev, R.M. Davydov, Processes of conformational relaxation in enzyme catalysis, in: E.R. Welch (Ed.), The Fluctuating Enzyme, Wiley, New York, 1986, pp. 369–402.

28. B.O. Boscovic, Carbon nanotubes and nanofibres, Nanotechnol. Percept. 3 (2007) 141–158.

29. E.A. Boucher, M.J.B. Evans, Pendent drop profiles and related capillary phenomena, Proc. R. Soc. A 346 (1975) 349–374.

30. D. Boyle, et al., Subtle alterations in swimming speed distributions of rainbow trout exposed to titanium dioxide nanoparticles are associated with gill rather than brain injury, Aquatic Toxicol. 126 (2013) 116–127.

31. A.M. Brintrup, et al., Evaluation of sequential, multi-objective, and parallel interactive genetic algorithms for multi-objective optimization problems, J. Biol. Phys. Chem. 6 (2006) 137–146.

32. R. Bruinsma, Physical aspects adhesion leukocytes, in: T. Riste, D. Sherrington (Eds.), Physics Biomaterials: Fluctuations, Self-Assembly and Evolution, Kluwer, Dordrecht, 1996, pp. 61–101.

33. L.E. Brus, Electron-electron and electron-hole interactions in small semiconductor crystallites: the size dependence of the lowest excited electronic state, J. Chem. Phys. 18 (1984) 4403–4409.

34. P.M. Budd, et al., Polymers of intrinsic microporosity (PIMs): robust, solution-processable, organic nanoporous materials, Chem. Commun. (2004) 230–231.

35. G.W. Burr, Phase change memory technology, J. Vac. Sci. Technol. B 28 (2010) 223–262.

36. S.Z. Butler, et al., Progress, challenges and opportunities in two-dimensional materials beyond graphene, ACS Nano 7 (2013) 2898–2926.

37. H. Cabral, et al., Systemic targeting of lymph node metastasis through the blood vascular system by using size-controlled nanocarriers, ACS Nano 9 (2015) 4957–4967.

38. M. Calis, M.P.Y. Desmulliez, Haptic sensing technologies for a novel design methodology in micro/nanotechnology, Nanotechnol. Percept. 1 (2007) 141–158.

39. C. Calonder, J. Talbot, J.J. Ramsden, Mapping the electron donor/acceptor potentials on protein surfaces, J. Phys. Chem. B 105 (2001) 725–729.

40. G. Carturan, et al., Inorganic gels for immobilization of biocatalysts: inclusion of invertase-active whole cells of yeast (*Saccharomyces cerevisiae*) into thin layers of SiO_2 gel deposited on glass sheets, J. Molec. Catal. 57 (1989) L13–L16.

41. G. Catalan, et al., Domains in three-dimensional ferroelectric nanostructures: theory and experiment, J. Phys.: Condens. Matter. 19 (2007) 132201.

42. G. Catalan, et al., Domain wall nanoelectronics, Rev. Modern Phys. 84 (2012) 119–156.

43. R.K. Cavin, V.V. Zhirnov, Generic device abstractions for information processing technologies, Solid State Electron. 50 (2006) 520–526.

44. X. Chen, et al., Probing the electron states and metal–insulator transition mechanisms in molybdenum disulphide vertical heterostructures, Nature Commun. 6 (2015) 6088.

45. X. Chen, et al., High-quality sandwiched black phosphorus heterostructure and its quantum oscillations, Nature Commun. 6 (2015) 7315.

46. C. Chiutu, et al., Precise orientation of a single C_{60} molecule on the tip of a scanning probe microscope, Phys. Rev. Lett. 108 (2012) 268302.

47. Y.K. Cho, et al., Self-assembling colloidal-scale devices: selecting and using short-range surface forces between conductive solids, Adv. Funct. Mater. 17 (2007) 379–389.

48. A.V. Chumak, et al., Magnon spintronics, Nature Phys. 11 (2015) 453–461.

49. S. Chumakov, et al., The theoretical basis of universal identification systems for bacteria and viruses, J. Biol. Phys. Chem. 5 (2005) 121–128.

50. S.A. Claridge, et al., Cluster-assembled materials, ACS Nano 3 (2009) 244–255.

51. Concise Oxford Dictionary, tenth ed., University Press, Oxford, 1999.

52. J. Corbett, et al., Nanotechnology: international developments and emerging products, Ann. CIRP 49 (2) (2000) 523–545.

53. A.M. Cormack, Representation of function by its line integrals, with some radiological applications, J. Appl. Phys. 34 (1963) 2722–2727.

54. K. Cottier, R. Horvath, Imageless microscopy of surface patterns using optical waveguides, Appl. Phys. B 91 (2008) 319–327.

55. G. Csúcs, J.J. Ramsden, Interaction of phospholipid vesicles with smooth metal oxide surfaces, Biophys. Biochim. Acta 1369 (1998) 61–70.

56. A. Dér, et al., Interfacial water structure controls protein conformation, J. Phys. Chem. B 111 (2007) 5344–5350.

57. R.H. Dennard, et al., Design of ion-implanted MOSFETs with very small physical dimensions, IEEE J. Solid State Circuits SC-9 (1974) 256–268.

58. M. Dettenkofer, R.C. Spencer, Importance of environmental decontamination—a critical view, J. Hospital Infect. 65 (2007) 55–57.

59. R. Dingle, C.H. Henry, Quantum effects in heterostructure lasers, US Patent 3,982,207, 1976.

60. P.A.M. Dirac, The Principles of Quantum Mechanics, fourth ed., Clarendon Press, Oxford, 1967, (Chapter 1).

61. R. Dobbs, Multi-carbide material manufacture and use as grinding media, US patent no 7,140,567, 2006.

62. K.E. Drexler, Molecular engineering: an approach to the development of general capabilities for molecular manipulation, Proc. Natl. Acad. Sci. USA 78 (1981) 5275–5278.

63. X. Duan, C.M. Lieber, General synthesis of compound semiconductor nanowires, Adv. Mater. 12 (2000) 298–302.

64. X. Duan, et al., Single nanowire electrically driven lasers, Nature 421 (2003) 241–245.

65. Y. Dubi, M. Di, Ventra, Colloquium: heat flow and thermoelectricity in atomic and molecular junctions, Rev. Modern Phys. 83 (2011) 131–155.

66. D. Duonghong, J.J. Ramsden, M. Grätzel, Dynamics of interfacial electron transfer processes in colloidal semiconductor systems, J. Am. Chem. Soc. 104 (1982) 2977–2985.

67. B. van Duuren-Stuurman, et al., Stoffenmanager Nano version 1.0: a web-based tool for risk prioritization of airborne manufactured nano objects, Ann. Occupational Hygiene (2012) 1–17.

68. M.H. Eckman, S.M. Greenburg, J. Rosand, Should we test for CYP2C19 before initiating anticoagulant therapy in patients with atrial fibrillation?, J. Gen. Intern. Med. 24 (2009) 543–549.

69. S.F. Edwards, R.D.S. Oakeshott, Theory of powders, Physica A 157 (1989) 1080–1090.

70. Al.L. Efros, A.L. Efros, Interband absorption of light in a semiconductor sphere, Soviet Phys. Semicond. 16 (1982) 772–775.

71. A.L. Efros, B.I. Shklovskii, Coulomb gap and low-temperature conductivity of disordered systems, J. Phys. C 8 (1975) L49–L51.

72. W.E. Fabb, Conceptual leaps in family medicine: are there more to come?, Asia Pacific Family Med. 1 (2002) 67–73.

73. Y. Fang, et al., Resonant waveguide grating biosensor for living cell sensing, Biophys. J. 91 (2006) 1925–1940.

74. M. Faraday, Experimental relations of gold (and other metals) to light (Bakerian Lecture), Phil. Trans. R. Soc. 147 (1857) 145–181.

75. A. Faulkner, W. Shu, Biological cell printing technologies, Nanotechnol. Percept. 8 (2012) 35–57.

76. X. Feng, et al., Converting ceria polyhedral nanoparticles into single-crystal nanospheres, Science 312 (2006) 1504–1507.

77. Y. Feng, et al., Nanopore-based fourth-generation DNA sequencing technology, Genom. Proteom. Bioinform. 13 (2015) 4–16.

78. I. Ferain, C.A. Colinge, J.-P. Colinge, Multigate transistors as the future of classical metal-oxide-semiconductor field-effect transistors, Nature 479 (2011) 310–316.

79. G. Férey, Hybrid porous solids: past present future, Chem. Soc. Rev. 37 (2008) 191–214.

80. G. Férey, C. Serre, Large breathing effects in three-dimensional porous hybrid matter: facts analyses rules and consequences, Chem. Soc. Rev. 38 (2009) 1380–1399.

81. A. Fernández, H. Cendra, In vitro RNA folding: the principle of sequential minimization of entropy loss at work, Biophys. Chem. 58 (1996) 335–339.

82. A. Fernández, A. Colubri, Microscopic dynamics from a coarsely defined solution to the protein folding problem, J. Math. Phys. 39 (1998) 3167–3187.

83. A. Fernández, R. Scott, Dehydron: a structurally encoded signal for protein interaction, Biophys. J. 85 (2003) 1914–1928.

84. A. Fernández, J.J. Ramsden, On adsorption-induced denaturation of folded proteins, J. Biol. Phys. Chem. 1 (2001) 81–84.

85. R. Feynman, There's plenty of room at the bottom, J. Microelectromech. Syst. 1 (1992) 60–66 (transcript of a talk given by the author on 26 December 1959 at the annual meeting of the American Physical Society at the California Institute of Technology).

86. H. von Foerster, On self-organizing systems their environments, in: M.C. Yorvitz, S. Cameron (Eds.), Self-Organizing Systems, Pergamon Press, Oxford, 1960, pp. 31–50.

87. S. Fordham, On the calculation of surface tension from measurements of pendant drops, Proc. R. Soc. A 194 (1948) 1–16.

88. R.A. Freitas Jr., Economic impact of the personal nanofactory, in: Nanotechnology Implications— More Essays, Collegium Basilea, Basel, 2006, pp. 111–126.

89. R.F. Frindt, Single crystals of MoS_2 several molecular layers thick, J. Appl. Phys. 37 (1966) 1928–1929.

90. A. Fujishima, K. Honda, Electrochemical photolysis of water at a semiconductor electrode, Nature 238 (1972) 37–38.

91. T. Funatsu, et al., Imaging of single fluorescent molecules and individual ATP turnovers by single myosin molecules in aqueous solution, Nature 374 (1995) 555–559.

92. S. Galam, A. Mauger, Universal formulas for percolation thresholds, Phys. Rev. E 53 (1996) 2177–2181.

93. Y. Gao, et al., Chemical activation of carbon nano-onions for high-rate supercapacitor electrodes, Carbon 51 (2013) 52–58.

94. Y. Gefen, Y. Meir, A. Aharony, Geometric implementation of hypercubic lattices with noninteger dimensionality by use of low lacunarity fractal lattices, Phys. Rev. Lett. 50 (1983) 145–148.

95. M. Gell-Mann, S. Lloyd, Information measures, effective complexity and total information, Complexity 2 (1996) 44–52.

96. L. Gerhard, et al., Magnetoelectric coupling at metal surfaces, Nature Nanotechnol. 5 (2010) 792–797.

97. I. Giaever, C.R. Keese, Micromotion of mammalian cells measured electrically, Proc. Natl. Acad. Sci. USA 81 (1984) 3761–3764.

98. I. Giaever, C.R. Keese, Micromotion of mammalian cells measured electrically, Proc. Natl. Acad. Sci. USA 88 (1991) 7896–7900.

99. M. Göppert-Mayer, Ueber Elementarakte mit zwei Quantensprüngen, Ann. Phys. 401 (Bd 9 der 5. Folge) (1931) 273–294.

100. R. Gogsadze, et al., Formulation and solution of the boundary value problem of viscous liquid flow in a nano tube taking external friction into account, Nanotechnol. Percept. 9 (2013) 57–69.

101. S.N. Gorb, Uncovering insect stickiness: structure and properties of hairy attachment devices, Am. Entomologist 51 (2005) 31–35.

102. S.N. Gorb, Biological fibrillar adhesives: functional principles and biomimetic applications, in: L.F.M. Da Silva, A. Oechsner, R.D. Adams (Eds.), Handbook of Adhesion Technology, Springer-Verlag, Berlin, 2011, pp. 1409–1436.

103. J. Gorelik, et al., Non-invasive imaging of stem cells by scanning ion conductance microscopy: future perspective, Tissue Engng. C 14 (2008) 311–318.

104. T. Graham, On liquid diffusion applied to analysis, J. Chem. Soc. 15 (1862) 216–255.

105. B.F. Gray, Reversibility and biological machines, Nature (Lond.) 253 (1975) 436–437.

106. M.J. Green, Analysis and measurement of carbon nanotube dispersions: nanodispersion *versus* macrodispersion, Polymer Intl. 59 (2010) 1319–1322.

107. P.A. Grigoriev, Electrical inductance properties of lipid bilayers with alamethicin channels, J. Biol. Phys. Chem. 4 (2004) 143–144.

108. P.A. Grigoriev, Memory effects in the fast substates system of alamethicin channels, J. Biol. Phys. Chem. 11 (2011) 6–10.

109. M. Groenendijk, J. Meijer, Microstructuring using femtosecond pulsed laser ablation, J. Laser Appl. 18 (2006) 227–235.

110. A. Yu Grosberg, T.T. Nguyen, B.I. Shklovskii, The physics of charge inversion in chemical and biological systems, Rev. Modern Phys. 74 (2002) 329–345.

111. M.S. Gudiksen, C.M. Lieber, Diameter-selective synthesis of semiconductor nanowires, J. Am. Chem. Soc. 122 (2000) 8801–8802.

112. R. Guo, et al., Nonvolatile memory based on the ferroelectric photovoltaic effect, Nature Commun. 4 (2013) 1990.

113. T. Gyalog, H. Thomas, Atomic friction, Z. Phys. B 104 (1997) 669–674.

114. K.C. Hall, et al., Nonmagnetic semiconductor spin transistor, Appl. Phys. Lett. 83 (2004) 2937–2939.

115. N. Hampp, Bacteriorhodopsin as a photochromic retinal protein for optical memories, Chem. Rev. 100 (2000) 1755–1776.

116. T. Hashimoto, H. Tanaka, H. Hasegawa, Ordered structure in mixtures of a block copolymer and Homopolymers 2. Effects of molecular weights of homopolymers, Macromolecules 23 (1990) 4378–4386.

117. D. Hanein, et al., Selective interactions of cells with crystal surfaces, J. Cell Sci. 104 (1993) 27–288.

118. M.B. Hastings, Superadditivity of communication capacity using entangled inputs, Nature Phys. 5 (2009) 255–257.

119. K. Hauffe, Fehlordnungsgleichgewichte in halbleitenden Kristallen vom Standpunkt des Massenwirkungsgesetzes, in: W. Schottky (Ed.), Halbleiterprobleme, Vieweg, Brunswick, 1954, pp. 107–127.

120. T.W. Healy, L.R. White, Ionizable surface group models of aqueous interfaces, Adv. Colloid Interface Sci. 9 (1978) 303–345.

121. S. Henke, A. Schneemann, R.A. Fischer, Massive anisotropic thermal expansion and thermoresponsive breathing in metalorganic frameworks modulated by linker functionalization, Adv. Funct. Mater. 23 (2013) 5990–5996.

122. H. Hidaka, et al., Photo-assisted dehalogenation and mineralization of chloro/fluorobenzoic acid derivatives in aqueous media, J. Photochem. Photobiol. A197 (2008) 150–123.

123. C. Hierold, From micro- to nanosystems: mechanical sensors go nano, J. Micromech. Microengng. 14 (2004) S1–S11.

124. H. Hillman, The Case for New Paradigms in Cell Biology and in Neurobiology, Edwin Mellen Press, Lewiston, 1991.

125. Y. Hiratsuka, et al., A microrotatory motor powered by bacteria, Proc. Natl. Acad. Sci. USA 103 (2006) 13618–13623.

126. A.O. Hirschman, The Strategy of Economic Development, University Press, Yale, 1958.

127. R. Hofmann, Surface science in photography, Nanotechnol. Percept. 11 (2015) 5–19.

128. B. Högberg, H. Olin, Programmable self-assembly—unique structures and bond uniqueness, J. Computat. Theor. Nanosci. 3 (2006) 391–397.

129. T. Hogg, Evaluating microscopic robots for medical diagnosis and treatment, Nanotechnol. Percept. 3 (2007) 63–73.

130. T.E. Holy, S. Leibler, Dynamic instability of microtubules as an efficient way to search in space, Proc. Natl. Acad. Sci. USA 91 (1994) 5682–5685.

131. W. Hoppe, et al., Three-dimensional electron microscopy of individual biological objects. Part I. Methods, Z. Naturforschung 31a (1976) 645–655.

132. P. Horcajada, et al., Metal–organic–framework nanoscale carriers as a potential platform for drug delivery and imaging, Nature Mater. 9 (2010) 172–178.

133. S. Horike, S. Shimomura, S. Kitagawa, Soft porous crystals, Nature Chem. 1 (2009) 695–704.

134. R. Horvath, J.J. Ramsden, Quasi-isotropic analysis of anisotropic thin films on optical waveguides, Langmuir 23 (2007) 9330–9334.

135. R. Horvath, L.R. Lindvold, N.B. Larsen, Reverse-symmetry waveguides: theory and fabrication, Appl. Phys. B 74 (2002) 383–393.

136. R. Horvath, H. Gardner, J.J. Ramsden, Apparent self-accelerating alternating assembly of semiconductor nanoparticles and polymers, Appl. Phys. Lett. 107 (2015) 041604.

137. R. Horvath, et al., Multidepth screening of living cells using optical waveguides, Biosensors Bioelectronics 24 (2008) 805–810.

138. S.M. Hsu, Nano-lubrication: concept and design, Tribol. Intl. 37 (2004) 537–545.

139. J. Hu, et al., Nonlinear thermal transport and negative differential thermal conductance in graphene nanoribbons, Appl. Phys. Lett. 99 (2011) 113101.

140. M.A. Hubbe, Adhesion and detachment of biological cells *in vitro*, Prog. Surf. Sci. 11 (1981) 65–138.

141. H.S. Hwang, Resistance RAM having oxide layer and solid electrolyte layer, and method for operating the same, US Patent No 8,116,116, 2012.

142. K. Ichikawa, et al., Nonvolatile thin film transistor memory with ferritin, J. Korean Phys. Soc. 54 (2009) 554–557.

143. I. Ichinose, T. Kawakami, T. Kunitake, Alternate molecular layers of metal oxides and hydroxyl polymers prepared by the surface sol-gel process, Adv. Mater. 10 (1998) 535–539.

144. S. Iijima, Direct observation of the tetrahedral bonding in graphitized carbon black by high resolution electron microscopy, J. Cryst. Growth 50 (1980) 675–683.

145. R.K. Iler, Multilayers of colloidal particles, J. Colloid Interface Sci. 21 (1966) 569–594.

146. A. Iles, Microsystems for the enablement of nanotechnologies, Nanotechnol. Percept. 5 (2009) 121–133.

147. A. Ishijima, et al., Simultaneous observation of individual ATPase and mechanical events by a single myosin molcule during interaction with actin, Cell 92 (1998) 161–171.

148. Y. Iwasaki, et al., Label-free detection of C-reactive protein using highly dispersible gold nanoparticles synthesized by reducible biomimetic block copolymers, Chem. Commun. 50 (2014) 5656–5658.

149. M. Jamali, et al., Spin wave nonreciprocity for logic device applications, Sci. Rep. 3 (2013) 3160.

150. I. Jäger, P. Fratzl, Mineralized collagen fibrils, Biophys. J. 79 (2000) 1737–1746.

151. M. Jaschke, H.-J. Butt, Deposition of organic material by the tip of a scanning force microscope, Langmuir 11 (1995) 1061–1064.

152. J.S. Jenness, Calculating landscape surface area from digital elevation models, Wildlife Soc. Bull. 32 (2004) 829–839.

153. F.O. Jones, K.O. Wood, The melting point of thin aluminium films, Br. J. Appl. Phys. 15 (1964) 185–188.

154. M. Kammerer, et al., Fast spin-wave-mediated magnetic vortex core reversal, Phys. Rev. B 86 (2012) 134426.

155. M.J. Kelly, Nanotechnology and manufacturability, Nanotechnol. Percept. 7 (2011) 79–81.

156. K. Kendall, The impossibility of comminuting samll particles by compression, Nature (Lond.) 272 (1978) 710–711.

157. A. Khitun, K.L. Wang, Non-volatile magnonic logic circuits engineering, J. Appl. Phys. 110 (2011) 034306.

158. A. Khitun, Multi-frequency magnonic logic circuits for parallel data processing, J. Appl. Phys. 111 (2012) 054307.

159. H.L. Khor, et al., Response of cells on surface-induced nanopatterns: fibroblasts and mesenchymal progenitor cells, Biomacromolecules 8 (2007) 1530–1540.

160. C.J. Kiely, et al., Ordered colloidal nanoalloys, Adv. Mater. 12 (2000) 640–643.

161. S. Kitagawa, K. Uemura, Dynamic porous properties of coordination polymers inspired by hydrogen bonds, Chem. Soc. Rev. 34 (2005) 109–119.

162. A.I. Kitaigorodskii, Organic Chemical Crystallography, especially Chs 3 and 4, Consultants Bureau, New York, 1961.

163. J. Kjelstrup-Hansen, C. Simbrunner, H.-G. Rubahn, Organic surface-grown nanowires for functional devices, Rep. Prog. Phys. 76 (2013) 126502.

164. E. Klavins, Directed self-assembly using graph grammars, in: Foundations of Nanoscience: Self-Assembled Architectures and Devices, Utah, Snowbird, 2004.

165. E. Klavins, Universal self-replication using graph grammars, in: Int. Conf. on MEMs, NANO and Smart Systems, Banff, Canada, 2004.

166. E. Klavins, R. Ghrist, D. Lipsky, Graph grammars for self-assembling robotic systems, in: Proc. Int. Conf. Robotics Automation, New Orleans, 2004, pp. 5293–5300.

167. W.D. Knight, et al., Electronic shell structure and abundances of sodium clusters, Phys. Rev. Lett. 52 (1984) 2141–2143.

168. A.N. Kolmogorov, A new invariant for transitive dynamical systems, Dokl. Akad. Nauk SSSR 119 (1958) 861–864.

169. V. Kondratyev, The Structure of Atoms and Molecules, Dover, New York, 1965, p. 485.

170. P. Kozma, et al., Grating coupled optical waveguide interferometer for label-free biosensing, Sensors Actuators B 155 (2011) 446–450.

171. Y. Kuang, Elongated nanostructures for radial junction solar cells, Rep. Progr. Phys. 76 (2013) 106502.

172. M. Kumar, Y. Ando, et al., Carbon nanotube synthesis and growth mechanism, Nanotechnol. Percept. 6 (2010) 7–28.

173. R. Kurzweil, The Singularity is Near, Viking Press, New York, 2005.

174. B. Laforge, Emergent properties in biological systems as a result of competition between internal and external dynamics, J. Biol. Phys. Chem. 9 (2009) 5–9.

175. Ph. Lavalle, et al., Direct observation of postadsorption aggregation of antifreeze glycoproteins on silicates, Langmuir 16 (2000) 5785–5789.

176. M. Lawo, et al., Simulation techniques for the description of smart structures and sensorial materials, J. Biol. Phys. Chem. 9 (2009) 143–148.

177. S.C.G. Leeuwenburgh, et al., Morphology of calcium phosphate coatings for biomedical applications deposited using electrostatic spray deposition, Thin Solid Films 503 (2006) 69–78.

178. P. Lehmann, G. Goch, Comparison of conventional light scattering and speckle techniques concerning an in-process characterization of engineered surfaces, Ann. CIRP 49 (1) (2000) 419–422.

179. P. Lehmann, Surface roughness measurement based on the intensity correlation function of scattered light under speckle-pattern illumination, Appl. Optics 38 (1999) 1144–1152.

180. P. Lehmann, S Patzelt, A. Schöne, Surface roughness measurement by means of polychromatic speckle elongation, Appl. Optics 36 (1997) 2188–2197.

181. X. Li, H. Gao, C.J. Murphy, K.K. Caswell, Nanoindentation of silver nanowires, Nano Lett. 3 (2003) 1495–4098.

182. K.K. Likharev, et al., Single-electron devices and their applications, Proc. IEEE 87 (1999) 606–632.

183. O.A. von Lilienfeld, Accurate ab initio energy gradients in chemical compound space, J. Chem. Phys. 131 (2009) 164102.

184. W.A. Lopes, H.M. Jaeger, Hierarchical self-assembly of metal nanostructures on diblock copolymer scaffolds, Nature 414 (2001) 735–738.

185. J.E. Lovelock, C.E. Giffin, Planetary atmospheres: compositional and other changes associated with the presence of life, Adv. Astronaut. Sci. 25 (1969) 179–193.

186. P.O. Luthi, et al., A cellular automaton model for neurogenesis in *Drosophila*, Physica D 118 (1998) 151–160.

187. P.O. Luthi, J.J. Ramsden, B. Chopard, The role of diffusion in irreversible deposition, Phys. Rev. E 55 (1997) 3111–3115.

188. Yu. Lvov, et al., Molecular film assembly via layer-by-layer adsorption of oppositely charged macromolecules (linear polymer, protein and clay) and concanavalin A and glycogen, Thin Solid Films 284–285 (1996) 797–801.

189. D.Q. Ly, et al., The Matter Compiler—towards atomically precise engineering and manufacture, Nanotechnol. Percept. 7 (2011) 199–217.

190. O. Lysenko, et al., Surface nanomachining using scanning tunnelling microscopy with a diamond tip, Nanotechnol. Percept. 6 (2010) 41–49.

191. J.K. Mcdonough, Y. Gogotsi, Carbon onions: synthesis and electrochemical applications, Interface (Fall 2013) 61–66.

192. P.A. Mckeown, et al., Ultraprecision machine tools—design principles and developments, Nanotechnol. Percept. 4 (2008) 5–14.

193. A. Maitra, Nanotechnology and nanobiotechnology—are they children of the same father?, Nanotechnol. Percept. 6 (2010) 197–204.

194. M. Malinauskas, et al., Ultrafast laser nanostructuring of photo polymers: a decade of advances, Phys. Rep. 533 (2013) 1–13.

195. A.G. Mamalis, A. Markopoulos, D.E. Manolakos, Micro and nanoprocessing techniques and applications, Nanotechnol. Percept. 1 (2005) 63–73.

196. S. Manghani, J.J. Ramsden, The efficiency of chemical detectors, J. Biol. Phys. Chem. 3 (2003) 11–17.

197. E.K. Mann, et al., Optical characterization of thin films: beyond the uniform layer model, J. Chem. Phys. 105 (1996) 6082–6085.

198. J.L. Mansot, et al., Nanolubrication, Braz. J. Phys. 39 (2009) 186–197.

199. A.M. Marconnet, M.A. Panzer, K.E. Goodson, Thermal conduction phenomena in carbon nanotubes and related nanostructured materials, Rev. Modern Phys. 85 (2013) 1295–1326.

200. C.P. Martin, et al., Controlling pattern formation in nanoparticle assemblies via directed solvent dewetting, Phys. Rev. Lett. 99 (2007) 116103.

201. M. Máté, J.J. Ramsden, Addition of particles of alternating charge, J. Chem. Soc. Faraday Trans. 94 (1998) 2813–2816.

202. A. Matsumoto, et al., A synthetic approach toward a self-regulated insulin delivery system, Angew. Chem. Intl Ed. 51 (2012) 2124–2128.

203. T. Matsunaga, et al., Photoelectrochemical sterilization of microbial cells by semiconductor powders, FEMS Microbiol. Lett. 29 (1985) 211–214.

204. J.W. May, Platinum surface LEED rings, Surf. Sci. 17 (1969) 267–270.

205. J. Mehnen, et al., Design for wire and arc additive layer manufacture, in: Proc. 20th CIRP Design Conference, Nantes, April 2010, pp. 19–21.

206. van der Merwe, F.C. Frank, Misfitting monolayers, Proc. Phys. Soc. A 62 (1949) 315–316.

207. J. Miao, et al., Nanomaterials applications in "green" functional coatings, Nanotechnol. Percept. 8 (2012) 181–189.

208. P.A. Midgley, et al., Nano tomography in the chemical, biological and materials sciences, Chem. Soc. Rev. 36 (2007) 1477–1494.

209. M.L. Minsky, Where is our microtechnology?, in: I.J. Good (Ed.), The Scientist Speculates, Heinemann, London, 1962, p. 139.

210. V.N. Mochalin, et al., The properties and applications of nanodiamonds, Nature Nanotechnol. 7 (2012) 11–23.

211. G. Möbus, R.C. Doole, B.J. Inkson, Spectroscopic electron tomography, Ultramicroscopy 96 (2003) 433–451.

212. A.M. Mood, The distribution theory of runs, Ann. Math. Statist. 11 (1940) 367–392.

213. M.K. Monastyrov, et al., Electroerosion dispersion-prepared nano- and submicrometre-sized aluminium and alumina powders as power-accumulating substances, Nanotechnol. Percept. 4 (2008) 179–187.

214. Y. Montelongo, et al., Plasmon nanoparticle scattering for color holograms, Proc. Natl. Acad. Sci. USA 111 (2014) 12679–12683.

215. T.L. Morkved, et al., Mesoscopic self-assembly of gold islands on diblock-copolymer films, Appl. Phys. Lett. 64 (1994) 422–424.

216. N.F. Mott, The electrical conductivity of transition metals, Proc. R. Soc. A 153 (1936) 699–717.

217. E.W. Müller, Das Auflösungsvermögen des Feldionenmikroskopes, Z. Naturforschg 11a (1956) 88–94.

218. M. Naruse, et al., Information physics fundamentals of nanophotonics, Rep. Progr. Phys. 76 (2013) 056401.

219. National Nanotechnology Initiative: The Initiative and its Implementation Plan, National Science and Technology Council, Committee on Technology,Subcommittee on Nanoscale Science, Engineering and Technology, Washington, D.C., 2000, pp. 19–20.

220. E. Nazaretski, et al., Pushing the limits: an instrument for hard X-ray imaging below 20 nm, J. Synchrotron Radiat. 22 (2015) 336–341.

221. K. Nikolić, A. Sadek, M. Forshaw, Architectures for reliable computing with unreliable nanodevices, in: Proc. IEEE-NANO 2001 M2.1 Nano-Devices II, pp. 254–259.

222. T. Nishizaka, et al., Unbinding force of a single motor molecule of muscle measured using optical tweezers, Nature 377 (1995) 251–254.

223. S. Oh, et al., Stem cell fate dictated solely by altered nanotube dimension, Proc. Natl. Acad. Sci. USA 106 (2009) 2130–2135.

224. C.J. van Oss, Forces interfaciales en milieux aqueux, Masson, Paris, 1996.

225. C.J. van Oss, R.F. Giese, Properties of two species of deadly nano-needles, Nanotechnol. Percept. 5 (2009) 147–150.

226. C.J. van Oss, et al., Impact of different asbestos species and other mineral particles on pulmonary pathogenesis, Clays Clay Minerals 47 (1999) 697–707.

227. N. Oyabu, et al., Mechanical vertical manipulation of selected single atoms by soft nanoindentation using near contact atomic force microscopy, Phys. Rev. Lett. 90 (2003) 176102.

228. N. Oyabu, et al., Single atomic contact adhesion and dissipation in dynamic force microscopy, Phys. Rev. Lett. 96 (2006) 106101.

229. T.O. Paine, L.I. Mendelsohn, F.E. Luborsky, Effect of shape anisotropy on the coercive force of elongated single-magnetic-domain iron particles, Phys. Rev. 100 (1955) 1055–1059.

230. P.C. Pandey, Bacteriorhodopsin—novel biomolecule for nano devices, Anal. Chim. Acta 568 (2006) 47–56.

231. K. Papageorgiou, G. Maistros, A. Koufaki, Modelling of dispersion quality of carbon nanotubes in thermosetting blends for capacitive behaviour enhancement of composite materials, Nanotechnol. Percept. 9 (2013) 147–158.

232. S.S.P. Parkin, M. Hayashi, L. Thomas, Magnetic domain-wall racetrack memory, Science 320 (2008) 190–194.

233. N.P. Paynter, et al., Cardiovascular disease risk prediction with and without knowledge of genetic variation at chromosome 9p21.3, Ann. Internal Med. 150 (2009) 65–72.

234. B.N.J. Persson, On the mechanism of adhesion in biological systems, J. Chem. Phys. 118 (2003) 7614–7621.

235. M. Planck, The concept of causality, Proc. Phys. Soc. 44 (1932) 529–539.

236. J. Platts, The offshore wind energy nano-industry, Nanotechnol. Percept. 9 (2013) 91–95.

237. M. Polanyi, Die Natur des Zerreissvorganges, Z. Phys. 7 (1921) 323–327.

238. A. Politi, J.L. O'Brien, Quantum computation with photons, Nanotechnol. Percept. 4 (2008) 289–294.

239. E.S. Polsen, et al., High-speed roll-to-roll manufacturing of graphene using a concentric tube CVD reactor, Sci. Rep. 5 (2015) 10257.

240. X.-L. Qi, S.-C. Zhang, Topological insulators and superconductors, Rev. Modern Phys. 83 (2011) 1057–1110.

241. J. Radon, Eine neue Transformation, Ber. Sächs Akad. Wiss Leipzig, Math. Phys. Kl. 69 (1917) 262–265.

242. L.V. Radushkevich, V.M. Luk'yanovich, The structure of carbon forming in thermal decomposition of carbon monoxide on an iron catalyst, Zh. Fiz. Khim. 26 (1952) 88–95 (in Russian).

243. H. Rafii-Tabar, Computational Physics of Carbon Nanotubes, University Press, Cambridge, 2008.

244. M.J. Rak, et al., Mechanosynthesis of ultra-small monodisperse amine-stabilized gold nanoparticles with controllable size, Green Chem. 16 (2014) 86–89.

245. J.J. Ramsden, The nucleation and growth of small CdS aggregates by chemical reaction, Surf. Sci. 156 (1985) 1027–1039.

246. J.J. Ramsden, Electronic processes in small semiconductor particles, Proc. R. Soc. A410 (1987) 89–103.

247. J.J. Ramsden, The stability of superspheres, Proc. R. Soc. Lond. A 413 (1987) 407–414.

248. J.J. Ramsden, Electron diffraction anomalies in small CdS clusters, J. Cryst. Growth 82 (1987) 569–572.

249. J.J. Ramsden, Impedance of pore-containing membranes, Stud. Biophys. 130 (1989) 83–86.

250. J.J. Ramsden, Molecular orientation in lipid bilayers, Phil. Mag. B 79 (1999) 381–386.

251. J.J. Ramsden, M. Grätzel, Formation and decay of methyl viologen radical cation dimers on the surface of colloidal CdS, Chem. Phys. Lett. 132 (1986) 269–272.

252. J.J. Ramsden, R. Horvath, Optical biosensors for cell adhesion, J. Recept. Signal Transduct. 29 (2009) 211–223.

253. J.J. Ramsden, G.I. Bachmanova, A.I. Archakov, Kinetic evidence for protein clustering at a surface, Phys. Rev. E. 50 (1994) 5072–5076.

254. J.J. Ramsden, Yu.A. Lvov, G. Decher, Optical and X-ray structural monitoring of molecular films assembled via alternate polyion adsorption, Thin Solid Films 254 (1995) 246–251 ibid. 261 (1995) 343–344.

255. J.J. Ramsden, et al., An optical method for the measurement of number and shape of attached cells in real time, Cytometry 19 (1995) 97–102.

256. J.J. Ramsden, et al., The design and manufacture of biomedical surfaces, Ann. CIRP 56/2 (2007) 687–711.

257. M. Reibold, et al., Carbon nanotubes in an ancient Damascus sabre, Nature 444 (2006) 286.

258. J. Ren, et al., One-pot synthesis of carbon nanofibers from CO_2, Nano Lett. 15 (2015) 6142–6148.

259. A. Rényi, Kémiai reakciók tárgyalása a sztochasztikus folyamatok elmélete segítségével, Magy. Tud. Akad. Mat. Kut. Int. Közl. 2 (1953) 83–101.

260. P.A. Revell, The biological effects of nanoparticles, Nanotechnol. Percept. 2 (2006) 283–298.

261. K. Ritz, Soil as a paradigm of a complex system, in: J.J. Ramsden, P.J. Kervalishvili (Eds.), Complexity and Security, IOS Press, Amsterdam, 2008, pp. 103–119.

262. K. Sääskilahti, et al., Frequency-dependent phonon mean free path in carbon nanotubes from nonequilibrium molecular dynamics, Phys. Rev. B 91 (2015) 115426.

263. J.R. Sanchez-Valencia, et al., Controlled synthesis of single-chirality carbon nanotubes, Nature 512 (2014) 61–64.

264. G. Schwarz, Coöperative binding to linear biopolymers, Eur. J. Biochem. 12 (1970) 442–453.

265. U.D. Schwarz, et al., Tip artefacts in scanning force microscopy, J. Microsc. 173 (1994) 183–197.

266. E.K. Schweizer, D.M. Eigler, Positioning single atoms with a scanning tunneling microscope, Nature (Lond.) 344 (1990) 524–526.

267. M.M. Shulaker, et al., Carbon nanotube computer, Nature 501 (2013) 526–530.

268. O. Sinanoğlu, Microscopic surface tension down to molecular dimensions and micro thermodynamic surface areas of molecules or clusters, J. Chem. Phys. 75 (1981) 463–468.

269. G.L. Snider, et al., Quantum dot cellular automata: review and recent experiments, J. Appl. Phys. 85 (1999) 4283–4285.

270. G. Sommerhoff, Analytical Biology, Oxford University Press, London, 1950, pp. 124–126.

271. G. Sommerhoff, Design for a Brain, Clarendon Press, London, 1970.

272. J.P. Spatz, et al., A.R. Khokhlov, R.G. Winkler, P. Reineker.Order-disorder transition in surface-induced nanopatterns of block copolymer films, Macromolecules 33 (2000) 150–157.

273. O. Spengler, Der Mensch und die Technik, C.H. Beck, Munich, 1931.

274. K. Spratte, L.F. Chi, H. Riegler, Physisorption instabilities during dynamic Langmuir wetting, Europhys. Lett. 25 (1994) 211–217.

275. Y. Sugimoto, et al., Complex patterning by vertical interchange atom manipulation using atomic force microscopy, Science 322 (2008) 413–417.

276. S.M. Sze, Physics of Semiconductor Devices, second ed., Wiley, New York, 1981, Appendix G.

277. L. Takacs, The historical development of mechanochemistry, J. Mater. Sci. 39 (2004) 4987–4993.

278. N. Taniguchi, On the basic concept of nano-technology, in: Proc. Intl Conf. Prod. Engng, Part II (Jap. Soc. Precision Engng), Tokyo, 1983.

279. N. Taniguchi, Current status in future trends of ultraprecision machining and ultrafine materials processing, Ann. CIRP 32/2 (1983) 573–582.

280. N. Taniguchi, On the basic concept of nano-technology, Precision Engrg. 7 (1985) 145–155.

281. T. Tate, On the magnitude of a drop of liquid formed under different circumstances, Phil. Mag. (S. 4) 27 (1864) 176–180.

282. A.I. Teixeira, P.F. Nealey, C.J. Murphy, Responses of human keratinocytes to micro- and nanostructured substrates, J. Biomed. Mater. Res. 71A (2004) 369–376.

283. M. Tayyab, et al., Synthesis and characterization of mechanically milled nanocomposites—carbon nanotube-reinforced aluminium, Nanotechnol. Percept. 10 (2014) 54–60.

284. E.C. Theil, Ferritin protein nanocages—the story, Nanotechnol. Percept. 8 (2012) 7–16.

285. P.J. Thomas, et al., The extent of regulatory consensus on health and safety expenditure. Part 1: Development of the J-value technique and evaluation of regulators' recommendations, Trans. IChemE B 84 (2006) 329–336.

286. S.R. Thomas, Modelling and simulation of the kidney, J. Biol. Phys. Chem. 5 (2005) 70–83.

287. G.G. Tibbetts, Lengths of carbon fibers grown from iron catalyst particles in natural gas, J. Cryst. Growth 73 (1985) 431–438.

288. K. Tiefenthaler, W. Lukosz, Sensitivity of grating couplers as integrated-optical chemical sensors, J. Opt. Soc. Am. B 6 (1989) 209–220.

289. T.T. Toth-Fejel, A few lesser implications of nanofactories, Nanotechnol. Percept. 5 (2009) 37–59.

290. H. Tributsch, J.C. Bennett, Electrochemistry and photochemistry of MoS2 layer crystals I, J. Electroanal. Chem. 81 (1977) 97–111.

291. H.N. Tsao, et al., Ultrahigh mobility in polymer field-effect transistors by design, J. Am. Chem. Soc. 133 (2011) 2605–2612.

292. A. Vanossi, et al., Modeling friction: From nanoscale to mesoscale, Rev. Modern Phys. 85 (2013) 529–552.

293. L. Vincent, L. Duchemin, E. Villermaux, Remnants from fast liquid withdrawal, Phys. Fluids 26 (2014) 031701. (6 pp).

294. Ph. Walter, et al., Early use of PbS nanotechnology for an ancient hair dying formula, Nano Lett. 6 (2006) 2215–2219.

295. F. Wang, et al., Hydrogel-retaining toxin-absorbing nanosponges for local treatment of methicillin-resistant *Staphylococcus aureus* infection, Adv. Mater. 27 (2015) 3437–3443.

296. G.S. Watson, B.W. Cribb, J.A. Watson, How micro/nanoarchitecture facilitates anti-wetting: an elegant hierarchical design of the termite wing, ACS Nano 4 (2010) 129–136.

297. R.A. Weale, The resistivity of thin metallic films, Proc. Phys. Soc. A 62 (1949) 135–136.

298. V.F. Weisskopf, Quality and quantity in quantum physics, Daedalus 88 (1959) 592–605.

299. T. West, The foot of the fly; its structure and action: elucidated by comparison with the feet of other insects, Trans. Linn. Soc. Lond. 23 (1862) 393–421.

300. R.A. Whiter, V. Narayan, S. Kar-Narayan, A scalable nanogenerator based on self-poled piezoelectric polymer nanowires with high energy conversion efficiency, Adv. Energy Mater. 4 (2014) 1400519.

301. Y. de. Wilde, et al., Thermal radiation scanning tunnelling microscopy, Nature (Lond.) 444 (2006) 740–743.

302. S.A. Wilson, et al., Enhanced dc conductivity of low volume-fraction nano-particulate suspensions in silicone and perfluorinated oils, J. Phys. D 42 (2009) 062003.

303. D. Winkel, Theoretical refinement of the pendant drop method for measuring surface tensions, J. Phys. Chem. 69 (1965) 348–350.

304. E.K. Wolff, A. Dér, All-optical logic, Nanotechnol. Percept. 6 (2010) 51–56.

305. S. Wood, R. Jones, A. Geldart, The Social and Economic Challenges of Nanotechnology, ESRC, Swindon, 2003.

306. A. Xu, et al., Van der Waals epitaxial growth of atomically thin Bi_2Se_3 and thickness-dependent topological phase transition, Nano Lett. 15 (2015) 2645–2651.

307. T. Xu, et al., The influence of molecular weight on nanoporous polymer films, Polymer 42 (2001) 9091–9095.

308. G.E. Yakubov, et al., Viscous boundary lubrication of hydrophobic surfaces by mucin, Langmuir 25 (2009) 2313–2321.

309. T. Yanagida, Y. Harada, A. Ishijima, Nanomanipulation of actomyosin molecular motors in vitro: a new working principle, Trends Biochem. Sci. (TIBS) 18 (1993) 319–323.

310. H. Yao, et al., Protection mechanisms of the iron-plated armor of a deep-sea hydrothermal vent gastropod, Proc. Natl. Acad. Sci. USA 107 (2010) 987–992.

311. A.D. Yoffe, Low-dimensional systems: quantum size effects and electronic properties of semiconductor microcrystallites (zero-dimensional systems) and some quasi-two-dimensional systems, Adv. Phys. 42 (1993) 173–266.

312. R. Young, et al., The Topografiner: an instrument for measuring surface microtopography, Rev. Sci. Instrum. 43 (1972) 999–1011.

313. L. Zhang, et al., Platinum-based nanocages with subnanometer-thick walls and well-defined, controllable facets, Science 349 (2015) 412–416.

314. M. Zhang, et al., Field-effect transistors based on a benzothiadiazole–cyclopentadithiophene copolymer, J. Am. Chem. Soc. 129 (2007) 3472–3473.

315. M. Zhang, K.R. Atkinson, R.H. Baughman, Multifunctional carbon nanotube yarns by downsizing an ancient technology, Science 306 (2004) 1358–1361.

316. X. Zhang, et al., Strong carbon-nanotube fibers spun from long carbon-nanotube arrays, Small 3 (2007) 244–248.

317. Y. Zhao, et al., Iodine doped carbon nanotube cables exceeding specific electrical conductivity of metals, Sci. Rep. 1 (2011) 83.

318. V.V. Zhirnov, et al., Limits to binary logic switch scaling—a gedanken model, Proc. IEEE 91 (2003) 1934–1939.

319. M. Zwolak, M. Di Ventra, Physical approaches to DNA sequencing and detection, Rev. Modern Phys. 80 (2008) 141–165.

Index

A

Abbe limit, 4, 19, 30, 32, 98
Abbott curve, 105
Absence of bulk criteria, 134
Accelerometers, 158
Acetylene, 237, 240
Acid–base (ab) interactions, 44, 45
Ackoff, R.L., 245
Actin, 264, 270, 271
Active sites, 269, 270
Actuator devices, 163
Adaptation (eukaryotic cells), 68, 70
Additives, 13, 14
Adenosine triphosphate (ATP), 270, 271
Adhesion, 42, 44, 45, 114–116
Adsorption,
    biomolecules, 71, 72
    impurity molecules, 48, 51
    nanotoxicology, 82, 83
    particle addition, 215, 216
AFGP, *see* Antifreeze glycoprotein
AFM, *see* Atomic force microscopy
Agglomerates, 72, 124, 125, 128, 130
Aggregates, 72, 124, 125, 130, 132, 219
Algorithmic information content (AIC), 110, 111
"all surface" matter, 25, 26
Alpha helices, 55
Alternating polyelectrolyte deposition (APED), 150, 152
Aluminium, 147
Amino acids,
    eukaryotic cells, 68, 70
    nanoscales, 19, 20
    three-body interactions, 55
    see also Proteins
Amorphous carbon, 231, 241
Amphibole asbestos needles, 83
Amphiphiles, 134, 135, 137
Anderson, P.W., 36
Anions, 249, 250
Anisotropic materials, 248
Anti-Frenkel type defects, 249
Anti-Schottky type defects, 249
Antiferromagnetic interactions, 178
Antifreeze glycoprotein (AFGP), 217
Antimony tin oxide (ATO), 145
Antistatic coatings, 145
APED, *see* Alternating polyelectrolyte deposition

APFIM, *see* Atom probe field ion microscopy
APM, *see* Atomically precise manufacturing
Apolar residues, 55, 222
APT, *see* Atomically precise technology
Aragonite, 141, 143, 145
Arginine-glycine-aspartic acid (RGD) triplets, 68, 70
Aromatic molecules, 141
Artificial composites, 143
Asbestos, 82, 83
Asexual regeneration, 258
Assemblers, 7
ATO, *see* Antimony tin oxide
Atom probe field ion microscopy (APFIM), 103
Atom sizes, 21
Atom, artificial, 29
Atom-by-atom assembly, 1, 2, 193, 195
Atomic force microscope, 94
Atomic force microscopy (AFM),
    chemography, 103, 104
    contact topography, 93–95
    nanomedicine, 76
Atomically precise manufacturing (APM), 4, *see also* Nanofacture
Atomically precise technology (APT), 1, 2
Atomistic simulation techniques, 252
ATP (adenosine triphosphate), 270, 271
Atypical char (carbon), 231
Aufbau principle, 141
Auger electron spectroscopy (AES), 101
Automata, 172
    see also Nanodevices,
Automated diagnosis, 290, 291
Automated reasoning, 285
Auxiliary concept systems, 74, 76
Available area function, 215

B

Bacteriophage, 12, 220, 221
Bacteriophage assembly, 220
Bacteriorhodopsin, 275
Bacteriorhodopsin (bR), 275–277
Ballistic deposition, 216
Ballistic transport, 157, 158, 165
Bases,
    bionanotechnology, 265, 274, 275
    folding, 221, 222

Printed in the United States
By Bookmasters